GLOBAL CLIMATE CHANGE AND EXTREME WEATHER EVENTS

Understanding the Contributions to Infectious Disease Emergence

WORKSHOP SUMMARY

Rapporteurs: David A. Relman, Margaret A. Hamburg, Eileen R. Choffnes, and Alison Mack

Forum on Microbial Threats
Board on Global Health

INSTITUTE OF MEDICINE
OF THE NATIONAL ACADEMIES

THE NATIONAL ACADEMIES PRESS
Washington, D.C.
www.nap.edu

THE NATIONAL ACADEMIES PRESS 500 Fifth Street, N.W. Washington, DC 20001

NOTICE: The project that is the subject of this report was approved by the Governing Board of the National Research Council, whose members are drawn from the councils of the National Academy of Sciences, the National Academy of Engineering, and the Institute of Medicine.

This project was supported by contracts between the National Academy of Sciences and the U.S. Department of Health and Human Services: National Institutes of Health, National Institute of Allergy and Infectious Diseases, Centers for Disease Control and Prevention, and Food and Drug Administration; U.S. Department of Defense, Department of the Army: Global Emerging Infections Surveillance and Response System, Medical Research and Materiel Command, and Defense Threat Reduction Agency; U.S. Department of Veterans Affairs; U.S. Department of Homeland Security; U.S. Agency for International Development; Lawrence Livermore National Laboratory; American Society for Microbiology; Sanofi Pasteur; Burroughs Wellcome Fund; Pfizer; GlaxoSmithKline; Infectious Diseases Society of America; and the Merck Company Foundation. Any opinions, findings, conclusions, or recommendations expressed in this publication are those of the author(s) and do not necessarily reflect the view of the organizations or agencies that provided support for this project.

International Standard Book Number-13: 978-0-309-12402-7
International Standard Book Number-10: 0-309-12402-6

Additional copies of this report are available from the National Academies Press, 500 Fifth Street, N.W., Lockbox 285, Washington, DC 20055; (800) 624-6242 or (202) 334-3313 (in the Washington metropolitan area); Internet, http://www.nap.edu.

For more information about the Institute of Medicine, visit the IOM home page at: **www.iom.edu.**

Printed in the United States of America

The serpent has been a symbol of long life, healing, and knowledge among almost all cultures and religions since the beginning of recorded history. The serpent adopted as a logotype by the Institute of Medicine is a relief carving from ancient Greece, now held by the Staatliche Museen in Berlin.

COVER: The cover image is a global anomaly mosaic of the combined normalized difference vegetation index (depicted over land surfaces) and sea surface temperatures (depicted over oceans) for January 2007 during the peak period of the 2006-2007 El Niño/Southern Oscillation warm event.
SOURCE: Data processing and analysis: Jennifer Small, Edwin Pak, Assaf Anyamba, Compton J. Tucker, GIMMS Group, NASA Goddard Space Flight Center. This image was provided by Dr. Assaf Anyamba of the University of Maryland Baltimore County and NASA Goddard Space Flight Center, GIMMS Group.

Suggested citation: IOM (Institute of Medicine). 2008. *Global climate change and extreme weather events: understanding the contributions to infectious disease emergence*. Washington, DC: The National Academies Press.

"Knowing is not enough; we must apply.
Willing is not enough; we must do."
—Goethe

INSTITUTE OF MEDICINE
OF THE NATIONAL ACADEMIES

Advising the Nation. Improving Health.

THE NATIONAL ACADEMIES

Advisers to the Nation on Science, Engineering, and Medicine

The **National Academy of Sciences** is a private, nonprofit, self-perpetuating society of distinguished scholars engaged in scientific and engineering research, dedicated to the furtherance of science and technology and to their use for the general welfare. Upon the authority of the charter granted to it by the Congress in 1863, the Academy has a mandate that requires it to advise the federal government on scientific and technical matters. Dr. Ralph J. Cicerone is president of the National Academy of Sciences.

The **National Academy of Engineering** was established in 1964, under the charter of the National Academy of Sciences, as a parallel organization of outstanding engineers. It is autonomous in its administration and in the selection of its members, sharing with the National Academy of Sciences the responsibility for advising the federal government. The National Academy of Engineering also sponsors engineering programs aimed at meeting national needs, encourages education and research, and recognizes the superior achievements of engineers. Dr. Charles M. Vest is president of the National Academy of Engineering.

The **Institute of Medicine** was established in 1970 by the National Academy of Sciences to secure the services of eminent members of appropriate professions in the examination of policy matters pertaining to the health of the public. The Institute acts under the responsibility given to the National Academy of Sciences by its congressional charter to be an adviser to the federal government and, upon its own initiative, to identify issues of medical care, research, and education. Dr. Harvey V. Fineberg is president of the Institute of Medicine.

The **National Research Council** was organized by the National Academy of Sciences in 1916 to associate the broad community of science and technology with the Academy's purposes of furthering knowledge and advising the federal government. Functioning in accordance with general policies determined by the Academy, the Council has become the principal operating agency of both the National Academy of Sciences and the National Academy of Engineering in providing services to the government, the public, and the scientific and engineering communities. The Council is administered jointly by both Academies and the Institute of Medicine. Dr. Ralph J. Cicerone and Dr. Charles M. Vest are chair and vice chair, respectively, of the National Research Council.

www.national-academies.org

FORUM ON MICROBIAL THREATS

DAVID A. RELMAN (*Chair*), Stanford University, Palo Alto, California
MARGARET A. HAMBURG (*Vice Chair*), Nuclear Threat Initiative/Global Health & Security Initiative, Washington, DC
DAVID W. K. ACHESON, Center for Food Safety and Applied Nutrition, Food and Drug Administration, Rockville, Maryland
RUTH L. BERKELMAN, Emory University, Center for Public Health Preparedness and Research, Rollins School of Public Health, Atlanta, Georgia
ENRIQUETA C. BOND, Burroughs Wellcome Fund, Research Triangle Park, North Carolina
ROGER G. BREEZE, Centaur Science Group, Washington, DC
STEVEN J. BRICKNER, Pfizer Global Research and Development, Pfizer Inc., Groton, Connecticut
GAIL H. CASSELL, Eli Lilly & Company, Indianapolis, Indiana
BILL COLSTON, Lawrence Livermore National Laboratory, Livermore, California
RALPH L. ERICKSON, Global Emerging Infections Surveillance and Response System, Department of Defense, Silver Spring, Maryland
MARK B. FEINBERG, Merck Vaccine Division, Merck & Co., West Point, Pennsylvania
J. PATRICK FITCH, National Biodefense Analysis and Countermeasures Center, Frederick, Maryland
DARRELL R. GALLOWAY, Medical S&T Division, Defense Threat Reduction Agency, Fort Belvoir, Virginia
S. ELIZABETH GEORGE, Biological and Chemical Countermeasures Program, Department of Homeland Security, Washington, DC
JESSE L. GOODMAN, Center for Biologics Evaluation and Research, Food and Drug Administration, Rockville, Maryland
EDUARDO GOTUZZO, Instituto de Medicina Tropical–Alexander von Humbolt, Universidad Peruana Cayetano Heredia, Lima, Peru
JO HANDELSMAN, College of Agricultural and Life Sciences, University of Wisconsin, Madison
CAROLE A. HEILMAN, Division of Microbiology and Infectious Diseases, National Institute of Allergy and Infectious Diseases, National Institutes of Health, Bethesda, Maryland
DAVID L. HEYMANN, Polio Eradication, World Health Organization, Geneva, Switzerland
PHIL HOSBACH, New Products and Immunization Policy, Sanofi Pasteur, Swiftwater, Pennsylvania

---

IOM Forums and Roundtables do not issue, review, or approve individual documents. The responsibility for the published workshop summary rests with the workshop rapporteur(s) and the institution.

---

*Deceased February 2, 2008.

IOM boards do not review or approve individual reports and are not asked to endorse conclusions and recommendations. The responsibility for the content of the report rests with the authors and the institution.

Reviewers

This report has been reviewed in draft form by individuals chosen for their diverse perspectives and technical expertise, in accordance with procedures approved by the National Research Council's Report Review Committee. The purpose of this independent review is to provide candid and critical comments that will assist the institution in making its published report as sound as possible and to ensure that the report meets institutional standards for objectivity, evidence, and responsiveness to the study charge. The review comments and draft manuscript remain confidential to protect the integrity of the deliberative process. We wish to thank the following individuals for their review of this report:

Ralph L. Erickson, DoD-Global Emerging Infections Surveillance and Response System, Walter Reed Army Institute of Research
Jonathan Patz, Center for Sustainability and the Global Environment, University of Wisconsin-Madison
Jeffrey Shaman, College of Oceanic and Atmospheric Sciences, Oregon State University
Mark Wilson, Department of Epidemiology, The University of Michigan
Mary Wilson, Department of Population and International Health, Harvard University

Although the reviewers listed above have provided many constructive comments and suggestions, they were not asked to endorse the final draft of the report before its release. The review of this report was overseen by **Dr. Melvin Worth**. Appointed by the Institute of Medicine, he was responsible for making certain that an independent examination of this report was carried out in accordance with institutional procedures and that all review comments were carefully considered. Responsibility for the final content of this report rests entirely with the authoring committee and the institution.

Preface

The Forum on Emerging Infections was created by the Institute of Medicine (IOM) in 1996 in response to a request from the Centers for Disease Control and Prevention (CDC) and the National Institutes of Health (NIH). The purpose of the Forum is to provide structured opportunities for leaders from government, academia, and industry to meet and examine issues of shared concern regarding research, prevention, detection, and management of emerging or reemerging infectious diseases. In pursuing this task, the Forum provides a venue to foster the exchange of information and ideas, to identify areas in need of greater attention, to clarify policy issues by enhancing knowledge and identifying points of agreement, and to inform decision makers about science and policy issues. The Forum seeks to illuminate issues rather than resolve them; for this reason, it does not provide advice or recommendations on any specific policy initiative pending before any agency or organization. Its value derives instead from the diversity of its membership and from the contributions that individual members make throughout the activities of the Forum. In September 2003, the Forum changed its name to the Forum on Microbial Threats.

ABOUT THE WORKSHOP

Long before the "germ theory" of disease was described, late in the nineteenth century,[1] humans have known that climatic conditions influence the appear-

[1]Pasteur, L. 1878. *Germ theory and its applications to medicine and surgery*. Read before the French Academy of Sciences, April 29, 1878. Published in Comptes rendus de l'Academie des Sciences, lxxxvi, pp. 1037-1043. Taken from *Scientific papers* (physiology, medicine, surgery, geology). New York: P. F. Collier and Son [c1910]. The Harvard classics v. 38. Modern History Sourcebook; http://www.fordham.edu/halsall/mod/1878pasteur-germ.html (accessed October 31, 2007).

ance and spread of epidemic diseases. As was pointed out in the report *Under the Weather: Climate, Ecosystems, and Infectious Disease*, "Since the dawn of medical science, people have recognized connections between a change in the weather and the appearance of epidemic disease. Roman aristocrats retreated to hill resorts each summer to avoid malaria. South Asians learned early that, in high summer, strongly curried foods were less likely to cause diarrhea."[2]

Ancient notions about the effects of weather and climate on disease remain in the medical and colloquial lexicon, in terms such as "cold" for rhinovirus infections; "malaria," derived from the Latin for "bad air"; and the common complaint of feeling "under the weather." Today, the evidence is mounting that Earth's climate is changing,[3] leading researchers to view the long-standing relationships between climate and disease from a global perspective.

Variations in climate may affect the health of humans, animals, and plants through direct impacts such as extreme heat or cold, or indirectly, by changing environments—in ways that may, for example, alter the geographic distribution or transmission dynamics of infectious diseases. The most recent report of the IPCC's Working Group II, whose members studied the influence of climate change on biological and social systems, stated with "very high confidence" that "climate change currently contributes to the global burden of disease and premature deaths." However, "at this early stage the effects are small but are projected to progressively increase in all countries and regions."[4]

The warming of the Earth is already contributing to the worldwide burden of disease and premature deaths, and is anticipated to influence the transmission dynamics and geographic distribution of malaria, dengue, tick-borne diseases, cholera, and other diarrheal diseases.[5] In the specific case of the relationship between climate and infectious diseases, it is important to recognize that a complex "web of causation" determines the distribution and transmission of infectious disease agents.[6] In addition to climate, factors influencing the geographic distribution and transmission of disease agents include: land-use patterns; a variety of social, demographic, and geographical variables; trade and transportation; human and animal migration; and public health interventions. Some of these factors are closely interrelated and influenced—directly or indirectly—by local, regional, or global variations in climate.

---

[2]NRC (National Research Council). 2001. *Under the weather: climate, ecosystems, and infectious disease.* Washington, DC: National Academy Press.

[3]IPCC (Intergovernmental Panel on Climate Change). 2007a. *Climate change 2007: the physical science basis.* Working Group I contribution to the Fourth Assessment Report of the IPCC. Cambridge, UK: Cambridge University Press.

[4]IPCC. 2007b. *Climate change 2007: climate change impacts, adaptation, and vulnerability.* Contribution of Working Group II to the Fourth Assessment Report of the IPCC. Cambridge, UK: Cambridge University Press.

[5]IPCC (2007b).

[6]NRC (2001).

The heating of the planet is also accelerating the hydrological cycle, increasing the likelihood of extreme weather events such as droughts, heavy precipitation, heat waves, hurricanes, typhoons, and cyclones. The projected health impacts of climate change and extreme weather events are predominately negative, with the most severe impacts in low-income countries where the capacity to adapt is weakest. Developed countries are also vulnerable to the health effects of extreme temperatures, as was demonstrated in 2003 when tens of thousands of Europeans died as a result of record-setting summer heat waves.[7] Climate change is expected to converge with, and intensify, additional contributors to infectious disease emergence and reemergence including global trade and transportation, land use, and human migration.[8]

The Forum on Microbial Threats hosted a public workshop in Washington, DC, on December 4 and 5, 2007, to consider the possible infectious disease impacts of global climate change and extreme weather events on human, animal, and plant health, as well as their implications for global and national security. Through invited presentations and discussions, participants explored a range of topics related to climate change and infectious diseases, including the ecological and environmental contexts of climate and infectious diseases; direct and indirect influences of extreme weather events and climate change on infectious diseases; environmental trends and their influence on the emergence, reemergence, and movement of vector- and non-vector-borne infectious diseases; opportunities and challenges for the surveillance, prediction, and early detection of climate-related outbreaks of infectious diseases; and the international policy implications of the potentially far-reaching health impacts of climate change.

ACKNOWLEDGMENTS

The Forum on Microbial Threats and the IOM wish to express their warmest appreciation to the individuals and organizations that gave their valuable time to provide information and advice to the Forum through their participation in this workshop. A full list of presenters can be found in Appendix A.

The Forum is indebted to the IOM staff who contributed during the course of the workshop and the production of this workshop summary. On behalf of the Forum, we gratefully acknowledge the efforts led by Eileen Choffnes, director of the Forum, Kate Skoczdopole, senior program associate, and Sarah Bronko, senior program assistant, for dedicating much effort and time to developing this workshop's agenda and for their thoughtful and insightful approach and skill in planning for the workshop and translating the workshop's proceedings and

---

[7]Kovats, R. S., and A. Haines. 2005. Global climate change and health: recent findings and future steps. *Canadian Medical Association Journal* 172(4):501-502.

[8]IOM (Institute of Medicine). 2003. *Microbial threats to health: emergence, detection, and response*. Washington, DC: The National Academies Press.

discussion into this workshop summary. We would also like to thank Dr. Assaf Anyamba of the NASA Goddard Space Flight Center and the University of Maryland Baltimore County for his invaluable contributions to this volume. Special thanks to the following IOM staff and consultants for their valuable contributions to this activity: Alison Mack, Bronwyn Schrecker, Lara Andersen, and Florence Poillon.

Finally, the Forum wishes to recognize the sponsors that supported this activity. Financial support for this project was provided by the U.S. Department of Health and Human Services: National Institutes of Health, National Institute of Allergy and Infectious Diseases, Centers for Disease Control and Prevention, and Food and Drug Administration; U.S. Department of Defense, Department of the Army: Global Emerging Infections Surveillance and Response System, Medical Research and Materiel Command, and Defense Threat Reduction Agency; U.S. Department of Veterans Affairs; U.S. Department of Homeland Security; U.S. Agency for International Development; Lawrence Livermore National Laboratory; American Society for Microbiology; Sanofi Pasteur; Burroughs Wellcome Fund; Pfizer; GlaxoSmithKline; Infectious Diseases Society of America; and the Merck Company Foundation. The views presented in this workshop summary report are those of the workshop participants and rapporteurs and are not necessarily those of the Forum on Microbial Threats or its sponsors.

David A. Relman, *Chair*
Margaret A. Hamburg, *Vice Chair*
Forum on Microbial Threats

Contents

Tables, Figures, and Boxes

TABLES

FIGURES

BOXES

Summary and Assessment

GLOBAL CLIMATE CHANGE AND EXTREME WEATHER EVENTS: UNDERSTANDING THE CONTRIBUTIONS TO INFECTIOUS DISEASE EMERGENCE

Humans have long recognized that climatic conditions influence the appearance and spread of epidemic diseases (NRC, 2001). Hippocrates' observations of seasonal illnesses, in the fifth century B.C.E., formed the basis for his treatise on epidemics. Hippocratic medicine, which attempted to predict the course and outcome of an illness according to its symptoms, also considered winds, waters, and seasons as diagnostic factors. Ancient notions about the effects of weather and climate on disease remain in the medical and colloquial lexicon, in terms such as "cold" for rhinovirus infections; "malaria," derived from the Latin for "bad air"; and the common complaint of feeling "under the weather."

Today, evidence that the Earth's climate is changing (IPCC, 2007b) is leading researchers to view the long-standing relationships between climate and disease from a global perspective. Increased atmospheric and surface temperatures are already contributing to the worldwide burden of disease and premature deaths, and are anticipated to influence the transmission dynamics and geographic distribution of malaria, dengue fever, tick-borne diseases, and diarrheal diseases such as cholera (IPCC, 2007a). Global warming is also accelerating the worldwide

The Forum's role was limited to planning the workshop, and the workshop summary has been prepared by the workshop rapporteurs as a factual summary of what occurred at the workshop.

hydrological cycle, increasing the intensity, frequency, and duration of droughts; heavy precipitation events; and flooding (IPCC, 2007a). Such extreme weather events have been increasing (IPCC, 2007a) and have been linked to global warming (Hoyos et al., 2006). These weather events may, in turn, contribute to and increase the risk for a wide range of vector- and non-vector-borne diseases in humans, plants, and animals (IPCC, 2007b).

The projected health consequences of future climate change and extreme weather events are predominantly negative.[1] The most severe impacts are expected to occur in low-income countries where adaptive capacity is weakest. Developed countries are also vulnerable to the health effects of weather extremes, as was demonstrated in 2003 when tens of thousands of Europeans died as a result of record-setting summer heat waves (Kovats and Haines, 2005). Climate change is expected to reinforce additional contributors to infectious disease emergence including global trade and transportation, land use, and human migration (IOM, 2003).

The Forum on Microbial Threats of the Institute of Medicine (IOM) held a public workshop in Washington, DC, on December 4 and 5, 2007, to explore the anticipated direct and indirect effects of global climate change and extreme weather events on infectious diseases of humans, animals, and plants and the implications of these health impacts for global and national security. Through invited presentations and discussions, invited speakers considered a range of topics related to climate change and infectious diseases, including the ecological and environmental contexts of climate and infectious diseases; direct and indirect influences of extreme weather events and climate change on infectious diseases; environmental trends and their influence on the transmission and geographic range of vector- and non-vector-borne infectious diseases; opportunities and challenges for the surveillance, prediction, and early detection of climate-related outbreaks of infectious diseases; and the international policy implications of the potentially far-reaching impacts of climate change on infectious disease.

Organization of the Workshop Summary

This workshop summary report was prepared for the Forum membership in the name of the rapporteurs and includes a collection of individually-authored

[1]In a personal communication on June 11, 2008, Diarmid Campbell-Lendrum (WHO) stated: "Some benefits undoubtedly exist, for some populations. But I don't know of any papers in the health literature, WHO, or otherwise that specifically focus on reviewing the benefits separate from the damages. These are usually referred to in reviews that look at the health effects overall. The health chapter of the IPCC refers to both harms and benefits, and I think this would be the best citation, and source for other studies. In IPCC (2007a), Confalonieri et al. note that the most important benefits are likely to be reduced deaths in winter at high latitudes, increased food production in high latitudes (for moderate climate change), and disruption of transmission cycles of some infectious disease in some places (e.g., where it may become too hot or dry for malaria transmission in some locations)."

papers and commentary. Sections of the workshop summary not specifically attributed to an individual reflect the views of the rapporteurs and not those of the Forum on Microbial Threats, its sponsors, or the IOM. The contents of the unattributed sections are based on presentations and discussions at the workshop.

The workshop summary is organized into chapters as a topic-by-topic description of the presentations and discussions that took place at the workshop. Its purpose is to present lessons from relevant experience, to delineate a range of pivotal issues and their respective problems, and to offer potential responses as discussed and described by workshop participants.

Although this workshop summary provides an account of the individual presentations, it also reflects an important aspect of the Forum philosophy. The workshop functions as a dialogue among representatives from different sectors and allows them to present their beliefs about which areas may merit further attention. The reader should be aware, however, that the material presented here expresses the views and opinions of the individuals participating in the workshop and not the deliberations and conclusions of a formally constituted IOM study committee. These proceedings summarize only the statements of participants in the workshop and are not intended to be an exhaustive exploration of the subject matter or a representation of consensus evaluation.

Workshop Context and Scope

Encouraged by opening remarks from the Forum's chair, David Relman, and Harvey Fineberg, President of the IOM, workshop presenters and discussants attempted to identify scientific questions that must be answered in order to discern—and, ultimately, to predict—the effects of a changing climate on specific infectious diseases, as well as the technical means to tackle these issues. At the same time, workshop participants grappled with an overarching question: What degree of scientific certainty that global climate change threatens human, animal, and plant health must be achieved before taking actions to mitigate these effects?

The National Research Council (NRC) report *Under the Weather: Climate, Ecosystems, and Infectious Diseases* (2001) has served as both a springboard and a resource for many discussions, including this workshop. The meeting began with a keynote address by Donald Burke of the University of Pittsburgh, who chaired the interdisciplinary committee that produced that influential report (see Burke in Chapter 1). Its key findings, summarized in Box SA-1, reflect considerable scientific uncertainty regarding the causal relationship between global climate change and infectious disease emergence.[2]

[2]Emerging infectious diseases are caused by pathogens that (1) have increased in incidence, geographical, or host range; (2) have altered capabilities for pathogenesis; (3) have newly evolved; or (4) have been discovered or newly recognized (Anderson et al., 2004; Daszak et al., 2000; IOM, 1992).

BOX SA-1
***Under the Weather* Key Findings:**
Linkages Between Climate and Infectious Diseases

- Weather fluctuations and seasonal-to-interannual climate variability influence many infectious diseases.
- Observational and modeling studies must be interpreted cautiously.
- The potential disease impacts of global climate change remain highly uncertain.
- Climate change may affect the evolution and emergence of infectious diseases.
- There are potential pitfalls in extrapolating climate and disease relationships from one spatial or temporal scale to another.
- Recent technological advances will aid efforts to improve modeling of infectious disease epidemiology.

SOURCE: NRC (2001).

This nuanced assessment has endured, as demonstrated in the 2007 report of Working Group II of the Intergovernmental Panel on Climate Change (IPCC), whose members studied the influence of climate change[3] on biological and social systems (IPCC, 2007a). The report states with "very high confidence" that "climate change currently contributes to the global burden of disease and premature deaths," but notes that "at this early stage the effects are small but are projected to progressively increase in all countries and regions."

Physical Evidence of Climate Change

There is little doubt that Earth's climate is changing as a result of human activities. The IPCC's Working Group I, which assessed the physical science of climate change, concluded that the "warming of the climate system is unequivocal, as is now evident from observations of increases in global average air and ocean temperatures, widespread melting of snow and ice, and rising global average sea level" and that "most of the observed increase in global average temperatures since the mid-twentieth century is very likely due to the observed increase in anthropogenic greenhouse gas concentrations" (IPCC, 2007b). A more detailed discussion of these findings appears in Appendix SA-1 (see page 43), "A Brief History of Climate Change," and in Chapter 1.

---

[3]Climate change in IPCC usage, and in this document as well, refers to any change in climate over time, whether due to natural variability or as a result of human activity.

Several workshop participants remarked on the IPCC's conclusions and called attention to the following general observations suggestive of the broad, profound, and rapidly accelerating impacts of climate change on Earth's physical systems:

- *Oceans as heat sinks*: Energy from global warming has been absorbed almost entirely by ocean waters, and relatively little has contributed to the melting of glacial ice or increases in air temperatures (Barnett et al., 2005; Levitus et al., 2005). To date, the thermal expansion of seawater accounts for about half of the observed rise in sea level. As sea levels rise, coastal flooding occurs more frequently and groundwater becomes increasingly saline.
- *Warming at high latitudes*: Warming is occurring fastest in boreal and arctic regions,[4] where its effects are amplified by the melting of snow, ice, and tundra (which also releases methane, a greenhouse gas), according to speaker Paul Epstein of the Harvard Medical School. Measurements by speaker Compton Tucker of the National Aeronautics and Space Administration (NASA) reveal that Greenland (which he described as a "canary for climate change") is melting at an accelerating pace that currently results in a net loss of approximately 160 km^3 of ice per year (see Tucker in Chapter 3).
- *Heat waves*: Epstein observed that climate change is not only associated with increases in the extent, breadth, intensity, and frequency of heat waves, but also with disproportionately elevated nighttime temperatures, which have increased twice as fast as average ambient temperatures since 1970. He also noted that as warming increases the levels of atmospheric water vapor, heat waves are more likely to be accompanied by increased humidity (IPCC, 2007b; see also Milly et al., 2005).
- *Dwindling freshwater supplies*: Warmer temperatures mean less water stored in glaciers and snow cover, which yield freshwater for approximately one-sixth of the world's population (IPCC, 2007a), according to presenter Sir Andrew Haines of the London School of Hygiene and Tropical Medicine. By 2050, he said, annual river runoffs are predicted to decrease by 10 to 30 percent in midlatitude dry regions and in the dry tropics (Milly et al., 2005).
- *Hydrological extremes*: Warming of the global climate system accelerates the hydrological cycle, producing more droughts, floods, and other extreme weather events. Warming-induced evaporation causes drought in some places, while higher atmospheric water content leads to more intense downpours elsewhere (Karl and Trenberth, 2003).

Epstein remarked that the confluence of trends toward increased interannual variability in precipitation (IPCC, 2001, 2007b), heavier precipitation events (Groisman et al., 2004), and more winter precipitation falling as rain rather than snow (Frederick and Gleick, 1999; Gleick, 2004; Levin et al., 2002) reflects the

[4]North and South Poles.

overall increase in seasonal (and, apparently, day-to-day) hydrological variability. He also noted that successive droughts punctuated by heavy rains not only favor flooding, but may also destabilize ecosystems, creating conditions that may be associated with clusters of mosquito-, rodent-, and water-borne disease outbreaks.[5]

- *Higher winds*: Circumpolar westerly winds are accelerating, particularly in the Southern Hemisphere (Gillett and Thompson, 2003; IPCC, 2007a), an effect Epstein described as a key sign of climatic instability.[6] Moreover, he said, as temperatures rise and pressure gradients build, winds can be expected to increase in intensity, generating stronger windstorms and altering the movement of weather fronts.

In June 2008, the U.S. Climate Change Science Program and the Subcommittee on Global Change Research released a report entitled *Weather and Climate Extremes in a Changing Climate*. While the IPCC (2007) report looked at the global effects of climate change on biological and social systems, this report focuses on the effects of climate change in North America, Hawaii, the Caribbean, and the U.S. Pacific Islands. Table SA-1 illustrates observed climate phenomena in the last 50 years and projects the likelihood of continued changes in North America. These phenomena include warmer days and nights, increased precipitation, more intense hurricanes, and larger areas affected by drought.

Over the last two decades, hydrometeorological disasters (e.g., hurricanes, droughts, floods) have affected a steadily increasing number of people living in vulnerable areas, most of them in developing countries, as shown in Figure SA-1. This development might be more accurately described as "global weirding," Burke said, in order to capture both the severity and the unpredictability of weather events spawned by global warming. As discussed in subsequent sections of this summary and in Chapter 1, extreme weather conditions increase the risk of transmission for a variety of infectious diseases, including diarrheal diseases, vector-borne diseases, and respiratory infections. Following a weather disaster such as a hurricane, affected areas must often cope with multiple infectious disease outbreaks.

Coincident Changes in Climate and Infectious Diseases

There are no appropriate, independent controls for the study of global climate change on Earth, Epstein observed. A wide range of methodologies must be har-

[5]In some cases, however, flooding may be associated with the destruction of vector breeding sites.

[6]Findings indicative of climate instability include (1) increasing rates of change, (2) wider fluctuations from norms, and (3) the appearance of major outliers (several standard deviations from the norm; Epstein and McCarthy, 2004).

TABLE SA-1 Observed Changes in North American Extreme Events, Assessment of Human Influence for the Observed Changes, and Likelihood That the Changes Will Continue Through the Twenty-first Century[a]

Phenomenon and direction of change	Where and when these changes occurred in past 50 years	Linkage of human activity to observed changes	Likelihood of continued future changes in this century
Warmer and fewer cold days and nights	Over most land areas, the last 10 years had lower numbers of severe cold snaps than any other 10-year period	Likely warmer extreme cold days and nights and fewer frosts[b]	Very likely[d]
Hotter and more frequent hot days and nights	Over most of North America	Likely for warmer nights[b]	Very likely[d]
More frequent heat waves and warm spells	Over most land areas, most pronounced over northwestern two-thirds of North America	Likely for certain aspects, e.g., night-time temperatures; and linkage to record high annual temperature[b]	Very likely[d]
More frequent and intense heavy downpours and higher proportion of total rainfall in heavy precipitation events	Over many areas	Linked indirectly through increased water vapor, a critical factor for heavy precipitation events[c]	Very likely[d]
Increases in area affected by drought	No overall average change for North America, but regional changes are evident	Likely, southwest USA.[c] Evidence that 1930s and 1950s droughts were linked to natural patterns of sea surface temperature variability	Likely in Southwest USA, parts of Mexico, and Carribean[d]
More intense hurricanes	Substantial increase in Atlantic since 1970; likely increase in Atlantic since 1950s; increasing tendency in W. Pacific and decreasing tendency in E. Pacific (Mexico West Coast) since 1980[e]	Linked indirectly through increasing sea surface temperature, a critical factor for intense hurricanes;[e] a confident assessment requires further study[c]	Likely[d]

[a]Based on frequently used family of IPCC emission scenarios.
[b]Based on formal attribution studies and expert judgment.
[c]Based on expert judgment.
[d]Based on model projections and expert judgment.
[e]As measured by the Power Dissipation Index (which combines storm intensity, duration, and frequency).
SOURCE: U.S. Climate Change Science Program and the Subcommittee on Global Change Research (2008).

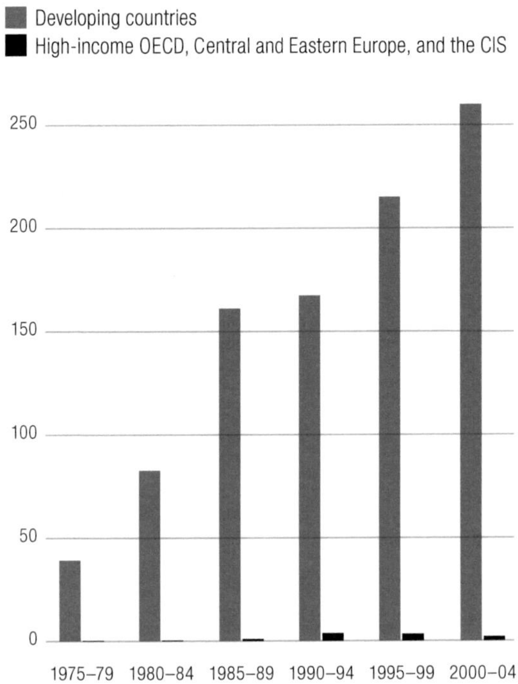

FIGURE SA-1 People affected by hydrometeorological disaster (millions per year).
SOURCE: Reproduced from United Nations Development Programme (2007) with permission of Palgrave Macmillan.

nessed, therefore, in order to assess changes in biological variables—including the geographic range and incidence of diseases—in relation to changes in temperature and precipitation (see Chapter 1). Information obtained from a variety of monitoring and mapping techniques can be integrated into geographic information systems (GISs) and used to identify and compare physical and biological phenomena. By enabling the overlay of multiple sets of data, GISs also provide contributions to descriptive and mathematical models that may be used to project the biological impacts of various climate change scenarios. Additional methods are used to analyze data gathered across scientific disciplines in order to reveal patterns and emerging trends associated with climate change, calculate rates of change (i.e., in the geographic range, prevalence, and incidence of infectious diseases), and compare these observations with predicted outcomes.

Many of the methodologies used to study the effects of climate change yield correlations, rather than proof of causation, Epstein acknowledged, but he argued that when observational data from multiple sources (1) match model projections,

(2) are consistent with each other, and (3) can be explained by plausible biological mechanisms, the preponderance of the evidence warrants further attention and exploration. Moreover, he added, models could be used to test such associations and their apparent underlying mechanisms (see Chapter 1).

In particular, Epstein identified three outcome variables as central to understanding the effect of climate change on the distribution of infectious diseases: shifts in altitude (and latitude), changes in seasonality, and responses to increased weather variability.

Shifts in altitude Many animal and plant species are adapted to specific habitats that occupy a narrow range along altitudinal and latitudinal climatic gradients.[7] Increasing temperatures not only melt alpine glaciers and drive the upward migration of plant communities, but also enable insects and other species that serve as infectious disease vectors to occupy higher altitudes (Epstein et al., 1998).[8] Such changes in conditions—which are conducive to changes in the ranges of disease agents and vectors—are occurring at high-altitude locations across the globe: in the Andes, the Sierra Nevada, the East African highlands, the European Alps, and the mountainous regions of India, Nepal, and Papua New Guinea, Epstein observed.

Seasonal shifts Climatic warming is expected to lengthen seasonal activity periods for mosquitoes and other insect vectors, thereby increasing opportunities for exposure to infectious diseases such as malaria (Tanser et al., 2003; van Lieshout et al., 2004). Ecological opportunists—including insects and rodents that serve as vectors of, and reservoirs for, infectious diseases—tend to proliferate rapidly in disturbed environments, while large predator species (infectious disease hosts) suffer under unstable environmental conditions, Epstein said.

Responses to increased weather variability Increased climate variability, along with habitat fragmentation and pollution, is likely to alter predator-prey relationships, which in turn influence infectious disease transmission dynamics. Such disequilibrium is thought to have precipitated the 1993 outbreak of a rodent-borne infection, hantavirus pulmonary syndrome, in the Four Corners region of the southwestern United States. That year, early, heavy rains ended an intense drought (during which predator populations declined) and provided new food for rodents, whose populations then expanded rapidly (Calisher et al., 2005; Patz et al., 1996).

---

[7]Plant and animal species first adapt to temperature changes by shifting their elevational ranges. A 1 km change in altitude is estimated to correspond to a geographic shift of 600 km north or south (Peters and Lovejoy, 1994). Highlands are considered sentinel regions for monitoring the biological response to global climate change.

[8]While some vectors may already be present at higher altitudes, higher temperatures may shorten the extrinsic incubation period, allowing the vector to transmit disease.

While it is anticipated that climate change will influence infectious disease emergence, several workshop participants emphasized that direct causal connections have yet to be established between climate change and infectious diseases, and that accurate predictions of infectious disease behavior cannot yet be made on the basis of climate projections alone.

Climate and Health

"Climate change will affect the health of humans as well as the ecosystems and species on which we depend, and . . . these health impacts will have economic consequences," predicts a recent report published by the Center for Health and the Global Environment (2005), edited by Epstein and Evan Mills (see Chapter 1 for the executive summary of this report, *Climate Change Futures: Health, Ecological and Economic Dimensions*). The report highlights a broad range of known and anticipated health consequences of climate change for humans, animals, and plants. In addition to influencing the location and frequency of infectious disease emergence and outbreaks, these effects include increased pest damage of crop plants, which in turn could contribute to human malnutrition; greater concentrations of pollen and fungi in the air, raising the risk of allergic symptoms and asthma; and higher rates of injury and death due to weather disasters and fires. Indeed, as Epstein (2005) has concluded, "it would appear that we may be underestimating the breadth of biologic responses to changes in climate."

Figure SA-2 illustrates the multiple pathways by which variations in climate affect the health of humans, animals, and plants. Direct influences include long-term regional changes in average temperature and precipitation, as well as extreme weather events such as floods, droughts, or violent storms. Climate change may also exert health effects indirectly, by altering ecosystems in ways that, for example, affect the geographic distribution or transmission dynamics of infectious diseases.

Direct and Indirect Effects of Climate on Infectious Diseases

Climate exerts both direct and indirect influences on the transmission and geographic distribution of infectious diseases, such as those shown in Table SA-2 (NRC, 2001). **Direct** effects of climate on infectious disease occur through the following mechanisms:

- Pathogen replication rate. This is particularly true of vector-borne diseases of warm-blooded animals, due to the exposure of pathogens to ambient weather conditions for part of their life cycle.
- Pathogen dissemination. This occurs when floods contaminate drinking water reservoirs, resulting in diarrheal diseases, and also when dry winds distribute soil-borne pathogens.

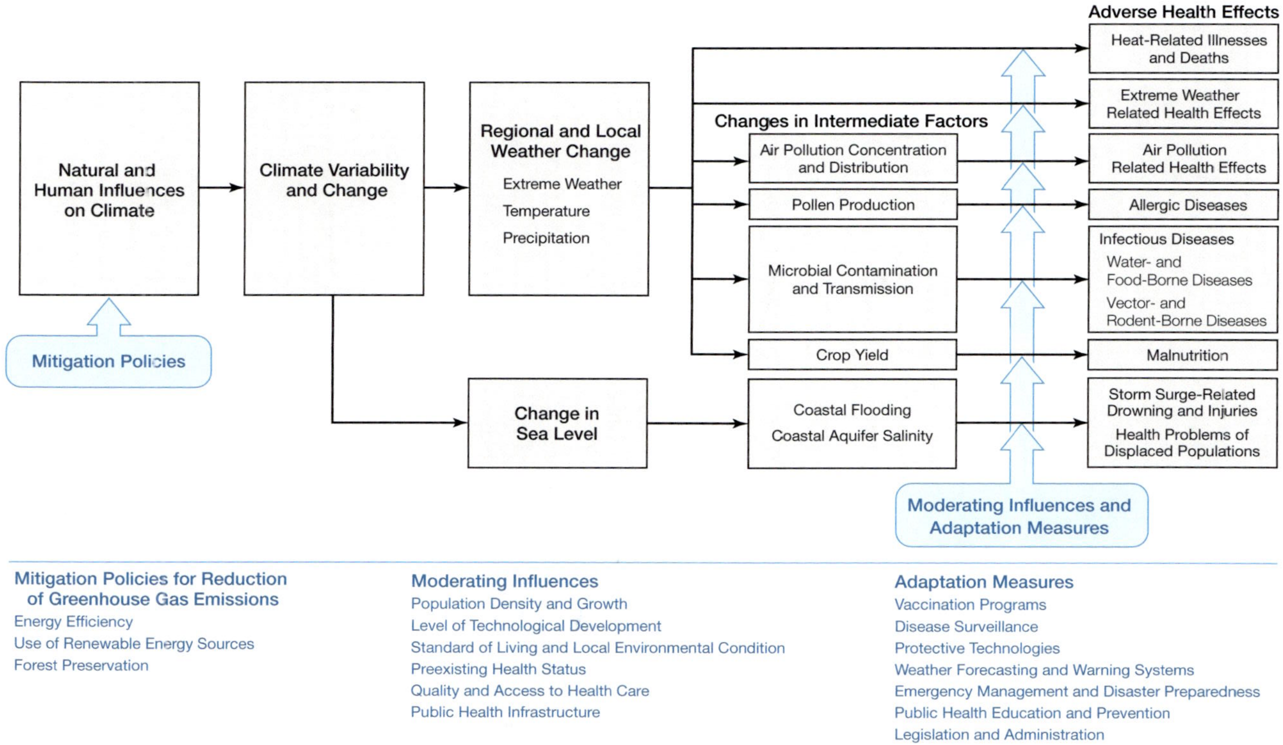

FIGURE SA-2 Potential health effects of climate variability and change.
SOURCE: Reprinted with permission from the American Medical Association from Haines and Patz (2004). Copyright 2004. All rights reserved; adapted from Patz et al. (2000).

TABLE SA-2 Examples of Diseases Influenced by Environmental Conditions

Environmental Condition	Disease Favored	Evidence
Warm	Malaria, dengue	Primarily tropical distribution, seasonal transmission pattern
Cold	Influenza	Seasonal transmission pattern
Dry	Meningococcal meningitis, coccidioidomycosis	Associated with arid conditions, dust storms
Wet	Cryptosporidiosis, Rift Valley fever	Associated with flooding

SOURCE: NRC (2001).

• Movement and replication of vectors and abundance of animal hosts. These include reservoir species for infectious diseases, such as migratory birds that carry avian influenza.

Climate also influences the distribution and transmission of infectious diseases through **indirect** effects on local ecosystems and human behavior. For example, abundant precipitation provides more and better breeding sites for vector species such as mosquitoes, ticks, and snails, while increasing the density of vegetation beneficial to these organisms (Githeko et al., 2000). Drought, on the other hand, may prompt people to store water in open containers, which also provide ideal breeding environments for mosquitoes.

Climate influences each component of the epidemiological triad of host-vector (see Figure SA-3), pathogen, and environment, which intersect to produce infectious disease. The complex ecologies of vector-borne diseases render them particularly sensitive to variations in temperature, which can alter patterns of disease incidence, seasonal transmission, and geographic range (McMichael et al., 2006; Sutherst, 2004). Some scientists predict that the effects of climate change and variability on vector-borne diseases are likely to be expressed in the form of short-term epidemics, as well as through gradual changes in disease trends (Githeko et al., 2000).

Climate's Role in Context

Climate interacts with a range of factors that shape the course of infectious disease emergence, including host, vector, and pathogen population dynamics; land use, trade, and transportation; social, political, and economic systems; human and animal migration; and interventions that control or prevent disease. These interdependent influences—or web of causation—can act together, resulting in outbreaks or epidemics of infectious disease; for example, people and animals (both domesticated and wild), if forced by climate disasters to migrate,

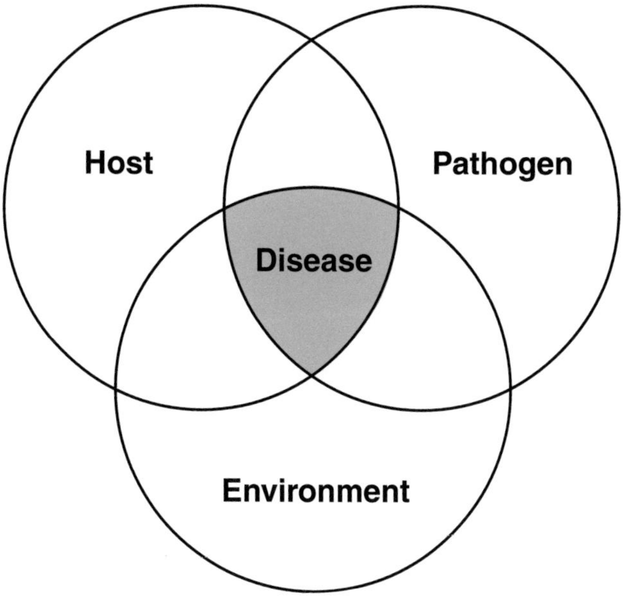

FIGURE SA-3 The epidemiological triad.
SOURCE: Reprinted from Snieszko (1974) with permission from Blackwell Publishing Ltd. Copyright 1974.

may introduce pathogens, parasites, and disease vectors into novel environments. The intersection of human, livestock, and wildlife movements and migration with climate change is discussed in greater detail later in this summary (see "Policy Implications") and in Chapter 4. An even broader view of disease emergence, the "Convergence Model" (see Figure SA-4), places climate among other physical environmental factors in disease emergence that intersect with biological and socioeconomic factors, as well as with host (human) and microbe (IOM, 2003).

Observed Effects of Climate Variation on Infectious Disease Range and Transmission Dynamics

The many factors confounding the interrelationships between climate change and infectious disease emergence vastly complicate attempts to investigate causality. As Haines and coauthors note, "Empirical observation of the health consequences of recent climate change, followed by formulation, testing, and then modification of hypotheses would require long time-series (probably several decades) of careful monitoring" (Haines et al., 2006). To inform health policy in the immediate future, risk assessments will need to be developed from short-

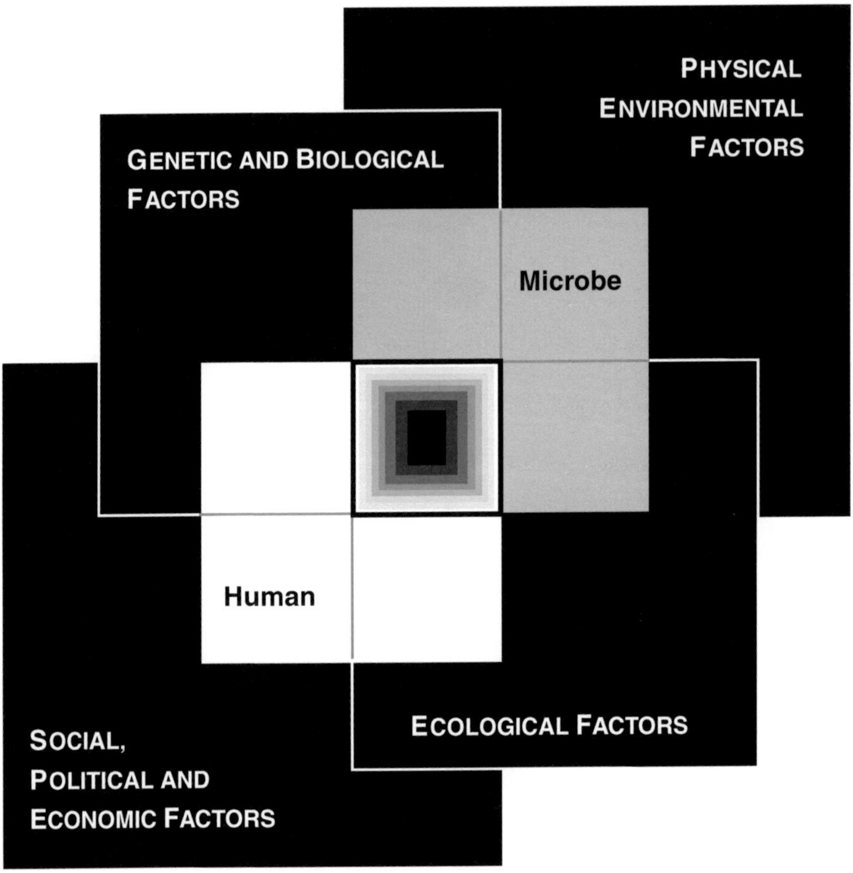

FIGURE SA-4 The Convergence Model. At the center of the model is a box representing the convergence of factors leading to the emergence of an infectious disease. The interior of the box is a gradient flowing from white to black; the white outer edges represent what is known about the factors in emergence, and the black center represents the unknown (similar to the theoretical construct of the "black box" with its unknown constituents and means of operation). Interlocking with the center box are the two focal players in a microbial threat to health—the human and the microbe. The microbe-host interaction is influenced by the interlocking domains of the determinants of the emergence of infection: genetic and biological factors; physical environmental factors; ecological factors; and social, political, and economic factors.
SOURCE: IOM (2003).

term observations of the effects of climate variation on infectious disease, taking into account the influence of confounding factors. Existing observations of these effects fall into two main categories: (1) climate-associated shifts in the geographical ranges of pathogens and vectors, and (2) studies of infectious disease transmission dynamics spanning relatively short periods of climatic variation.

Infectious Diseases in New Places

The following illustrative examples suggest that climate change has contributed to recent shifts in the geographic distribution of certain vector-borne diseases. In each case, additional factors may also contribute to the emergence and spread of these diseases.

- Bluetongue, a midge-borne viral disease of ruminant animals, emerged for the first time in northern Europe in 2006, during the hottest summer on record for that region and following nearly a decade of anomalously warm years. In the summer of 2007, the disease was reported in nine European countries, including the United Kingdom[9] and Denmark, during a massive outbreak that affected tens of thousands of farms (Enserink, 2008; IOM, 2008; ProMed Mail, 2007a,b, 2008; see Figure SA-5).
- Ticks that carry viruses known to be associated with encephalitis have been found at increasingly higher latitudes in northern Europe. A recent study in Denmark reveals a marked shift in the distribution of the tick-borne encephalitis virus as predicted by climate change models (IOM, 2008; Skarphedinsson et al., 2005).
- A 2004 outbreak of *Vibrio parahaemolyticus* gastroenteritis, associated with human consumption of raw oysters taken from Alaskan waters, extended the northernmost documented source of shellfish carrying this pathogen by 1,000 km. *Vibrio parahaemolyticus* had not been found in oyster beds in this region before 2004 (McLaughlin et al., 2005).
- In South Africa, the spread of wheat stripe rust has accompanied changes in rainfall patterns (Garrett et al., 2006), while needle blight of pine trees caused by *Dothistroma septosporum,* formerly a concern only in the Southern Hemisphere, is causing massive defoliation and mortality in the forests of British Columbia following climate change-associated increases in summer precipitation (Woods et al., 2005).
- Malaria incidence in the highlands of East Africa has risen since the late 1970s. The specific influence of rising temperatures on disease incidence has been a subject of considerable debate. Recent analyses employing a dynamical model

[9]It is believed that bluetongue was carried in a cloud of midges blown by warm winds across the English Channel from France, the Low Countries, or Germany, who, at the time, had similar outbreaks. The first case in the United Kingdom was discovered at a farm near Ipswich, Suffolk (BBC News, 2007; McKie and Revill, 2007).

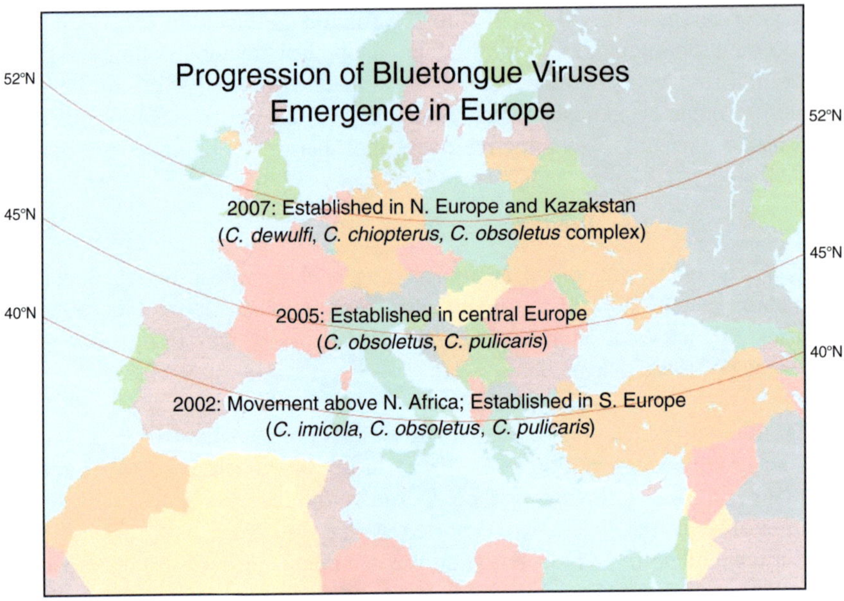

FIGURE SA-5 Progression of bluetongue viruses emergence in Europe.
SOURCE: Figure updated from Osburn (2008) and created by Rick Hayes, School of Veterinary Medicine, University of California, Davis.

suggest that a significant warming trend in this region has amplified mosquito population dynamics so as to contribute, along with drug resistance and land-use patterns, to the increased incidence of malaria (Harrus and Baneth, 2005; IOM, 2008; Pascual et al., 2006).

Climate Variation and Infectious Disease Transmission

Several recent studies have examined the relationship between short-term climatic variation and the occurrence of infectious diseases, in particular the influence of the El Niño/Southern Oscillation (ENSO) on the transmission of such vector- and non-vector-borne diseases as malaria, dengue fever, cholera, Rift Valley fever (RVF), and hantavirus pulmonary syndrome (Anyamba et al., 2006; McMichael et al., 2006; see Figure SA-6). ENSO, the irregular cycling between warm (El Niño) and cool (La Niña) phases of surface water temperatures across the central and east-central equatorial Pacific, is a well-known source of climate variability (see Haines in Chapter 1 and Chretien in Chapter 2). ENSO-associated shifts in ocean surface temperatures influence temperature and precipitation patterns throughout the global tropics, simultaneously producing excessive rainfall in some areas and drought in others (Kovats et al., 2003).

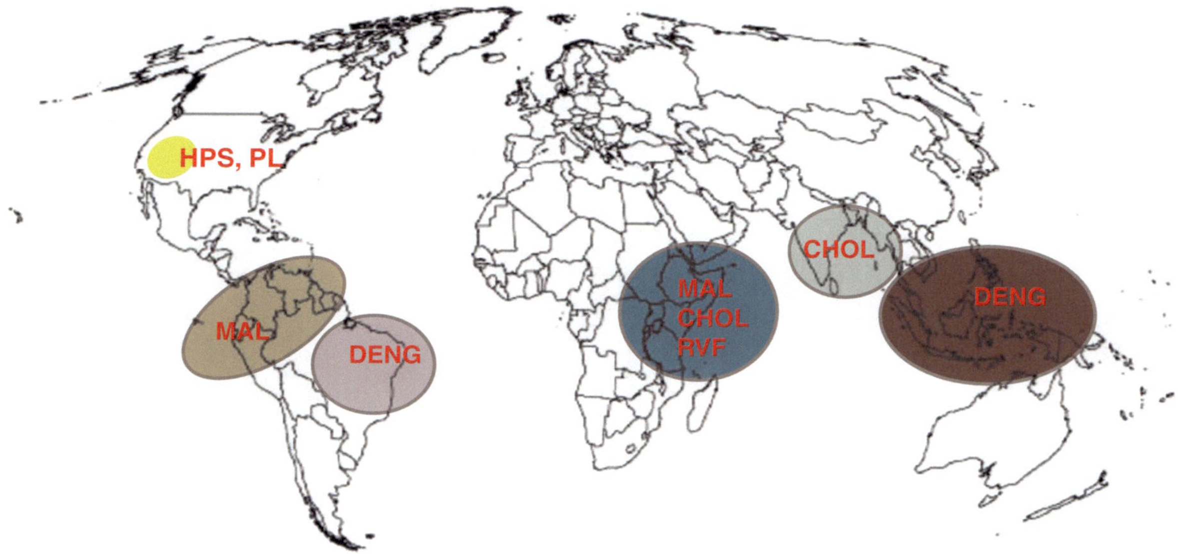

DENG Dengue Fever
CHOL Cholera
MAL Malaria
RVF Rift Valley Fever
HPS Hanta Virus Pulmonary Syndrome
PL Plague

FIGURE SA-6 Hot spots of potential elevated risk for disease outbreaks under El Niño conditions, 2006-2007.
SOURCE: Anyamba et al. (2006).

Global climate change is expected to intensify ENSO-related climate variability (WHO et al., 2003), which in turn offers a means to study the effects of climate variability on infectious disease (see Haines in Chapter 1). In his workshop presentation, Jean-Paul Chretien, of the U.S. Department of Defense, described key examples of such research, which examined connections between ENSO-related weather extremes and two infectious diseases: RVF and chikungunya fever (see Chretien in Chapter 2).

El Niño and Rift Valley fever An acute mosquito-borne viral disease, RVF primarily affects livestock (e.g., cattle, buffalo, sheep, goats) but can also be transmitted to humans through direct contact with the tissue or blood of infected animals, as well as by mosquito bites. Outbreaks of RVF among animals can spread to humans. The largest reported human outbreak, which occurred in Kenya during 1997-1998, resulted in an estimated 89,000 infections and 478 deaths (CDC, 2007b). For decades, RVF outbreaks have been associated with periods of heavy rainfall, which occur during El Niño; this observation led researchers to develop an operational model for RVF risk based on vegetation density (a marker for rainfall) as measured by satellite (see Figure SA-7A; Linthicum et al., 1999).

During the El Niño event of 2006-2007, above normal rainfall resulted in anomalous vegetation growth in East Africa, northern Australia, and parts of eastern China, and drought and diminished vegetation growth in southeastern Australia and northern South America. Above normal rainfall and anomalous vegetation growth in eastern Africa created ideal ecological conditions for the emergence of mosquito vectors of RVF, resulting in an outbreak of the disease in East Africa from December 2006 to May 2007 (see Figure SA-7B; A. Anyamba, personal communication, April 2008).[10]

Throughout the autumn of 2006, this model identified high risk for RVF in the same area affected by the 1997 epidemic, leading the U.S. Army Medical Research Unit (USAMRU) in Kenya to intensify its surveillance of local mosquitoes. Positive results provided early warning of a pending epidemic, enabling the Kenyan government—in concert with international partners including the Centers for Disease Control and Prevention (CDC) and the World Health Organization (WHO)—to mount a timely and targeted response, Chretien said (see CDC, 2007b).

La Niña and chikungunya fever Another mosquito-borne viral disease, chikungunya fever, is rarely fatal, but can cause severe joint pain, prolonged disability, and complications including protracted fatigue and arthritis (CDC, 2007b). A

[10]Moisture is required for egg development. Flooding often occurs following periods of heavy precipitation, enabling full development of the larvae and an increase in the mosquito population, thus spreading the virus during their next bloodmeal (WHO, 2007).

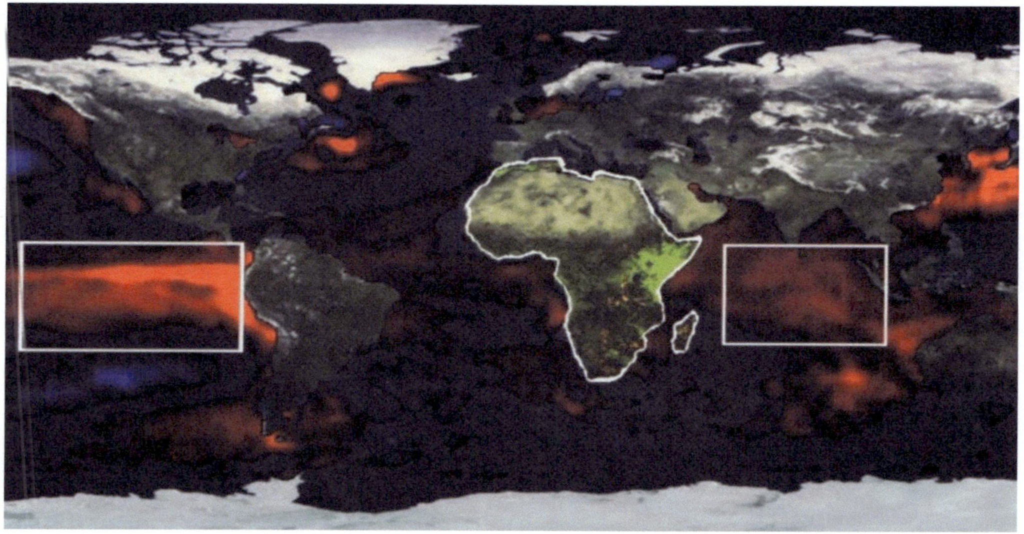

FIGURE SA-7A Using satellites to track Rift Valley fever.
NOTE: Scientists have discovered that the combination of warmer-than-normal equatorial Pacific Ocean temperatures associated with El Niño and rising sea surface temperatures in the western equatorial Indian Ocean can trigger outbreaks of Rift Valley fever in eastern Africa. This February 1998 image of sea surface temperature and vegetation (above), from the Advanced Very High Resolution Radiometer (AVHRR) onboard the National Oceanic and Atmospheric Administration's (NOAA) polar-orbiting weather satellites, illustrates the close relationship between ocean temperature (warmer-than-normal ocean temperatures are shown in red, cooler-than-normal temperatures shown in blue), rainfall, and their impacts on land vegetation (greener-than-normal vegetation shown in light green). The two warm pools of water (highlighted in the boxes) affect atmospheric circulation patterns such that there is an increase in rainfall over a large area of eastern Africa, which can lead to large-scale outbreaks of mosquito-borne diseases (NASA Goddard Space Flight Center, 2000).
SOURCE: NASA Goddard Space Flight Center, Scientific Visualization Studio (2000).

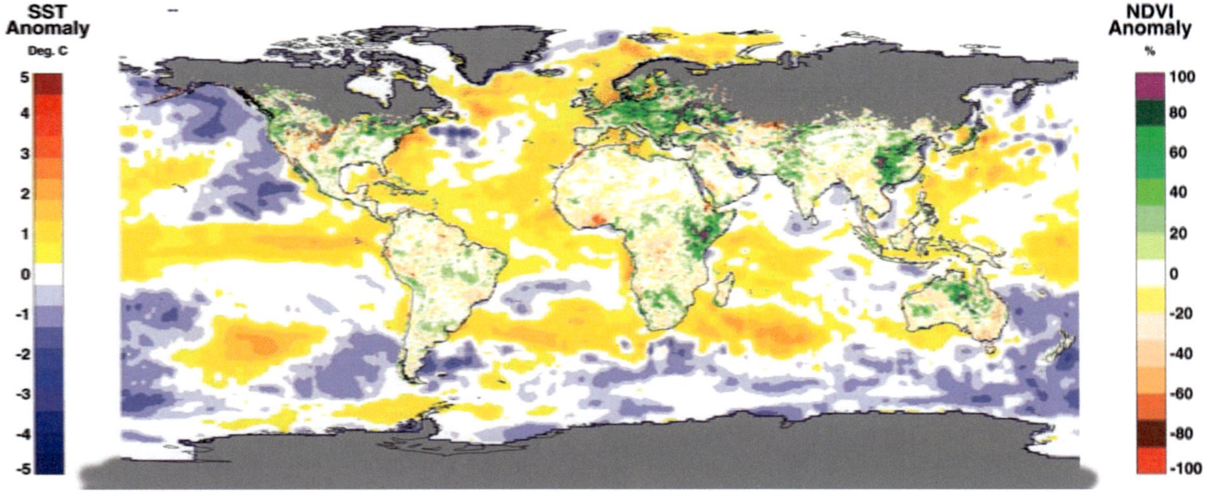

FIGURE SA-7B January 2007 combined global Normalized Difference Vegetation Index (NDVI) (depicted over land surfaces) and sea surface temperature (SST) (depicted over oceans) anomaly mosaic. NDVI and SST data are collected daily by several satellites in an ongoing fashion as part of NASA's and NOAA's global climate observing efforts. According to the Global Inventory Modeling and Mapping Studies (GIMMS) Group at the NASA Goddard Space Flight Center, the El Niño event of 2006-2007 was manifest by anomalous warming (~+2°C) of SSTs in the equatorial eastern Pacific Ocean with corresponding anomalous warming (~+1°C) in the equatorial western Indian Ocean. Such El Niño events result in anomalous displacement of global tropical precipitation yielding regional patterns with above normal rainfall in some areas and severe drought in other areas. These anomalies in precipitation are illustrated through biospheric response patterns represented by satellite derived vegetation index anomalies over land surfaces.

SOURCE: Data processing and analysis: Jennifer Small, Edwin Pak, Assaf Anyamba, Compton J. Tucker, GIMMS Group, NASA Goddard Space Flight Center.

string of outbreaks along the Kenyan coast in 2004 apparently spread to several western Indian Ocean islands and to India, resulting in the largest chikungunya fever epidemic on record (Chretien et al., 2007). Upon investigation, Chretien and coworkers discovered that at the time of the initial outbreaks in Kenya, a regional drought—corresponding to the La Niña phase of ENSO—had gripped the region. "There is some evidence that suggests that there may be a connection [between the drought and the chikungunya fever epidemic]," Chretien observed. "We know from the outbreak investigations in [Kenya], that domestic water wasn't being changed as frequently as usual because of the drought, and it wasn't being protected properly from the peridomestic mosquitoes that transmit chikungunya virus." Also, he noted, previous experimental studies in Kenya found that warm conditions can accelerate viral development within the mosquito (Chretien et al., 2007).

In addition to ENSO-associated weather anomalies, other short-term variations in climate, including drought, temperature, and wind patterns, have also been linked with changes in infectious disease incidence and geographic range:

Drought and diarrheal disease While diarrheal disease is frequently associated with periods of heavy rainfall and flooding and the subsequent contamination of water supplies with fecal bacteria (NRC, 2001), Haines described findings from a recent review of cross-sectional studies from 36 low- and middle-income countries that correlate increased incidence of diarrhea in young children with decreased rainfall (Lloyd et al., 2007). Because the vast majority of freshwater is used for irrigation, rather than for personal consumption, the relationship between these variables is unclear. Haines noted that handwashing behavior has been shown to decline when freshwater is less available (Curtis and Cairncross, 2003).

Temperature and food poisoning Comparing data from 16 sites in industrialized countries, investigators examined the incidence of sporadic cases of food poisoning (rather than outbreaks, which tend to be triggered by specific contamination incidents) attributed to the bacterium *Salmonella*. They found that such cases rose in a linear relationship to the previous week's temperature (Kovats et al., 2004). The lag in time suggests that temperature exerts this effect by accelerating bacterial replication in prepared food, Haines observed. Similar patterns of seasonal incidence also occur in cases of gastroenteritis caused by another bacterial agent *Campylobacter* (Kovats et al., 2005; Louis et al., 2005; Tam et al., 2006). However, unlike salmonellosis, seasonal patterns of *Campylobacter* infection in humans are not completely attributable to food-borne transmission of the pathogen, according to speaker Rita Colwell of the University of Maryland.

In a study conducted in England and Wales, Colwell and colleagues found that an increased incidence of *Campylobacter* gastroenteritis was associated with higher temperatures in districts supplied primarily with surface water, while those with the lowest incidence received mainly groundwater (Louis et al., 2005). The researchers therefore hypothesized that water ingested by poultry was the source of the seasonal increase in cases of human *Campylobacter* gastroenteritis and noted that surface water may be especially prone to contamination with the pathogen in the spring, when cattle and sheep give birth and are put out to pasture.

Wind-borne disease The annual arrival of dry, dust-laden winds—thought to render mucosal membranes vulnerable to infection—heralds the onset of epidemic meningococcal meningitis in West Africa (Sultan et al., 2005). There is some evidence that the geographical distribution of meningococcal meningitis in West Africa has expanded in the recent past, possibly as a result of changes in land use and climate (Molesworth et al., 2003; see Haines in Chapter 1).

Coccidioidomycosis—a fungal disease caused by inhaling the spores of *Coccidioides immitis*—along with meningococcal meningitis, can travel across continents in spore-laden desert dust clouds (Flynn et al., 1979; Garrison et al., 2003; NRC, 2001). The winds pick up these spores, along with dry, dusty soils, and transport them hundreds of miles (NRC, 2001; Schneider et al., 1997).

High winds and extreme weather have also been linked to the emergence, reemergence, and long-distance transport of vector-borne pathogens such as bluetongue and the citrus tristeza virus (IOM, 2003, 2008; NRC, 2001). Asian soybean rust, a pathogenic fungus, was apparently blown into the United States from South America by Hurricane Ivan in 2004 (Schneider et al., 2005). This nonnative plant pathogen has now become established in soybean-growing areas of the United States and Canada.

Synergies and Threshold Effects of Climate Change on Infectious Disease Emergence

In addition to the short-term observations of the effects of climate variation on the range and transmission of infectious disease described in the previous section, workshop participants considered the apparent near-term and long-range impacts of climate change on infectious diseases in several illustrative contexts: plant communities and crops; aquatic and marine environments; the Arctic; and central Asian ecosystems that have long served as incubators for plague epidemics. A common theme uniting these diverse accounts was the recognition that climate does not act gradually or entirely predictably upon ecosystems, but combines with other influences to produce threshold effects. Although typically expressed in terms of population dynamics (e.g., explosions, migrations, extinctions), such threshold effects also include the emergence of infectious diseases.

Plant Disease

According to speaker Karen Garrett of Kansas State University, climate change has the potential to produce huge—and largely unanticipated—impacts on agricultural and natural systems by altering patterns of plant infections. These effects include the direct consequences of crop diseases, such as declining food supplies; indirect effects on agricultural productivity, such as reduced soil formation (and thereby lower crop yields) resulting from more frequent tillage to remove infected plant residue; and health risks associated with increased pesticide usage. While efforts to understand these potential impacts typically focus on ecosystems, populations, and communities, Garrett and coworkers study plant responses to infectious disease at the molecular level, in order to understand and model genetic constraints for pathogen and plant adaptation to climate change (see Garrett in Chapter 2 for specific examples of these studies in various crop plants and plant communities).

Extending such observations to predict the repercussions of climate change on plant disease at the ecosystem level requires consideration of a broad range of influences on each member of the disease triad. Moreover, Garrett explained, any such perturbation may cross a threshold to an unexpectedly dramatic response. Many diseases, such as potato late blight, the disease that caused the Irish potato famine in the mid-nineteenth century, exhibit compound interest increases during a growing season, so that a slightly longer growing season can result in much higher regional inoculum loads. "The effects of climate change will be most important when there are thresholds and interactions that produce unanticipated large responses, and one of the most important effects might be that the systems will change more rapidly than in the past," Garrett observed.

Considerable resource investments will be needed to improve our understanding of the various and interacting factors that influence plant disease, she said. These include long-term, large-scale records of pathogen and host distributions (currently lacking even for agriculturally-important diseases); models of regional processes that incorporate disease dynamics; data and models that describe the dispersal of pathogens and vectors; and integrated, multidisciplinary, international collaborative networks for data collection and synthesis.

Research is also needed to identify and improve the introduction of disease resistance genes, a proven and promising strategy for responding to changes in disease threats to crops. In the tropics, where climate change is viewed as a considerable threat to food security due to the likelihood of greater climate variability, and where resources for crop protection are limited, efforts to characterize genetic resources are especially important. The Consultative Group for International Agricultural Research (CGIAR, 2008) currently undertakes such efforts on a "shoestring budget," Garrett reported.

Aquatic and Marine Environments

Two speakers at this workshop offered different perspectives on the direct and indirect influences of climate change in aquatic ecosystems. Leslie Dierauf of the U.S. Geological Survey (USGS) described the apparent impacts of climate and disease trends for a broad cross-section of aquatic and marine species and ecosystems, while Colwell discussed ecological and climatological factors that influence cholera, a water-borne infectious disease of considerable public health significance.

Aquatic and marine wildlife Marine life has suffered significant increases in the frequency and number of novel disease epidemics over the past few decades due to a variety of factors including, but not limited to, the disruption of ocean ecosystems by climate variability and warming water temperatures (Harvell et al., 1999). Much like their human counterparts in drought- or storm-stricken areas, marine mammals are being forced out of their home ranges by warming-induced population declines in plankton. As they follow their food to new territories, migrant marine mammals both encounter and introduce novel disease agents. Mass die-offs of certain species (e.g., seals, dolphins, porpoises) have occurred when these animals were exposed to morbilliviral diseases, such as distemper, for the first time during their annual migrations. Phocine distemper virus, identified as the cause of a die-off of harbor and gray seals in northern European coastal waters, is thought to have been transmitted to these species by harp seals that migrated to this region in response to overfishing-induced food shortages around their native Greenland in the late 1980s (Harvell et al., 1999).

The effects of climate variability on the health and disease of aquatic (freshwater-dwelling) and marine (ocean-dwelling) organisms are frequently exerted through the food web, as shown in Figure SA-8. In addition to these relationships, Dierauf emphasized that because aquatic and marine ecosystems are interconnected, infectious diseases of fish and wildlife may have the opportunity to move from freshwater sources to intertidal zones to marine environments, affecting species that may have not encountered these disease agents before. She also noted the particular vulnerability of coastal and intertidal zones to the effects of extreme weather, both directly as a result of damaging winds and water and indirectly though runoff from inland floods. On the U.S. Gulf Coast where, she said, "two hurricanes can turn an intertidal seagrass area into a mudflat," a majority of such areas—which act as buffers between ocean and land, and between fresh and marine waters—have been lost in recent years.

Examples of emerging infectious diseases along the aquatic-marine continuum, and their potential links to climate change, are presented in Box SA-2. Noting the lack of evidence-based literature on the effects of climate change and wildlife health, Dierauf joined the chorus of workshop participants calling for

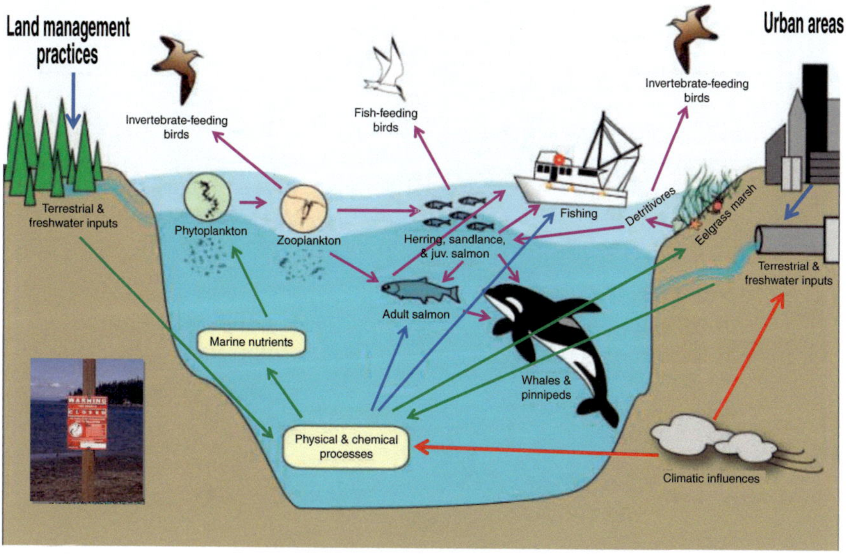

FIGURE SA-8 Interconnectedness of terrestrial, aquatic, and marine food webs.
SOURCE: Figure courtesy of Mary Ruckelshaus, NOAA Fisheries.

greater investments in collaborative efforts to monitor, model, and research these connections.

Water-borne human disease The incidence and distribution of food- and water-borne diseases are shaped by numerous factors, including climate variation, water temperature, precipitation patterns, and/or water salinity. Extreme weather events, including heavy rainfall and flooding, are associated with outbreaks of several important water-borne diseases (NRC, 2001). These include cholera, an acute diarrheal illness caused by the bacterium *Vibrio cholerae* (CDC, 2005); cryptosporidiosis, one of the most common water-borne diseases in the United States, caused by microscopic parasites of the genus *Cryptosporidium* (CDC, 2007a); and giardiasis, another diarrheal illness common in the United States, which is caused by the single-celled parasite *Giardia intestinalis* (CDC, 2004). Soil washed into coastal waters by floods contains animal wastes (and, therefore, fecal bacteria) as well as other organic matter and nutrients that promote rapid growth, or "blooms," of certain toxic algae species. These harmful algal blooms (also known as "red tides") produce neurotoxins that can be transferred through the marine web—killing some marine animals along the way—to seafood-consuming humans (Woods Hole Oceanographic Institution, 2008).

BOX SA-2
Emerging Infectious Diseases in the Aquatic-Marine Continuum

The following infectious diseases, described by workshop speaker Leslie Dierauf of the U.S. Geological Survey's National Wildlife Health Center, are of considerable concern in freshwater, intertidal, and marine wildlife, due to recent increases in incidence and/or geographic range, as well their potential to disrupt aquatic and marine ecosystems.

Freshwater Zone

- *Ranavirus*, within the family *Iridoviridae*, is a skin-destroying viral pathogen that infects North American amphibians (see Figure SA-9);

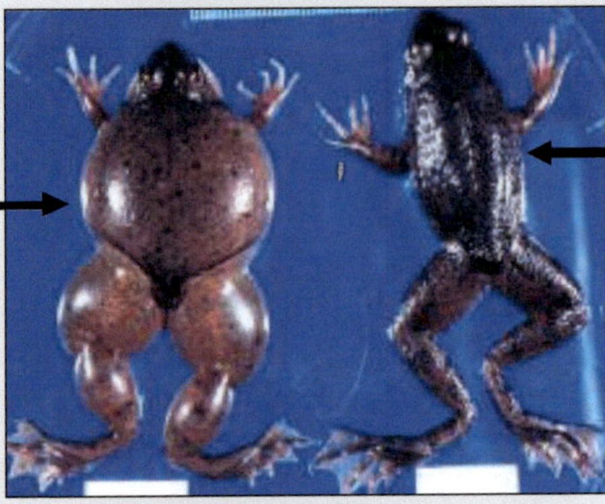

FIGURE SA-9 *Ranavirus*-associated disease in frogs.
SOURCE: USGS; Dierauf (2007).

- Viral hemorrhagic septicemia (*Rhabdoviridae novirhabdovirus*) is a newly discovered viral disease associated with large-scale mortality of many common fish species. The virus is able to survive in warm and cold waters and in estuarine and marine waters, as well as in freshwater systems (see Figure SA-10);

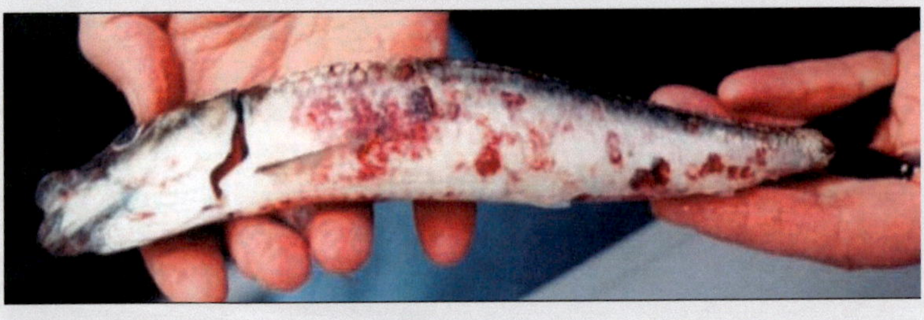

FIGURE SA-10 Viral hemorrhagic septicemia (VHS) *Rhabdoviridae novirhabdovirus.*
SOURCE: USGS; Dierauf (2007).

- Chytridiomycosis (*Batrachochytrium dendrobatidis*) is a fungal infection of not only North American frogs, but is now being detected worldwide (see Figure SA-11); and
- A *Perkinsus*-like protozoal organism has been identified as causing fatalities in tadpoles of several North American frog species (see Figure SA-12). Discovered in 1999, this pathogen decimates frog populations and is thought to be adapted to warmer temperature waters.

FIGURE SA-11 Chytridiomycosis (*Batrachochytrium dendrobatidis*) in Chiricahua leopard frog (New Mexico).
SOURCE: USGS; Dierauf (2007).

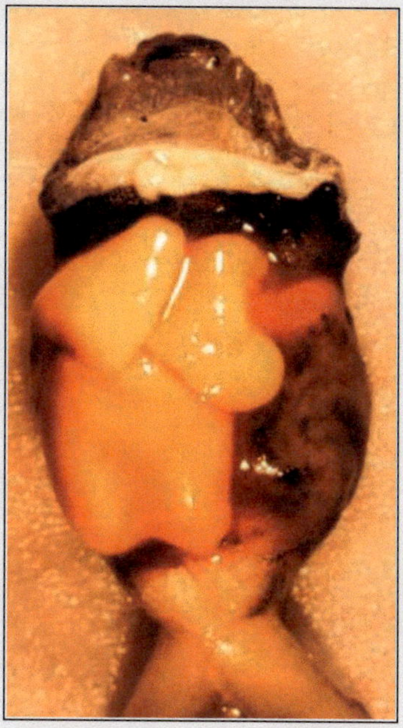

FIGURE SA-12 *Perkinsus*—wood frog (*Rana sylvatica*) tadpole with massively enlarged yellow liver.
SOURCE: USGS; Dierauf (2007).

Intertidal Zone

- Another *Perkinsus*-like protozoan affecting oysters—may be related to the species that infects frogs.
- Various species of *Vibrio* bacteria that infect shellfish and are transmitted up the food chain to birds and mammals, including humans.

Marine Zone

- The acceleration of coral bleaching by opportunistic infections during periods of elevated temperature (Harvell et al., 2002). Coral bleaching occurs when, under extreme environmental stress, corals expel their symbiotic algae. In 1997-1998, a dramatic global increase in the severity of coral bleaching coincided with El Niño (Harvell et al., 1999).

A complex web of ecological relationships is involved in the incidence and prevalence of cholera, which Colwell estimated affects 100,000 people per year and kills 10,000 (see Colwell in Chapter 2). Over the course of three decades of study, she and coworkers have determined that this water-borne disease, although caused by a bacterium (*V. cholerae*), is actually transmitted by the plankton species with which it associates (Colwell, 2004). *Vibrio cholerae* is a natural inhabitant of aquatic environments of appropriate salinity, but remains quiescent except when temperatures rise above 15°C, and an influx of nutrients causes the plankton to bloom, increasing *V. cholerae* concentrations to levels capable of causing disease when water is consumed. This relationship is sufficiently robust to permit the use of remote sensing data—incorporating sea surface temperature, sea surface height, and chlorophyll levels (an indicator of phytoplankton bloom) observed in the Bay of Bengal—to predict the onset of cholera epidemics in the Ganges delta region of Bangladesh, known as the "home of cholera" due to its long history of epidemic disease.

Primarily confined to the Indian subcontinent, cholera was spread by the shipping trade from India to Europe and the Americas in the early nineteenth century (Colwell, 2004). Subsequent improvements in sanitation drastically reduced cholera incidence in the West, but the disease reemerged in Peru in 1991, after being absent from that country for nearly a century. Although initially attributed to contaminated ballast water from a foreign ship, cholera's return to Peru was eventually linked to elevated sea surface temperature, coincident with El Niño (Lipp et al., 2003).

The Arctic

The physical effects of climate change are dramatically apparent in the Arctic, where temperatures have increased at nearly twice the global average over the past century, causing widespread melting of land and sea ice (see Figure SA-13; Borgerson, 2008; IPCC, 2007b). This trend is expected to continue and intensify, resulting in warmer winters, increased annual precipitation, more frequent extreme weather events, and—as the ice continues to melt—greater river discharge and increased sea height, according to workshop speaker Alan Parkinson, of the CDC's Arctic Investigations Program in Anchorage, Alaska.

These rapidly changing environmental conditions are ripe for infectious disease emergence on several fronts, Parkinson observed (see Chapter 2). Higher temperatures at these latitudes permit the survival and replication of cold-sensitive pathogens such as *Vibrio parahaemolyticus*, as previously noted (McLaughlin et al., 2005), or increase the prevalence of existing pathogens such as *Clostridium botulinum* (a particular concern for indigenous peoples, who traditionally preserve food by fermentation). Preliminary studies suggest that warmer ambient temperatures, which would be predicted to occur with climate change, may result in higher rates of food-borne botulism associated with the consumption

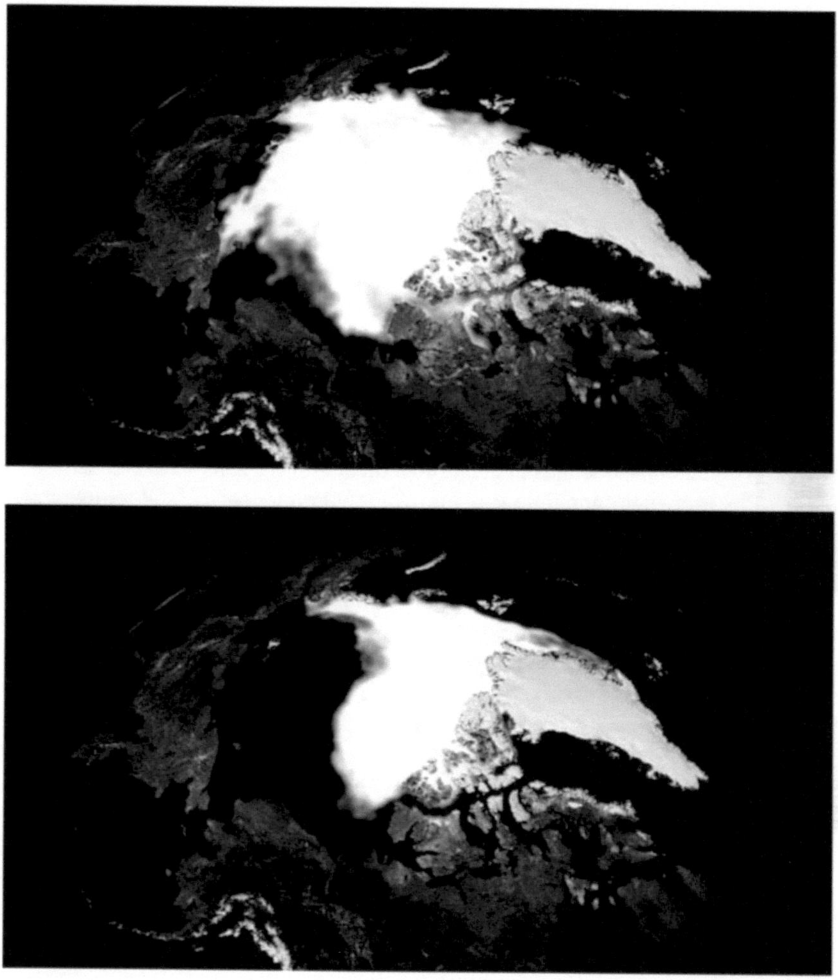

FIGURE SA-13 The Arctic ice cap, September 2001 (Top) and September 2007 (Bottom).
SOURCE: NASA, as printed in Borgerson (2008).

of fermented seal meat (Leclair et al., 2004). In addition to fermentation, many Arctic residents store fish and meat by air-drying or by burying it on or near the permafrost; changes in climate may therefore result in higher rates of spoilage of food preserved by either method (see Parkinson in Chapter 2).

As temperatures increase, reservoir species for zoonotic diseases may survive winters in larger numbers, increase in population, or expand their geographic ranges. Beavers, common hosts for the water-borne protozoan *Giardia intestinalis*, are migrating northward in Alaska, into areas that have become more habitable due to changes in vegetation and habitat (Parkinson and Butler, 2005). Similarly, climate conditions that favor range expansion by foxes or rodents that carry alveolar echinococcosis—a lethal zoonotic infection caused by the larval stage of the tapeworm *Echinococcus multilocularis*—may increase the human incidence of this disease (Holt et al., 2005; Parkinson and Butler, 2005; Schweiger et al., 2007).

Climate change may enable mosquito-borne diseases such as the West Nile virus (WNV) to move into the Arctic by increasing vector survival and disease transmission rates, as well as by altering migration patterns of birds and other reservoir species. WNV has already reached Canadian provinces adjacent to Alaska at a latitude of 57°N, and its mosquito vector, *Aedes albopictus,* is present in the state, Parkinson reported. Climate change is also projected to shift the range of the tick vector of Lyme disease northward (Ogden et al., 2005), but as with WNV, the consequences of such movements for human disease depend on a range of factors, including land use, human population density, and temperatures warm enough for pathogens to reach an infective dose in the vector.

Public health challenges Together with a catalog of health impacts attributable to climate change in the Arctic, Parkinson noted two indirect effects that appear especially favorable to infectious disease transmission: (1) damage to the sanitation infrastructure resulting from the melting of permafrost (upon which many Arctic communities are built) and from flooding, and (2) the opening of the Northwest Passage.

Inadequate housing and sanitation are already important determinants of infectious disease transmission in many Arctic regions, Parkinson observed. In a recent study conducted in western Alaska, Parkinson and coworkers found significantly higher rates of hospitalization for young children with pneumonia, influenza, and respiratory syncytial virus (RSV) and for people of all ages with outpatient *Staphylococcus aureus* infections and hospitalization for skin infections, in communities without in-house piped water service, compared to communities with in-house piped water service (Hennessey et al., 2008). This suggests that the loss of existing basic sanitation services, through climate change-related infrastructural damage, may raise infectious disease rates in Arctic populations. Furthermore, Parkinson noted, sewage leaking from pipes ruptured by melting permafrost contains water-borne pathogens such as *Giardia, Cryptosporidium,* and the hepatitis A virus.

A second potential route for infectious disease emergence in the Arctic is being cleared along with the sea ice. With the opening of the Northwest Passage—and perhaps, eventually, the Northeast Passage—more ships will take this shorter route as a sea lane alternative to the Panama Canal when crossing between the Atlantic and Pacific Oceans (see Figure SA-14). Increased maritime shipping in the Arctic is expected to bring many economic benefits to these north-

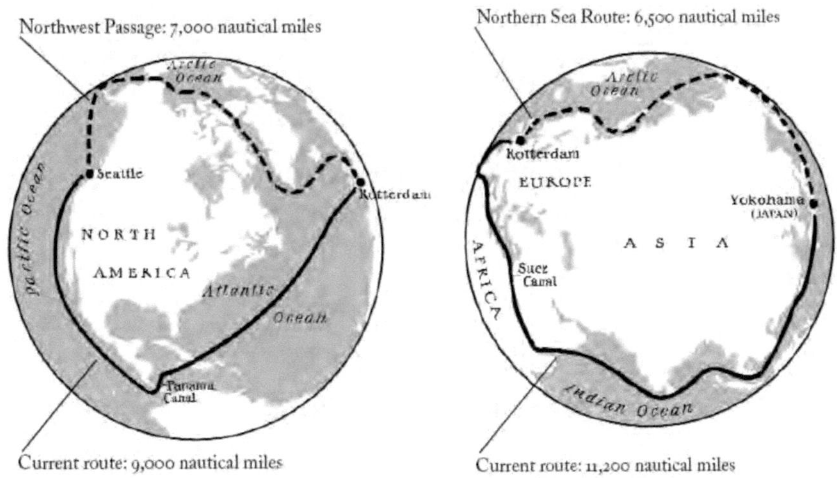

FIGURE SA-14 Arctic shipping shortcuts.
SOURCE: Reprinted with permission from *Foreign Affairs* (Borgerson, 2008). Copyright 2008 by the Council on Foreign Relations, Inc.

ernmost communities, Parkinson noted, but it is also likely to expose the region's inhabitants and ecosystems to invasive species of all kinds, including potentially pathogenic microbes and their vectors. In the face of these challenges, Parkinson echoed the suggestions of many other speakers at this workshop to enhance the surveillance and monitoring of climate-sensitive infectious diseases on a global basis and to establish international networks to share such information.

Plague Dynamics

Throughout human history the various forms of plague, caused by the bacterium *Yersinia pestis* and transmitted by fleas among a wide range of hosts, have caused both endemic and epidemic disease. "Plague is a highly variable disease," explained speaker Nils Christian Stenseth of the University of Oslo, Norway. "It is a complex system with complex temporal, seasonal, interannual dynamics." In his contribution to Chapter 2, Stenseth describes these intricate relationships and his approach to modeling plague dynamics based on long-term monitoring of pathogen prevalence in central Asian rodent populations. These studies have led him to conclude that although relatively few cases of plague are currently reported, the disease poses a significant and imminent threat to human populations due, in part, to the influence of climate change.

Using longitudinal data collected over 50 years in Kazakhstan, a focal region for plague where cases are regularly reported, Stenseth and colleagues determined that *Y. pestis* prevalence increases dramatically in its primary host, the great gerbil (*Rhombomys opimus*), during warmer springs and wetter summers (Stenseth et al., 2006). Rodent populations also tend to increase under these conditions and, along with them, the possibility that plague will be transmitted to humans. These conditions apparently existed in central Asia during the onset of the Black Death in the fourteenth century, as well as in the years preceding a mid-nineteenth-century plague pandemic. As Earth's climate warms, warmer springs and wet summers are expected to become increasingly common in the region, and also in North America.

Historical, Scientific, and Technological Approaches

Several workshop presentations described methods used to identify, measure, analyze, and predict the direct and indirect effects of climate change on infectious disease emergence. Each of the topics discussed below represents part of an interdisciplinary approach that, participants agreed, must continue to expand in order to pursue a common goal. As Colwell observed, understanding complex interactions between biological and physical environments paves the way for the development of predictive models and, thereby, for early and efficient responses to infectious disease threats.

Analysis of Historical Data

Historical analysis provides a perspective on climate and infectious disease far more sweeping than can be obtained from scientific monitoring. When speaker Rodolfo Acuña-Soto of the Universidad Nacional Autónoma de México and coworkers chronicled major epidemics (based on historical accounts) and climate conditions (as reflected in the width of tree rings) over the last millennium in the Valley of Mexico, they revealed an association between severe and prolonged droughts, catastrophic epidemics, and societal collapse (see Acuña-Soto et al. in Chapter 3).

Amid one such "megadrought" during the sixteenth century, hemorrhagic fever appears to have killed an estimated 80 percent of the indigenous people of the Valley of México; survivors mated primarily with Spanish colonists, repopulating the region with predominantly Mestizo offspring. Droughts also accompanied each of a series of 22 typhus epidemics that occurred between 1655 and 1915.

Drought is still a major problem for Mexico and is expected to continue to burden the country in the future, Acuña-Soto noted. In addition, contemporary increases in human connectivity and infectious disease emergence resemble the circumstances of past regional epidemics, such as those that followed the Spanish conquest of the region more than 400 years ago.

Wildlife Monitoring

Emerging infectious diseases of wildlife, such as those described in Box SA-2, arise from a disturbance in a delicate balance of host, pathogen, and environment. For this reason—and also because wild animals often serve as reservoir species for zoonotic threats to human health—they represent a critical target for infectious disease monitoring efforts of all sorts, including those that seek to track the influence of climate change, according to speaker William Karesh of the Wildlife Conservation Society (see Karesh in Chapter 3).

Wild animals also offer a number of advantages for disease monitoring programs, Karesh explained. Their comparatively short generation times reflect environmental changes more quickly than do humans; the great variety of wild species offers an equally wide range of life histories from which researchers can choose to model disease scenarios at different generational rates; and they provide sensitive sentinels for changes in the environments to which they are specifically adapted. Karesh observed that fish, bird, and marine mammal populations in South America declined dramatically during the El Niño event that occurred there in 1991-1992. In the case of Ebola hemorrhagic fever, Karesh observed that gorilla die-offs have preceded human outbreaks of Ebola virus by several weeks.

Highly pathogenic avian influenza can move between wild birds, domesticated poultry, and people, resulting in an increased risk of disease in cohabitated populations. Although wild birds cannot predict efficient human-to-human transfer of H5N1 avian influenza, the Global Avian Influenza Network for Surveillance (GAINS) gathers data in 23 developing countries—largely through the efforts of volunteers—on wild bird diseases; disseminates information to governments, international organizations, the private sector, and the general public; and helps to develop appropriate responses before outbreaks occur (GAINS, 2008).

Long-term monitoring of infectious diseases in wildlife also made possible the previously described model of plague dynamics and climate by Stenseth and coworkers (Stenseth et al., 2006). Under the Soviet regime, scientists began surveying rodent populations in Kazakhstan and testing them for plague in 1949; the practice continued through 1995, providing a wealth of data for statistical analysis. Such long-term studies are crucial to the prevention of human epidemics of plague and other zoonotic diseases that cannot be eradicated because they persist in a vast range of wildlife species, Stenseth said.

Remote Sensing

Satellite imagery is used to measure environmental variables over time, including land cover (a proxy for rainfall) and surface air temperature and humidity. Trends in these conditions, when compared with epidemiological data, reveal relationships between climate and infectious disease transmission and geographic distribution—for example, the previously discussed link between vegetation density and risk for epidemic RVF in humans (Linthicum et al., 1999).

In his workshop presentation, speaker Compton Tucker of the National Aeronautics and Space Administration (NASA) described the collection and analysis of remote sensing data and presented two examples of its use in examining links between climate and infectious disease (see Tucker in Chapter 3). The first involved a search for significant environmental factors common to sporadic outbreaks of Ebola hemorrhagic fever. Analyzing satellite data collected continuously since 1981, he and coworkers found an apparent "trigger event" that occurred prior to each outbreak: a period of drought, followed by a sudden return to very wet conditions (Pinzon et al., 2004). Today, satellite data from eastern equatorial Africa are screened routinely for this weather pattern; the results guide targeted testing for Ebola virus in local primates, which may provide an early warning of future outbreaks in humans.

Tucker and colleagues have also used satellite imagery to investigate an outbreak of RVF in Yemen, which seemed suspicious because of its proximity in location and time to a terrorist attack on a U.S. Navy ship, the *USS Cole*. Records of a satellite-derived index of photosynthetic capacity in local vegetation (another rainfall indicator) suggested that significant precipitation had fallen in the region prior to the outbreak, so the researchers concluded that it probably arose naturally.

Predictive Models

Several models for predicting the onset or prevalence of infectious diseases based on climatic indicators have been discussed in previous sections of this chapter (see also contributions to Chapter 2 by Chretien, Colwell, and Stenseth). Remote sensing of sea surface temperature and height, along with vegetation indices, are used to anticipate ENSO effects on a variety of diseases (Anyamba et al., 2006), to identify areas at risk for RVF outbreaks (Linthicum et al., 1999), and to provide early warning of epidemic cholera in Bangladesh (Gil et al., 2004; Speelmon et al., 2000). Stenseth suggested that statistical models capable of predicting past plague epidemics in central Asia (from tree-ring-derived measures of temperature and humidity) (Stenseth et al., 2006) could anticipate the influence of current climate conditions on population density and disease prevalence in rodent reservoirs of plague.

Climate-driven predictive models of mosquito-borne encephalitis transmission are also used by the State of California to estimate disease risk and inform public health interventions. Speaker William Reisen of the University of California, Davis, described the ongoing development of these models and their use in targeting surveillance to support integrated vector management (see Reisen and Barker in Chapter 3). The goal of these efforts, Reisen said, is to limit local population sizes of mosquitoes in order to prevent these vectors from amplifying

West Nile and related viruses to levels that put humans at risk for infection and to do so as efficiently as possible.

Reisen and coworkers found that although regional mosquito abundance was positively correlated with antecedent (January-February) temperatures and with precipitation levels, and inversely correlated with summer temperatures, climate measures alone explained only a fraction of the variability in mosquito populations. They discovered that climate variation produced very different responses (in both mosquito population size and viral amplification) in different environments. "One model doesn't fit all," Reisen concluded. "These relationships are very complex and have to be developed for specific biomes." Thus, the California Mosquito-borne Encephalitis Virus Surveillance and Response Plan currently incorporates measures of climate variation; however, the researchers are refining their models with the goal of using climate forecasts to provide earlier warning of transmission risk.

Challenges

As they explored the various routes by which climate variability and extreme weather events influence infectious disease emergence, workshop participants identified a range of challenges inherent to research on this topic. Many of these considerations were also discussed in *Under the Weather* (NRC, 2001), as noted in the Executive Summary of that report:

> There are many substantial research challenges associated with studying linkages among climate, ecosystems, and infectious diseases. For instance, climate-related impacts must be understood in the context of numerous other forces that drive infectious disease dynamics, such as rapid evolution of drug- and pesticide-resistant pathogens, swift global dissemination of microbes and vectors through expanding transportation networks, and deterioration of public health programs in some regions. Also, the ecology and transmission dynamics of different infectious diseases vary widely from one context to the next, thus making it difficult to draw general conclusions or compare results from individual studies. Finally, the highly interdisciplinary nature of this issue necessitates sustained collaboration among disciplines that normally share few underlying scientific principles and research methods, and among scientists that may have little understanding of the capabilities and limitations of each other's fields.

Consistent with these prior findings, workshop participants noted the following challenges intrinsic to the tasks of detecting, predicting, and mitigating infectious disease threats associated with climate change:

- Complexity of disease transmission patterns
- Global inequalities
- Varying space and time scales

- Establishing causation
- Lack of scientific certainty versus need for action

Complexity of Disease Transmission Patterns

Studies of influenza and dengue fever, as well as theoretical models, reveal that oscillations in disease incidence may occur even in the absence of seasonal changes in person-to-person transmissibility (see Burke in Chapter 1). Depending on parameters such as human birth rate, disease duration, and length of immunity, different epidemic viruses can display different intrinsic epidemic oscillatory frequencies. Burke observed that if such intrinsic epidemic frequency oscillations coincide with (resonate with) the annual seasonal changes in environmental conditions, then even very small annual environmentally-driven changes in transmissibility may, under some circumstances, drive very large seasonal changes in disease incidence (Dushoff et al., 2004). The impact of a given increment of change in climate upon the future transmission of a given disease cannot be determined without understanding the particular relationship between two oscillating patterns—the intrinsic incidence oscillation and the seasonally-driven oscillation. Resonance can raise the magnitude of seasonally epidemic disease. "Every effort should be made to isolate and thereby understand these component subsystems if we are to explain and predict epidemic patterns," Burke concluded.

Global Inequalities

The effects of climate change are likely to be far greater in the tropics, where the majority of the world's poorest people live, than in the wealthier temperate zones. As Haines observed, in the areas where links between climate and disease may best be studied, people are least able to investigate them. Similarly, a predictive model that highlights regions at higher risk for infectious disease emergence (see Figure SA-15) suggests that such "hot spots" are concentrated in equatorial developing countries, where opportunities for monitoring and research are severely limited (Jones et al., 2008). The model's developers conclude that "[t]he global effort for [emerging infectious disease] surveillance and investigation is poorly allocated, with the majority of our scientific resources focused on places [such as North America, Europe, and Australia] from where the next important emerging pathogen is least likely to originate" (Jones et al., 2008). They argue, instead, that the resources for emerging infectious disease surveillance should target hot spots in tropical Africa, Latin America, and Asia, and populations at greatest risk for infection, in order to detect outbreaks of emergent diseases at the earliest possible stage.

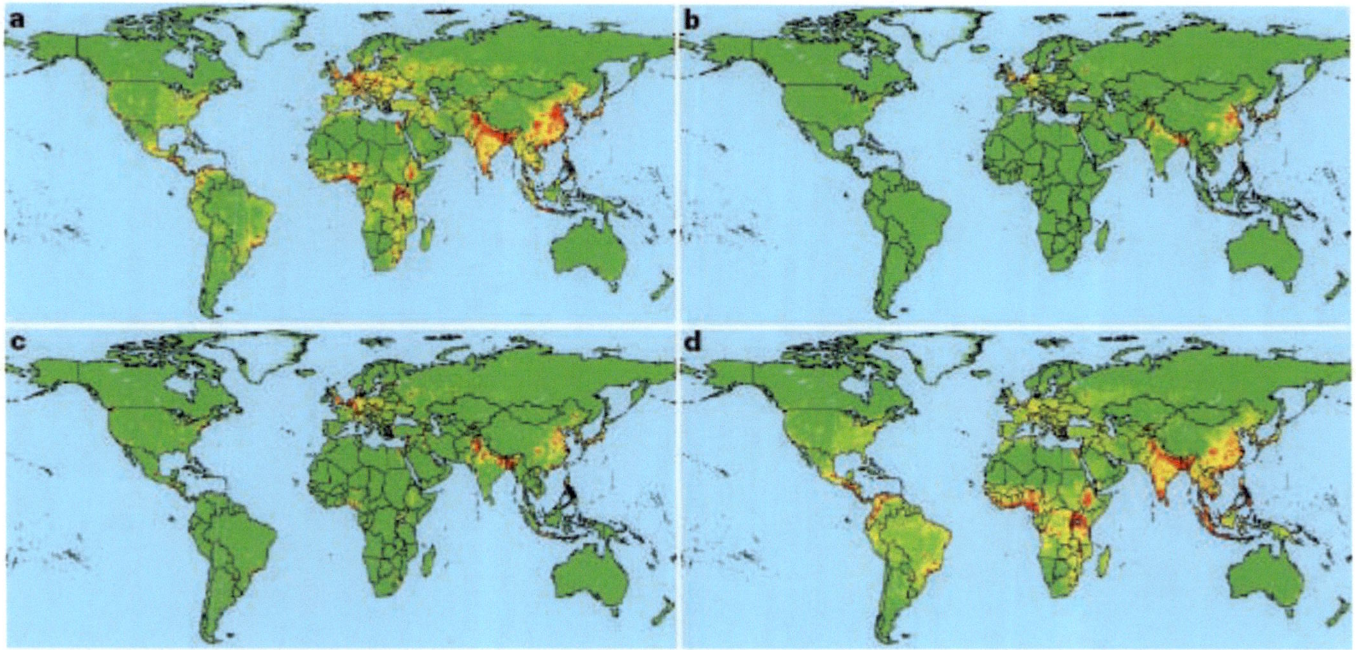

FIGURE SA-15 Global distribution of relative risk of an emerging infectious disease (EID) event. Maps depict predicted hot spots for EID events caused by (a) zoonotic pathogens from wildlife; (b) zoonotic pathogens from non-wildlife; (c) drug-resistant pathogens; and (d) vector-borne pathogens.
SOURCE: Reprinted with permission from Macmillian Publishers Ltd: *Nature*, Jones et al. (2008).

Varying Space and Time Scales

The influence of climate on infectious diseases is often highly dependent on local-scale parameters. It is sometimes impossible to extrapolate these relationships meaningfully to broader spatial scales; likewise, examples of seasonal or interannual climatic variability, such as ENSO, may not always provide a useful analog for the impacts of long-term trends in climate. Ecological responses on the timescale of an El Niño event, for example, may differ significantly from ecological responses and social adaptations that occur over the course of long-term climate change. Conversely, it is difficult to predict the influence of long-term climate change on regional patterns of climate variability, and even the effects of regional climate may be modified by landcover features.

Establishing Causation

In order to establish that a pattern of climatic variability or an extreme weather event *caused* a change in the transmission or geographic range of a particular infectious disease, several requirements must be met. For example, Haines observed, to infer a causal relationship between an El Niño cycle and a given health outcome, three elements are necessary: climate data, preferably local; a plausible biological relationship between a particular disease outcome and climate data; and a relatively long time series (e.g., decades) that can be analyzed statistically and adjusted for potentially confounding relationships.

Lack of Scientific Certainty Versus Need for Action

Because, as Haines observed, health lies at the end of a long chain of causality, several participants warned that by waiting to act on the potential adverse health impacts posed by climate change until their inevitability is scientifically confirmed, the world will lose the opportunity to prevent, and possibly to mitigate, these threats. "There is going to be a great deal of continuing debate about the precise magnitude and effects of climate change on health," Haines conceded, "but I think in view of the potential for very major impacts, these uncertainties don't justify inaction. We certainly need to adapt more effectively to a changing climate." Indeed, as Haines noted, it would be unethical for scientists to observe the emergence of infectious diseases, whether or not this trend was caused by climate change, and not intervene.

Implications for Public Health Policy and Global Security

Several workshop discussions raised the urgent question of how to act on what is known—or even suspected—about the potential health consequences of climate change. At the same time, participants supported the continuation and expansion of research on the significance of climate to the health of organisms and ecosystems. The need for action, as well as for knowledge, underscored sessions devoted to scientific observations and technical approaches and was addressed directly in presentations that focused on issues of public health policy and of national and global security.

Research Framework

Haines described three basic tasks for researchers studying the potential health impacts of climate change: (1) to examine past associations between climate variability and health; (2) to determine climate's role in present-day trends in disease transmission and geographic range; and (3) to create predictive models of future disease that account for a changing climate, as well as for other influential factors.

A fourth task was proposed by speaker Douglas MacPherson of McMaster University and Migration Health Consultants, Incorporated: to recognize and address the contribution of human behavior to global climate change and its further effects on infectious disease emergence. In his coauthored contribution to Chapter 4, MacPherson describes the complex, two-way association between climate change and human mobility—"a determinant of health that is directly linked to globalization of microbial disease threats and risks."

Connecting Climate and Migration

The immense contribution of human mobility and migration to infectious disease emergence is often illustrated in terms of annual global statistics such as these presented by MacPherson:

- 802 million international arrivals (2006)
- 200 million permanent residents outside of their country of birth
- 32.9 million "persons of concern," as defined by the office of the United Nations High Commissioner for Refugees (UNHCR)[11]
- 50,000 smuggled and trafficked persons
- 15,000 unaccompanied or separated children

Human mobility also includes the movement of animals, plants, and microbes, as well as trade in goods that include living organisms and their byproducts—topics that will be taken up in a future workshop of the Forum on Microbial Threats in calendar year 2008.

[11]The UNHCR collectively refers to people who have been forcibly uprooted from their homes as "persons of concern." They include asylum-seekers, refugees, stateless persons, the internally displaced, and returnees (UNHCR, 2006).

Interactions between human mobility, climate change, and infectious disease dynamics take several different forms, MacPherson explained. Extreme weather events create opportunities for infectious disease outbreaks; these conditions may force people and animals (including disease vectors) to migrate to different ecosystems, where they may both encounter and introduce novel pathogens. Slowly evolving impacts of climate change (such as sea level elevation, reduced freshwater availability, or increased average temperatures) may force or attract human population movements. Patterns of temporary migration (e.g., travel), and thereby disease transmission, are also likely to shift with climate change.

Calling the policy implications of these complex associations "daunting," MacPherson observed that in many cases, preventing the global spread of an infectious disease is not feasible. In his view, health officials should focus their energies on anticipating and mitigating problems that will doubtless arrive, rather than waiting to react.

Addressing Climate's Contribution to the Global Disease Burden

Certain high-profile outbreaks in developed areas of the world—such as shellfish poisoning (*Vibrio parahaemolyticus*) in Alaska, WNV in the United States, and bluetongue in northern Europe—have raised attention regarding the potential effects of climate change on infectious disease emergence and spread. However, as speaker Diarmid Campbell-Lendrum of WHO noted, climate change is expected to exact its most profound toll on the health of the world's poor, through increased rates of malaria, diarrheal diseases, and malnutrition. "We shouldn't lose sight of the fact that these [common and preventable] killers are also highly sensitive to climatic conditions," he cautioned.

Separating the effects of climate variability and change from the context of other determinants, or assessing the influence of climate versus other factors as a mutually exclusive "either/or" debate, is unproductive. Instead, considering climate as one important determinant of health risks, mediated by other contextual determinants, is more likely to lead to sound health policy, according to Campbell-Lendrum. This is the approach that WHO has taken since 1990 with its Programme on Climate Change, as evidenced in various projects described in Chapter 4. Although initially involved in the issue of climate change from the perspective of risk assessment, WHO has recently assumed an operational role, organizing activities that include advocating for climate change as a health security issue, generating evidence for action, and monitoring and evaluating the health effects of climate change.

The most effective available protective measures against the adverse health effects of climate change are basic public health interventions, Campbell-Lendrum said. "If we did a better job of controlling dengue now, or malaria now," he said, "we would have less to worry about from climate change."

Considerations for National and Global Security

At its summit in March 2008, a paper presented to the European Union included a grim catalog of threats to international security posed by climate change: conflicts over water, energy, and other increasingly scarce resources; loss of infrastructure and territory; border disputes; environmentally-induced migration; and political tension at all levels of governance (European Commission, 2008). Acknowledging that climate change may pose new challenges for national security, the National Intelligence Council (NIC) of the U.S. government is in the process of preparing a National Intelligence Assessment (NIA) forecasting these potential impacts over the next two decades.

In his workshop presentation, Major General Richard Engel (U.S. Air Force, retired), NIC deputy national intelligence officer for science and technology, described work in progress on this NIA, which is intended to inform decision making at the highest levels of the U.S. government (see Chapter 4). The NIC has chosen to evaluate the potential impacts of climate change on four essential components of national power: geopolitical power, military power, economic power, and social cohesion. "When we talk about climate change impacting the United States, we talk about it impacting one of those four classical elements of national power," Engel explained. Inherent uncertainties in predicting the course of climate change prompted the NIC to consider a "system vulnerability approach" for this assessment, which identifies existing internal vulnerability of states or regions of interest to U.S. security, then examines how the added stress of climate change could affect these states or regions (see Chapter 4). To date, the NIC has received considerable nongovernmental expert opinion on this issue, from which Engel crafted his remarks; however, the NIC and the intelligence community have yet to complete their own analysis and interpretation of these contributions.

In addition to this focused analysis of potential challenges, the NIC is contemplating the international political response to climate change. Engel noted that climate change has the potential to create geopolitical divisions, several of which have already been reported in the open press. "Developed countries want the developing countries to participate so they don't bear the full burden [of the cost of addressing climate change], and the developing countries want the developed countries to pay for it," he observed. Experts have reported to the NIC that a north-south division even exists within Europe, resulting from the varied effects of climate change along this axis. The differential effects of climate change in regions of Asia—where some areas may suffer droughts while others flood—may also prove a source of tension, particularly where water is concerned. The NIC's expert consultants agree that actions taken by the United States will profoundly influence the fate of a global consensus on climate change.

Needs and Opportunities

Two days of workshop discussion bore out Burke's claim, made during his keynote address, of the continuing relevance of the recommendations offered in *Under the Weather* (NRC, 2001), as summarized in Box SA-3.

Participants placed particular emphasis on the following considerations for studying the influence of climate change on infectious disease emergence:

- Developing a greater understanding of the interaction of climate with other major factors in disease emergence and resurgence, such as the globalization of travel and trade, population growth, urbanization, land-use patterns, and habitat destruction;
- Establishing long-term monitoring programs to simultaneously track climate and infectious disease dynamics, and optimizing instruments (many of which were designed for other purposes) for use in such programs;
- Devising metrics to relate changes in the physical environment to ecological and epidemiological trends and to evaluate potential adaptation and mitigation measures; and
- Continuing the development and refinement of predictive models of climate and infectious disease as the basis for early warning and public health response systems, and involving of stakeholders in the operation of such systems.

Many discussants urged that immediate action be taken to address the health effects of climate change—and indeed climate change itself—before irreversible harm is done to the Earth and its inhabitants. Several participants advocated the implementation of "win-win" adaptation strategies: improving access to clean water and sanitation; increasing the availability and uptake of immunization; and strengthening health systems. Many of the workshop's participants believed that these actions could produce near-term health benefits and might also improve the world's ability to withstand the potential stress of climate change.

While the scope of the workshop necessarily limited discussion of the larger issue of climate change itself and the potential to address it, this topic was the proverbial "elephant in the room." Despite the apparent gravity of the threats posed by climate change, some were able to view it as an opportunity, and one participant has characterized the demand for clean energy as "an engine of growth for the twenty-first century" (Epstein, 2005, 2007). Furthermore, Epstein said, by taking the "right" approach to addressing the consequences of climate change, "we can get a strong public health sector, and that will be good for security, good for the economy, and we certainly hope that it will stabilize the climate."

BOX SA-3
***Under the Weather* Recommendations for Future Research and Surveillance**

- Research on the linkages between climate and infectious diseases must be strengthened.
- Further development of disease transmission models is needed to assess the risks posed by climatic and ecological changes.
- Epidemiological surveillance programs should be strengthened.
- Observational, experimental, and modeling activities are all highly interdependent and must progress in a coordinated fashion.
- Research on climate and infectious disease linkages inherently requires interdisciplinary collaborations.

SOURCE: NRC (2001).

APPENDIX SA-1
A BRIEF HISTORY OF CLIMATE CHANGE

Long-Term Trends

As illustrated in Figure SA-16, human history spans several periods of climatic upheaval (WHO et al., 2003). However, the warmth of the last half-century is unusual; indeed, evidence suggests that the last time the polar regions remained significantly warmer than they are today—approximately 125,000 years ago—reductions in polar ice volume caused global sea levels to rise by 4 to 6 meters.

Recent Changes

Over the last century, global average temperatures and sea levels have risen significantly, while snow cover in the Northern Hemisphere has declined (see Figure SA-17; National Geographic Society, 2007). The total temperature increase from 1850-1899 to 2001-2005, estimated at 0.76°C (0.57°C to 0.95°C), occurred during a warming trend that appears to be gaining momentum. The rate of warming for the last 50 years was double that during the previous half-century, and 11 of the last 12 years (1995-2006) rank among the 12 warmest years in the instrumental record of global surface temperature (since 1850). Over the last 50 years, cold days, cold nights, and frost have become less frequent, while hot days, hot nights, and heat waves have become more frequent.

Measurements conducted since 1961 show that the average temperature of the global ocean has increased to depths of at least 3,000 meters and that oceans have absorbed more than 80 percent of the heat added to the climate system.

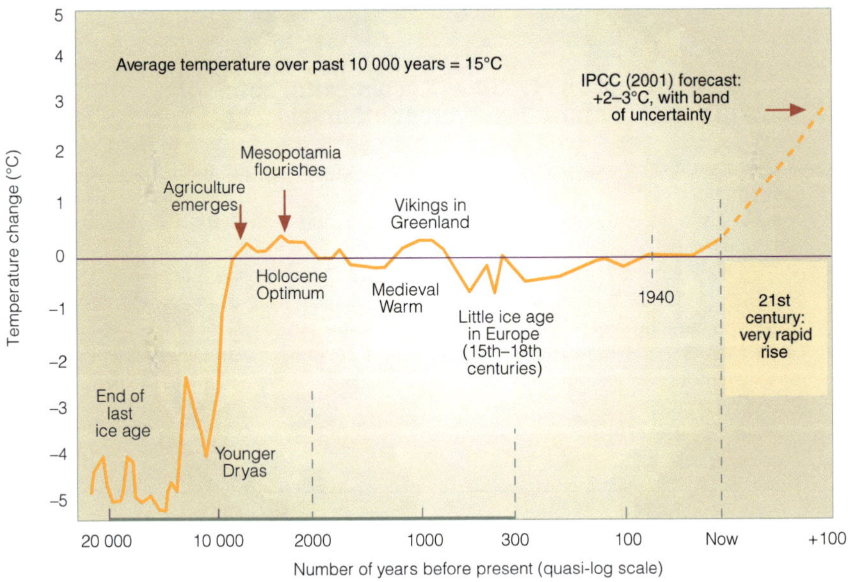

FIGURE SA-16 Variation in Earth's average surface temperature over the past 20,000 years.
SOURCE: Reprinted from WHO et al. (2003) with permission from the World Health Organization. Copyright 2003.

Such warming causes seawater to expand, contributing to sea level rise, as have widespread decreases in glaciers and ice caps. Global average sea level rose at an average rate of 1.8 mm per year between 1961 and 2003 and at a rate of about 3.1 mm per year between 1993 and 2003. Current estimates indicate that sea levels rose 0.17 m over the course of the twentieth century (see Figure SA-18; IPCC, 2007).

Present Effects and Future Projections

A warmer global climate system accelerates the hydrological cycle, increasing the likelihood of extreme weather phenomena such as droughts, heavy precipitation, heat waves, hurricanes, typhoons, and cyclones (see Figure SA-19; National Geographic Society, 2007). More intense and longer droughts, which have been observed over wider areas since the 1970s and particularly in the tropics and subtropics, have been associated with higher global temperatures, but also

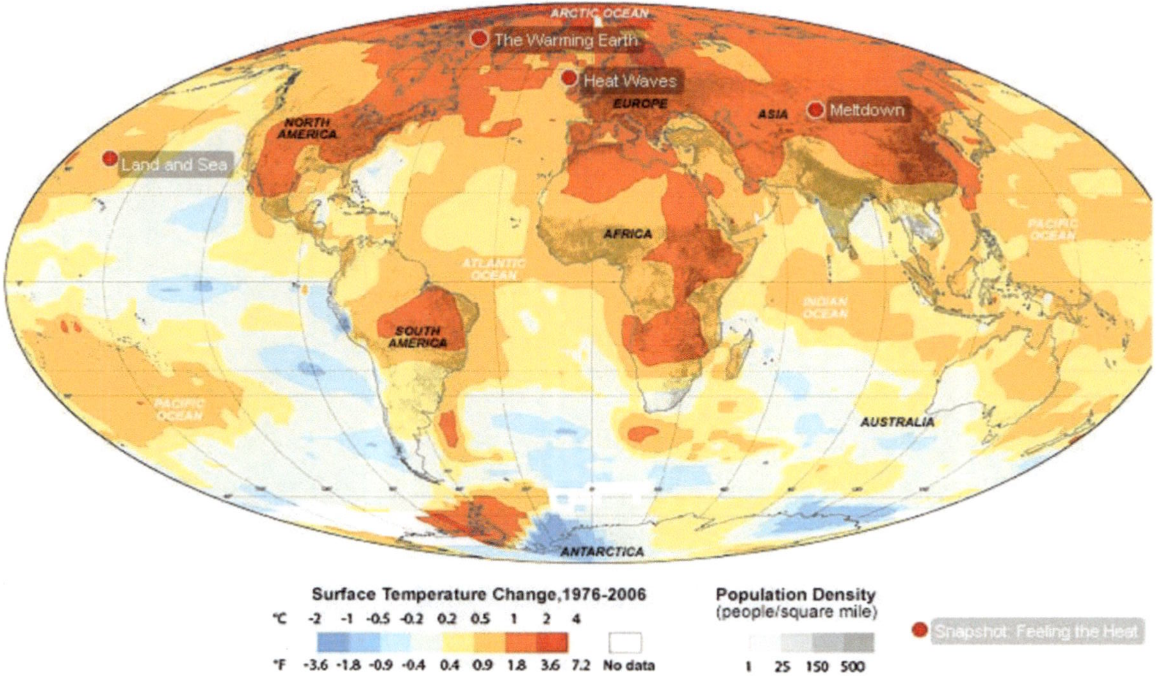

FIGURE SA-17 The Arctic is experiencing the fastest rate of warming as its reflective covering of ice and snow shrinks. In the midlatitudes, there are now fewer cold nights; heat waves are more common. The Indian Ocean and the western Pacific Ocean are warmer than at any point in the last 11,500 years. Against the trend: Pockets of the oceans are cooled by deepwater upwellings. Ozone loss over the South Pole may have cooled parts of Antarctica.

SOURCE: Reprinted from National Geographic Society (2007) with permission from the National Geographic Society.

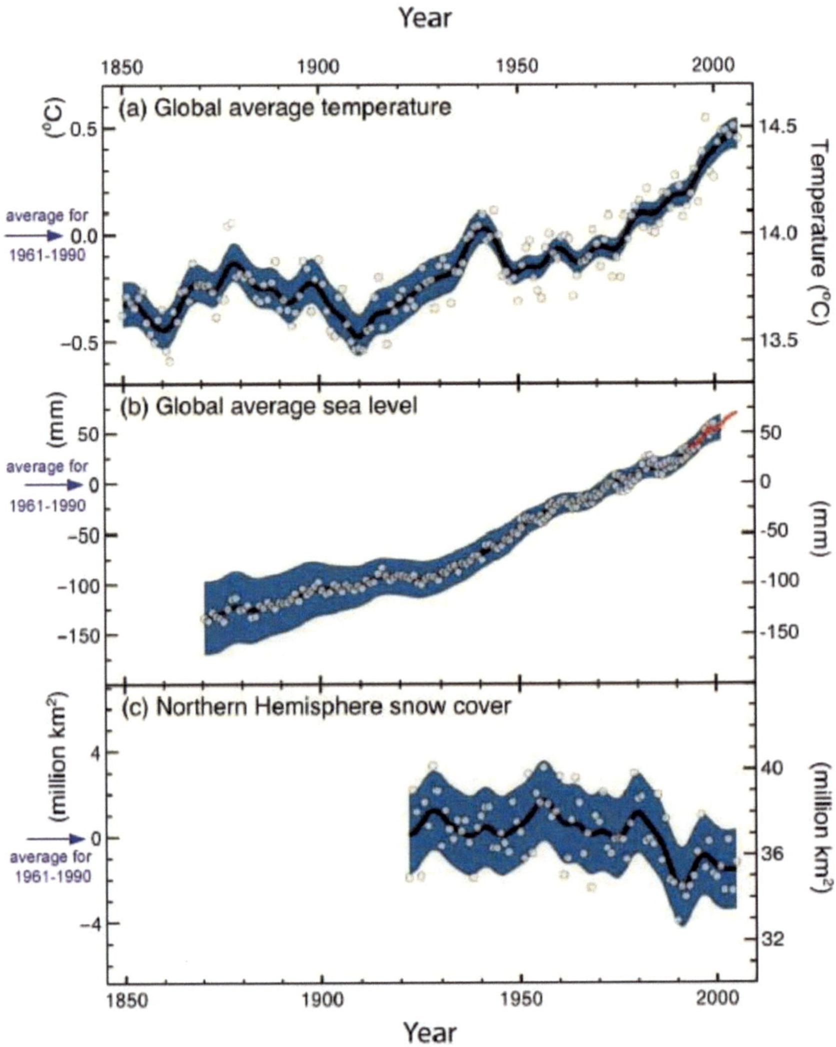

FIGURE SA-18 Observed changes in (A) global average surface temperature; (B) global average sea level rise from tide gauge (blue) and satellite (red) data; and (C) Northern Hemisphere snow cover for March-April. All changes are relative to corresponding averages for the period 1961-1990. Smoothed curves represent decadal averaged values while circles show yearly values. The shaded areas are the uncertainty intervals estimated from a comprehensive analysis of known uncertainties (A and B) and from the time series (C). SOURCE: Figure SPM.3 in IPCC (2007).

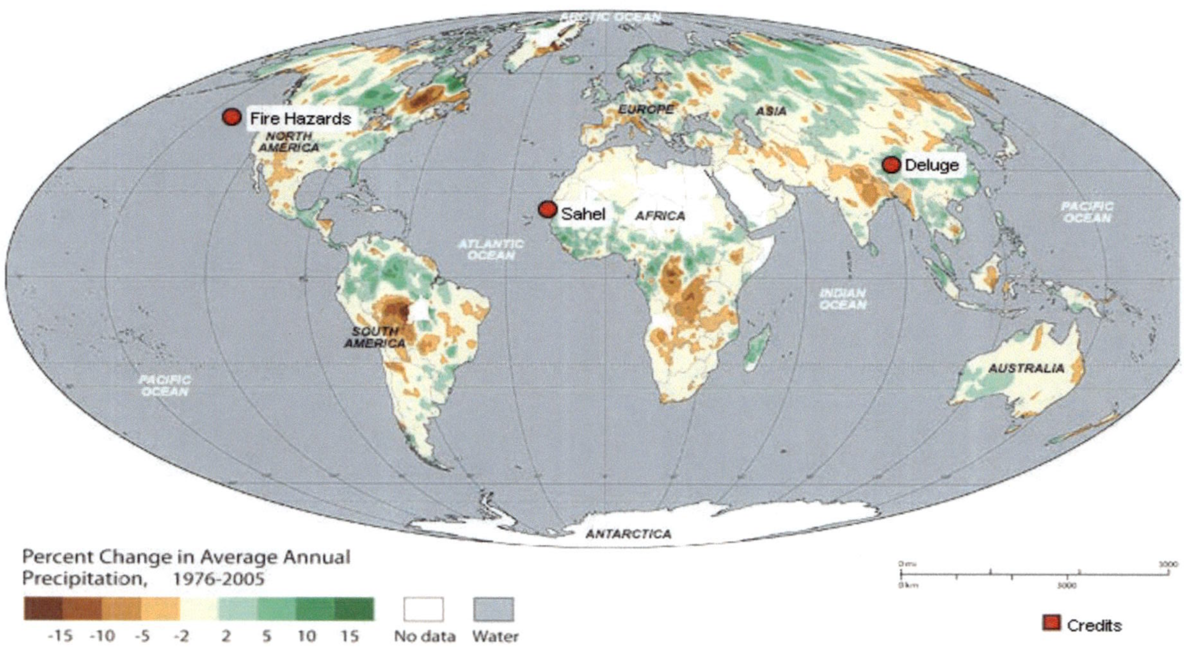

FIGURE SA-19 Drought is seizing more territory in the wake of mounting temperatures. Drying trends in the last 30 years are evident in the rain forests of Africa and South America and in already dry regions such as southern Europe and western North America. In wet areas, precipitation increasingly arrives in heavy downpours, raising the risk of flooding. Winter rain is replacing snow, an ominous development for hundreds of millions of people who depend on spring snowmelt for their water supply.

SOURCE: Reprinted from National Geographic Society (2007) with permission from the National Geographic Society.

with changes in wind patterns and decreases in snowpack and snow cover. Periods of heavy precipitation have occurred with greater frequency over most land areas in parallel with increases in atmospheric water vapor.

Over the next two decades, the Earth is expected to warm by an additional 0.2°C. Even if the concentrations of all greenhouse gases and aerosols (both of which cause the atmosphere to trap heat) could be kept the same levels as in 2000, warming would still be expected to proceed at about half the present rate. Continued greenhouse gas emissions at or above current rates are very likely to induce changes in the global climate system during the twenty-first century of even greater magnitude than has been observed during the late twentieth century.

REFERENCES

Summary and Assessment References

Anderson, P. K., A. A. Cunningham, N. G. Patel, F. J. Morales, P. R. Epstein, and P. Daszak. 2004. Emerging infectious diseases of plants: pathogen pollution, climate change and agrotechnology drivers. *Trends in Ecology and Evolution* 19(10):535-544.

Anyamba, A., J. P. Chretien, J. Small, C. J. Tucker, and K. J. Linthicum. 2006. Developing global climate anomalies suggest potential disease risks for 2006-2007. *International Journal of Health Geographics* 5:60.

Barnett, T. P., D. W. Pierce, K. M. AchutaRao, P. J. Gleckler, B. D. Santer, J. M. Gregory, and W. M. Washington. 2005. Penetration of human-induced warming into the world's oceans. *Science* 309(5732):284-287.

BBC News. 2007. The rise of bluetongue disease. *BBC News*, http://news.bbc.co.uk/2/hi/uk_news/7009288.stm (accessed June 20, 2008).

Borgerson, S. G. 2008. Arctic meltdown: the economic and security implications of global warming. *Foreign Affairs* 87(2):63-77.

Calisher, C. H., J. J. Root, J. N. Mills, J. E. Rowe, S. A. Reeder, E. S. Jentes, K. Wagoner, and B. J. Beaty. 2005. Epizootiology of Sin Nombre and El Moro Canyon hantaviruses, southeastern Colorado, 1995-2000. *Journal of Wildlife Diseases* 41(1):1-11.

CDC (Centers for Disease Control and Prevention). 2004. *Giardiasis fact sheet*, http://www.cdc.gov/ncidod/dpd/parasites/giardiasis/factsht_giardia.htm#child_diagnosed (accessed March 17, 2008).

———. 2005. *Cholera: frequently asked questions*, http://www.cdc.gov/ncidod/dbmd/diseaseinfo/cholera_g.htm (accessed March 17, 2008).

———. 2007a. *Cryptosporidium infection fact sheet*, http://www.cdc.gov/ncidod/dpd/parasites/cryptosporidiosis/factsht_cryptosporidiosis.htm (accessed March 17, 2008).

———. 2007b. Rift Valley fever outbreak—Kenya, November 2006-January 2007. *Morbidity and Mortality Weekly Report* 56(4):73-76.

Center for Health and the Global Environment. 2005. *Climate change futures: health, ecological and economic dimensions*. Cambridge, MA: Harvard Medical School.

CGIAR (Consultative Group for International Agricultural Research). 2008. http://www.cgiar.org/ (accessed March 16, 2008).

Chretien, J. P., A. Anyamba, S. A. Bedno, R. F. Breiman, R. Sang, K. Sergon, A. M. Powers, C. O. Onyango, J. Small, C. J. Tucker, and K. J. Linthicum. 2007. Drought-associated chikungunya emergence along coastal East Africa. *American Journal of Tropical Medicine and Hygiene* 76(3):405-407.

Colwell, R. R. 2004. Infectious disease and environment: cholera as a paradigm for waterborne disease. *International Microbiology* 7(4):285-289.

Curtis, V., and S. Cairncross. 2003. Effect of washing hands with soap on diarrhoea risk in the community: a systematic review. *Lancet Infectious Diseases* 3(5):275-281.

Daszak, P., A. A. Cunningham, and A. D. Hyatt. 2000. Emerging infectious diseases of wildlife—threats to biodiversity and human health. *Science* 287(5452):443-449.

Dushoff, J., J. B. Plotkin, S. A. Levin, and D. J. Earn. 2004. Dynamical resonance can account for seasonality of influenza epidemics. *Proceedings of the National Academy of Sciences* 101(48):16915-16916.

Enserink, M. 2008. Animal disease: exotic disease of farm animals tests Europe's responses. *Science* 319(5864):710-711.

Epstein, P. R. 2005. Climate change and human health. *New England Journal of Medicine* 353(14):1433-1436.

Epstein, P. R. 2007. Climate change: healthy solutions (Guest Editorial). *Environmental Health Perspectives* 115(4):A180.

Epstein, P. R., and J. J. McCarthy. 2004. Assessing climate stability. *Bulletin of the American Meteorological Society* 85(12):1863-1870.

Epstein, P. R., H. F. Diaz, S. Elias, G. Grabherr, N. E. Graham, W. J. M. Martens, E. Mosley-Thompson, and E. J. Susskind. 1998. Biological and physical signs of climate change: focus on mosquito-borne disease. *Bulletin of the American Meteorological Society* 78:409-417.

European Commission. 2008. *Climate change and international security*. Paper presented at European Union Summit, Brussels.

Flynn, N. M., P. D. Hoeprich, M. M. Kawachi, K. K. Lee, R. M. Lawrence, E. Goldstein, G. W. Jordan, R. S. Kundargi, and G. A. Wong. 1979. *New England Journal of Medicine* 301(7):358-361.

Frederick, K. D., and P. H. Gleick. 1999. *Water and global climate change: potential impacts on U.S. water resources*. Arlington, VA: Pew Center on Global Climate Change.

GAINS (Global Avian Influenza Network for Surveillance). 2008. *Global avian influenza network for surveillance*, http://www.gains.org/ (accessed June 20, 2008).

Garrett, K. A., S. P. Dendy, E. E. Frank, M. N. Rouse, and S. E. Travers. 2006. Climate change effects on plant disease: genomes to ecosystems. *Annual Review of Phytopathology* 44:489-509.

Garrison, V. H., E. A. Shinn, W. T. Foreman, D. W. Griffin, C. W. Holmes, C. A. Kellogg, M. S. Majewski, L. L. Richardson, K. B. Ritchie, and G. W. Smith. 2003. African and Asian dust: from desert soils to coral reefs. *BioScience* 53(5):469-480.

Gil, A. I., V. R. Louis, I. N. Rivera, E. Lipp, A. Huq, C. F. Lanata, D. N. Taylor, E. Russek-Cohen, N. Choopun, R. B. Sack, and R. R. Colwell. 2004. Occurrence and distribution of *Vibrio cholerae* in the coastal environment of Peru. *Environmental Microbiology* 6(7):699-706.

Gillett, N. P., and D. W. J. Thompson. 2003. Simulation of recent Southern Hemisphere climate change. *Science* 302(5643):273-275.

Githeko, A. K., S. W. Lindsay, U. E. Confalonieri, and J. A. Patz. 2000. Climate change and vector-borne diseases: a regional analysis. *Bulletin of the World Health Organization* 78(9):1136-1147.

Gleick, P. H. 2004. *Water conflict chronology*. Oakland, CA: The Pacific Institute.

Groisman, P. Y., R. W. Knight, T. R. Karl, D. R. Easterling, B. Sun, and J. H. Lawrimore. 2004. Contemporary changes of the hydrological cycle over the contiguous United States: trends derived from *in situ* observations. *Journal of Hydrometeorology* 5:64-85.

Haines, A., and J. A. Patz. 2004. Health effects of climate change. *Journal of the American Medical Association* 291(1):99-103.

Haines, A., R. S. Kovats, D. Campbell-Lendrum, and C. Corvalan. 2006. Climate change and human health: impacts, vulnerability, and public health. *Lancet* 367(9528):2101-2109.

Harrus, S., and G. Baneth. 2005. Drivers for the emergence and reemergence of vector-borne protozoal and bacterial diseases. *International Journal of Parasitology* 35(11-12):1309-1318.

Harvell, C. D., K. Kim, J. M. Burkholder, R. R. Colwell, P. R. Epstein, D. J. Grimes, E. E. Hofmann, E. K. Lipp, A. D. Osterhaus, R. M. Overstreet, J. W. Porter, G. W. Smith, and G. R. Vasta. 1999. Emerging marine diseases—climate links and anthropogenic factors. *Science* 285(5433):1505-1510.

Harvell, C. D., C. E. Mitchell, J. R. Ward, S. Altizer, A. P. Dobson, R. S. Ostfeld, and M. D. Samuel. 2002. Climate warming and disease risks for terrestrial and marine biota. *Science* 296(5576):2158-2162.

Hennessy, T., T. Ritter, R. C. Holman, D. L. Bruden, K. L. Yorita, L. Bulkow, J. E. Cheek, R. J. Singleton, and J. Smith. 2008. Relationship between in-home water service and the risk of respiratory tract, skin, and gastrointestinal tract infections among Alaska Natives. *American Journal of Public Health* 98(5):1-8.

Holt, D. W., C. Hanns, T. O'Hara, K. Burek, and R. Frantz. 2005. New distribution records of *Echinococcus multilocularis* in the brown lemming from Barrow, Alaska, USA. *Journal of Wildlife Diseases* 41(1):257-259.

Hoyos, C. D., P. A. Agudelo, P. J. Webster, and J. A. Curry. 2006. Deconvolution of the factors contributing to the increase in global hurricane intensity. *Science* 312(5770):94-97.

IOM (Institute of Medicine). 1992. *Emerging infections: microbial threats to health in the United States*. Washington, DC: National Academy Press.

———. 2003. *Microbial threats to health: emergence, detection, and response*. Washington, DC: The National Academies Press.

______. 2008. *Vector-borne diseases: understanding the environmental, human health and ecological connections*. Washington, DC: The National Academies Press.

IPCC (Intergovernmental Panel on Climate Change). 2001. *Climate Change 2001: the scientific basis*. Working Group I contribution to the third assessment report of the IPCC. New York: Cambridge University Press.

———. 2007a. *Climate change 2007: climate change impacts, adaptation and vulnerability*. Working Group II contribution to the fourth assessment report of the IPCC. Geneva, Switzerland: IPCC.

———. 2007b. *Climate change 2007: the physical science basis*. Working Group I contribution to the fourth assessment report of the IPCC. Geneva, Switzerland: IPCC.

Jones, K. E., N. G. Patel, M. A. Levy, A. Storeygard, D. Balk, J. L. Gittleman, and P. Daszak. 2008. Global trends in emerging infectious diseases. *Nature* 451(7181):990-993.

Karl, T. R., and K. E. Trenberth. 2003. Modern global climate change. *Science* 302(5651): 1719-1723.

Kovats, R. S., and A. Haines. 2005. Global climate change and health: recent findings and future steps. *Canadian Medical Association Journal* 172(4):501-502.

Kovats, R. S., M. J. Bouma, S. Hajat, E. Worrall, and A. Haines. 2003. El Niño and health. *Lancet* 362(9394):1481-1489.

Kovats, R. S., S. J. Edwards, S. Hajat, B. G. Armstrong, K. L. Ebi, and B. Menne. 2004. The effect of temperature on food poisoning: a time-series analysis of salmonellosis in ten European countries. *Epidemiology and Infection* 132(3):443-453.

Kovats, R. S., S. J. Edwards, D. Charron, J. Cowden, R. M. D'Souza, K. L. Ebi, C. Gauci, P. Gerner-Smidt, S. Hajat, S. Hales, G. Hernandez Pezzi, B. Kriz, K. Kutsar, P. McKeown, K. Mellou, B. Menne, S. O'Brien, W. van Pelt, and H. Schmid. 2005. Climate variability and campylobacter infection: an international study. *International Journal of Biometeorology* 49(4):207-214.

Leclair, D., J. W. Austin, J. Faber, B. Cadieux, and B. Blanchfield. 2004. *Toxicity of aged seal meat challenged with Clostridium botulinum type E. Federal Food Safety and Nutrition Research Meeting*. Ottawa, Ontario, October 4-5.

Levin, R. B., P. R. Epstein, T. E. Ford, W. Harrington, E. Olson, and E. G. Reichard. 2002. Drinking water challenges in the twenty-first century. *Environmental Health Perspectives* 110:43-52.

Levitus, S., J. I. Antonov, and T. Boyer. 2005. Warming of the world ocean, 1955-2003. *Geophysical Research Letters* 32:LO2604.

Linthicum, K. J., A. Anyamba, C. J. Tucker, P. W. Kelley, M. F. Myers, and C. J. Peters. 1999. Climate and satellite indicators to forecast Rift Valley fever epidemics in Kenya. *Science* 285(5426):397-400.

Lipp, E. K., I. N. Rivera, A. I. Gil, E. M. Espeland, N. Choopun, V. R. Louis, E. Russek-Cohen, A. Huq, and R. R. Colwell. 2003. Direct detection of *Vibrio cholerae* and *ctxA* in Peruvian coastal water and plankton by PCR. *Applied Environmental Microbiology* 69(6):3676-3680.

Lloyd, S. J., R. S. Kovats, and B. G. Armstrong. 2007. Global diarrhea morbidity, weather and climate. *Climate Research* 34:119-127.

Louis, V. R., I. A. Gillespie, S. J. O'Brien, E. Russek-Cohen, A. D. Pearson, and R. R. Colwell. 2005. Temperature-driven campylobacter seasonality in England and Wales. *Applied Environmental Microbiology* 71(1):85-92.

McKie, R., and J. Revill. 2007 (September 23). The first case of bluetongue disease found at U.K. farm. *Guardian*, http://www.guardian.co.uk/uk/2007/sep/23/ruralaffairs.robinmckie (accessed June 20, 2008).

McLaughlin, J. B., A. DePaola, C. A. Bopp, K. A. Martinek, N. P. Napolilli, C. G. Allison, S. L. Murray, E. C. Thompson, M. M. Bird, and J. P. Middaugh. 2005. Outbreak of *Vibrio parahaemolyticus* gastroenteritis associated with Alaskan oysters. *New England Journal of Medicine* 353(14):1463-1470.

McMichael, A. J., R. E. Woodruff, and S. Hales. 2006. Climate change and human health: present and future risks. *Lancet* 367(9513):859-869.

Milly, P. C. D., K. A. Dunne, and A. V. Vecchia. 2005. Global pattern of trends in streamflow and water availability in a changing climate. *Nature* 438(7066):347-350.

Molesworth, A. M., L. E. Cuevas, S. J. Connor, A. P. Morse, and M. C. Thomson. 2003. Environmental risk and meningitis epidemics in Africa. *Emerging Infectious Diseases* 9(10):1287-1293.

NASA (National Aeronautics and Space Administration) Goddard Space Flight Center. 2000. *Using satellites to track Rift Valley fever*, http://eospso.gsfc.nasa.gov/ftp_docs/lithographs/RVF.pdf (accessed April 16, 2008).

National Geographic Society. 2007 (October). The warming Earth and precipitation fallout figures. *National Geographic* (supplement).

NRC (National Research Council). 2001. *Under the weather: climate, ecosystems, and infectious disease*. Washington, DC: National Academy Press.

Ogden, N. H., A. Maarouf, I. K. Barker, M. Bigras-Poulin, L. R. Lindsay, M. G. Morshed, C. J. O'Callaghan, F. Ramay, D. Waltner-Toews, and D. F. Charron. 2005. Climate change and the potential for range expansion by the Lyme disease vector *Ixodes scapularis* in Canada. *International Journal for Parasitology* 36(1):63-70.

Osburn, B. I. 2008. Vector-borne zoonotic diseases and their ecological and economic implications: bluetongue disease in Europe. In: Institute of Medicine. *Vector-borne diseases: understanding the environmental, human health, and ecological connections*. Washington, DC: The National Academies Press.

Parkinson, A. J., and J. C. Butler. 2005. Potential impacts of climate change on infectious diseases in the arctic. *International Journal of Circumpolar Health* 64(5):478-486.

Pascual, M., J. A. Ahumada, L. F. Chaves, X. Rodo, and M. Bouma. 2006. Malaria resurgence in the East African highlands: temperature trends revisited. *Proceedings of the National Academy of Sciences* 103(15):5829-5834.

Patz, J. A., P. R. Epstein, T. A. Burk, and J. M. Balbus. 1996. Global climate change and emerging infectious diseases. *Journal of the American Medical Association* 275(3):217-223.

Patz, J. A., M. A. McGeehin, S. M. Bernard, K. L. Ebi, P. R. Epstein, A. Grambsch, D. J. Gubler, P. Reiter, I. Romieu, J. B. Rose, J. M. Samet, and J. Trtanj. 2000. The potential health impacts of climate variability and change for the United States: executive summary of the report of the health sector of the U.S. National Assessment. *Environmental Health Perspectives* 108(4):367-376.

Peters, R. L., and T. E. Lovejoy. 1994. *Global warming and biodiversity*. New Haven, CT: Yale University Press.

Pinzon, J. E., J. M. Wilson, C. J. Tucker, R. Arthur, P. B. Jahrling, and P. Formenty. 2004. Trigger events: enviroclimatic coupling of Ebola hemorrhagic fever outbreaks. *American Journal of Tropical Medicine and Hygiene* 71(5):664-674.

ProMed Mail. 2008. *Assessment of the outbreak of bluetongue serotype 8 in Europe, 2007 (July-December)*, http://www.promedmail.org (accessed February 29, 2008).

Schneider, E., R. A. Hajjeh, R. A. Spiegel, R. W. Jibson, E. L. Harp, G. A. Marshall, R. A. Gunn, M. M. McNeil, R. W. Pinner, R. C. Baron, R. C. Burger, L. C. Hutwagner, C. Crump, L. Kaufman, S. E. Reef, G. M. Feldman, D. Pappagianis, and S. B. Werner. 1997. A coccidioidomycosis outbreak following the Northridge, Calif, earthquake. *Journal of the American Medical Association* 277(11):904-908.

Schneider, W., C. A. Hollier, H. K. Whitam, M. E. Palm, J. M. Mckemy, J. Hernandez, L. Levy, and R. Devries-Paterson. 2005. First report of soybean rust caused by *Phakopsora pachyrhizi* in the continental United States. *Plant Disease* 89:774.

Schweiger, A., R. W. Ammann, D. Candinas, P. A. Clavien, J. Eckert, B. Gottstein, N. Halkic, B. Muellhaupt, B. M. Prinz, J. Reichen, P. E. Tarr, P. R. Torgerson, and P. Deplazes. 2007. Human alveolar echinococcosis after fox population increase, Switzerland. *Emerging Infectious Diseases* 13(6):878-882.

Skarphedinsson, S., P. M. Jensen, and K. Kristiansen. 2005. Survey of tickborne infections in Denmark. *Emerging Infectious Diseases* 11(7):1055-1061.

Snieszko, S. F. 1974. The effects of environmental stress on outbreaks of infectious diseases of fishes. *Journal of Fish Biology* 6(2):197-208.

Speelmon, E. C., W. Checkley, R. H. Gilman, J. Patz, M. Calderon, and S. Manga. 2000. Cholera incidence and El Niño-related higher ambient temperature. *Journal of the American Medical Association* 283(23):3072-3074.

Stenseth, N. C., N. I. Samia, H. Viljugrein, K. L. Kausrud, M. Begon, S. Davis, H. Leirs, V. M. Dubyanskiy, J. Esper, V. S. Ageyev, N. L. Klassovskiy, S. B. Pole, and K. S. Chan. 2006. Plague dynamics are driven by climate variation. *Proceedings of the National Academy of Sciences* 103(35):13110-13115.

Sultan, B., K. Labadi, J. F. Guegan, and S. Janicot. 2005. Climate drives the meningitis epidemics onset in West Africa. *PLoS Medicine* 2(1):e6.

Sutherst, R. W. 2004. Global change and human vulnerability to vector-borne diseases. *Clinical Microbiology Reviews* 17(1):136-173.

Tam, C. C., L. C. Rodrigues, S. J. O'Brien, and S. Hajat. 2006. Temperature dependence of reported *Campylobacter* infection in England, 1989-1999. *Epidemiology and Infection* 134(1):119-125.

Tanser, F. C., B. Sharp, and D. Le Sueur. 2003. Potential effect of climate change on malaria transmission in Africa. *Lancet* 362:1792-1798.

UNDP (United Nations Development Programme). 2007. *Human development report 2007/2008. Fighting climate change: human solidarity in a divided* world. New York: United Nations Development Programme.

UNHCR (Office of the United Nations High Commissioner for Refugees). 2006. *UNHCR and international protection: a protection induction programme.* Geneva, Switzerland: Office of the United Nations High Commissioner for Refugees.

U.S. Climate Change Science Program and the Subcommittee on Global Change Research. 2008. *Weather and climate extremes in a changing climate*. Synthesis and Assessment Product 3.3. Edited by T. R. Karl, G. A. Meehl, C. D. Miller, S. J. Hassol, A. M. Waple, and W. L. Murray. Washington, DC: U.S. Climate Change Science Program.

van Lieshout, M., R. S. Kovats, M. Livermore, and P. Martens. 2004. Climate change and malaria: analysis of the SRES climate and socio-economic scenarios. *Global Environmental Change* 14:87-99.

WHO (World Health Organization). 2007. *Rift Valley fever,* http://www.who.int/mediacentre/factsheets/fs207/en/ (accessed June 19, 2008).

WHO, World Meterological Organization, and United Nations Environment Programme. 2003. *Climate change and human health—risks and responses*. Geneva, Switzerland: World Health Organization.

Woods, A., K. D. Coates, and A. Hamann. 2005. Is an unprecedented dothistroma needle blight epidemic related to climate change? *Bioscience* 55(9):761-769.

Woods Hole Oceanographic Institution. 2008. *What are harmful algal blooms (HABs)?,* http://www.whoi.edu/redtide/whathabs/whathabs.html (accessed March 17, 2008).

Appendix SA-1 References

IPCC (Intergovernmental Panel on Climate Change). 2007. *Climate change 2007: the physical science basis. Summary for policymakers.* Working Group I contribution to the fourth assessment report of the IPCC. Geneva, Switzerland: IPCC. P. 17.

National Geographic Society. 2007 (October). http://ngm.nationalgeographic.com/climate connections/climate-map (accessed June 11, 2008).

WHO (World Health Organization), World Meterological Organization, and United Nations Environment Programme. 2003. *Climate change and human health—risks and responses*. Geneva, Switzerland: World Health Organization.

1

Climate Change Challenges

OVERVIEW

The contributions that comprise this chapter establish the context of workshop discussions and depict the "big picture" within which the papers collected in subsequent chapters are set. The three presenters represented herein—Sir Andrew Haines of the London School of Hygiene and Tropical Medicine; Paul Epstein of Harvard Medical School; and keynote speaker Donald Burke of the University of Pittsburgh—offer varied, and occasionally contrasting, perspectives on what is known, suspected, or unknown regarding the consequences of global climate change for health and, more specifically, for infectious disease emergence. At the same time, each of these contributors observes the accelerating pace of ecological upheaval and emphasizes the inherent complexity of biological responses to climate change and extreme weather events, which frequently involve nonlinear "tipping points." These characteristics inspire both uncertainty and urgency in the quest to better understand, anticipate, and respond to the potentially wide-ranging health effects of climate change.

In the chapter's first paper, Haines reviews the Intergovernmental Panel on Climate Change's (IPCC's) most recent findings on global climate change to date and the panel's predictive scenarios for the future. He then describes several approaches that have been taken to identify and model the potential health impacts of climate change, along with the methodological challenges presented by such studies. Several examples illustrate how variations in climate—in the form of floods, droughts, and other extreme weather events—influence the range and transmission dynamics of infectious diseases, and therefore suggest potential effects of climate change. Yet, as Haines observes, "in the case of infectious diseases there is still considerable controversy about the degree to which climate change has been responsible for changes in the incidence and distribution of disease. This is due to the potential contribution of other factors, such as changing land-use patterns, human behavior, and methodological issues including the use and analysis of appropriate climate data."

Haines also reviews several efforts to date to estimate the future impact of climate change on infectious diseases, which—although individually problematic—support his overall conclusion that "it is likely that the disease burden as a result of climate change will [increase] substantially over time and will be particularly concentrated in the poorer populations." Thus, his proposed strategies to address the negative health effects of climate change focus on the poor: first, by improving their access to basic public health services (clean water, sanitation, immunization); second, by providing them with cleaner fuels, which offer both immediate health benefits and long-term protection for the atmosphere. "Infectious diseases are one of a number of categories of health outcomes that are likely to be affected adversely by climate change," Haines concludes. "Public health policies should take into account the need to adapt to a changing climate, as well as the potential for near-term benefits to health from a range of policies to mitigate climate change."

Haines's paper is followed by two reprinted pieces, authored (in the first case) and coedited by Epstein (in the second), that illustrate the breadth of biological responses to climate change. The first manuscript is an essay, originally published in the *New England Journal of Medicine* in October 2005—weeks after Hurricane Katrina devastated the Gulf Coast—that focuses on the wide-ranging health effects of extreme weather. The second, an excerpt from the report *Climate Change Futures: Health, Ecological and Economic Dimensions* (Center for Health and the Global Environment, 2005), introduces a "multidimensional assessment" incorporating trend analysis, case studies, and scenarios that focus on health, ecological, and economic impacts of climate change. This project, undertaken in collaboration by the Center for Health and the Global Environment at Harvard Medical School, the United Nations Development Programme, and Swiss Re, a global reinsurance company, was designed to assess threats posed by climate change to the institution of insurance, a "time-tested method for adapting to change." Such threats go far beyond immediate property damage, and indeed even health consequences, to the social and political stability of regions affected by climate disasters (see also Chapter 4).

In his workshop presentation, Epstein emphasized the methodologies that make such threat assessments possible. He identified three phenomena that underlie climate- and weather-related changes in disease distribution:

1. Since 1950, nighttime and winter warming have occurred twice as fast as overall global warming.

2. The pace of warming in temperate, boreal, and polar latitudes is occurring faster than warming in the tropics.
3. Since the first International Geophysical Year in 1957, when many global measurements were initiated, the world's oceans have accumulated 22 times the amount of heat that the atmosphere has, accelerating the global hydrological cycle.

There are no appropriate, independent controls for the study of global climate change on Earth, and the experiment we are conducting (with an *n* of 1) cannot be repeated, Epstein observed. Therefore, he explained, a wide range of methodologies must be harnessed to assess changes in biological variables, such as the geographic range and incidence of diseases in relation to changes in temperature and precipitation. Monitoring and mapping produce data that can be integrated into geographic information systems (GISs) to identify and compare physical and biological phenomena. Further, GISs overlay multiple sets of data, providing input for descriptive and mathematical models that project the biological impacts of various climate change scenarios. Models are used for understanding dynamics, for predicting outcomes, and for decision making (Hilborn and Mangel, 1997).

In his presentation, Epstein also described methods for analyzing data gathered across scientific disciplines (e.g., diseases affecting a range of taxonomic groups) in order to reveal patterns and emerging trends associated with climate change, calculate rates of change (i.e., in the geographic range, prevalence, and incidence of infectious diseases), and compare observations with predicted outcomes. Researchers also conduct experiments, called "fingerprint" studies, to compare data with model projections (such as those undertaken by the IPCC to analyze climate models and energy fluxes driven by increases in heat-trapping, greenhouse gases).

Many of the methodologies used to study the effects of climate change yield correlations, rather than proof of causation, Epstein acknowledged. However, he added, such associations and their plausible mechanisms can be then tested via qualitative, schematic, and quantitative models. Bayesian methods of assessing causality based on prior probabilities and prior knowledge via first physical principles can also be used to analyze the effects of global climate change, he said.

Moreover, Epstein asserted, when observational data from multiple sources match model projections (i.e., the findings are internally consistent) and can be explained by plausible biological mechanisms (e.g., changes in observed altitude ranges in tandem with observed temperature changes), the composite pattern warrants further attention. This may take the form of analyzing attribution probabilities for anomalous events; for example, such an assessment indicates that global warming increased the likelihood of the European heat wave of 2003 two- to fourfold (Stott et al., 2004). Extreme conditions that favor infectious disease outbreaks via multiple pathways may be revealed by "cluster analyses" or characterized through the use of "principal component analyses," which identify spatial and temporal associations among variables.

Despite the existence of such methodologies, "our current understanding of the relationships between climate and weather, and epidemic infectious diseases, is insufficient to make credible predictions about future threats posed by infectious diseases under various global change scenarios," Burke argues in the chapter's final paper. To support this contention, he presents detailed analyses of the transmission dynamics of two infectious diseases: influenza and dengue fever. These studies reveal that oscillations in disease incidence occur in the absence of seasonal transmission effects; if these patterns coincide with seasonal variation, small changes in transmissibility may, under some circumstances, produce considerable variability from year to year in epidemic disease occurrence (Dushoff et al., 2004).

Burke notes approvingly that conclusions by "respected scientific bodies" regarding the probable impact of global climate change on epidemic infectious diseases remain measured since the publication of the landmark report *Under the Weather: Climate, Ecosystems, and Infectious Diseases* (NRC, 2001) by the interdisciplinary committee that he chaired. "This caution honestly reflects the uncertainties involved, which in turn reflect the difficulty of the underlying scientific problems," he states. Calling readers' attention to the recommendations for future research and surveillance made in that report (see also Box SA-3), Burke concludes that "it is safe to say that [these recommendations] continue to be relevant."

CLIMATE CHANGE, EXTREME EVENTS, AND HUMAN HEALTH

Andy Haines, M.B.B.S., M.D.
London School of Hygiene and Tropical Medicine

Greenhouse gases are now accumulating in the atmosphere at unprecedented rates. The annual growth rate of carbon dioxide (CO_2) concentration was highest over the last 10 years since the beginning of continuous direct atmospheric measurements (IPCC, 2007a). The atmospheric concentration of CO_2 now greatly exceeds the natural range over the last 650,000 years. CO_2 is the most important greenhouse gas produced by humankind and accounts for around 77 percent of the total. The concentrations of all three major greenhouse gases: CO_2, methane, and nitrous oxide, are at the highest levels for at least 10,000 years and have resulted in clear changes in the Earth's climate. Of the last 12 years (1995-2006), 11 are among the 12 warmest years since 1850 when instrumental records began. Over past 100 years (1906-2005), global surface temperature has increased by 0.74°C (90 percent, uncertainty interval 0.56-0.92) with warming faster over land than over oceans. Of course climate will still vary (e.g., in February 2008, NOAA [National Oceanic and Atmospheric Administration] predicted La Niña conditions—the cold phase of the El Niño Southern Oscillation [ENSO]—to continue throughout spring 2008; however, by May La Niña began to transition to more ENSO-neutral conditions [NOAA Climate Prediction Center, 2008]). But after accounting for these known climate fluctuations, the long-term warming trend will continue.

Figure 1-1 shows the trends for temperature, global average sea level, and observed decreases in snow and ice. Sea level rise is due to a combination of thermal expansion of the oceans together with increased melting of glaciers and polar ice sheets. Other significant changes in climate include declines in precipitation from 1900 to 2005 in the Sahel, Mediterranean, southern Africa, and parts of southern Asia. More intense and longer droughts have occurred over increasing areas since the 1970s, particularly in the tropics and subtropics. The IPCC concludes that increased drying as a result of temperature increases and reductions in precipitation has contributed to changes in drought.

In contrast, precipitation has increased significantly in eastern areas of North and South America, as well as northern Europe and northern and central Asia. There is an apparent increase in intense tropical cyclone activity since 1970 in the North Atlantic.

Over the next two decades, warming of about 0.2°C per decade has been projected for a range of emission scenarios according to the IPCC. After that, different scenarios of socioeconomic development and the use of mitigation strategies result in markedly different trajectories for greenhouse gas emissions (see Figure 1-2). The different scenarios result in best estimates for temperature change in 2090-2099 relative to 1980-1999, ranging between 1.8°C for B1 scenario and 4°C for A1F1 scenario, but the likely range of estimates is even wider, extending to 6.4°C for the latter scenario. Regional-scale changes are outlined in Box 1-1.

Climate Variability, Climate Change, and Health

It has been known for thousands of years, at least since the time of Hippocrates, that climatic variations can influence health, particularly through changes in temperature and precipitation, as well as extreme weather events. Growing scientific consensus about the existence of global climate change has rekindled interest in linkages between climate and health. The potential range of impacts is wide and they have been reviewed extensively (IPCC, 2007b).

There are a number of approaches to studying the potential health impacts of climate change. These include studies of the associations between past climate variability and disease; of the associations between trends in climatic variables over recent decades and the epidemiology of diseases; and of the response of vector species to changes in temperature and rainfall. In addition, there have been a number of approaches to modeling the potential future impacts of climate change on health.

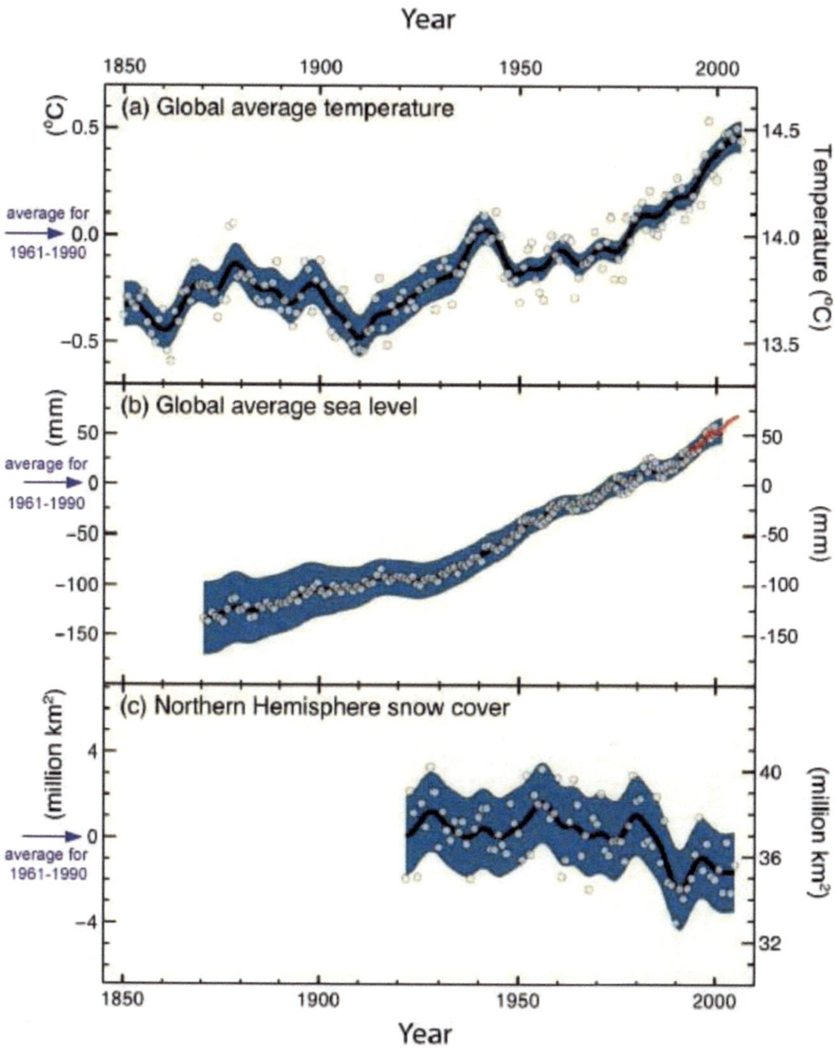

FIGURE 1-1 Observed changes in (A) global average surface temperature; (B) global average sea level rise from tide gauge (blue) and satellite (red) data; and (C) Northern Hemisphere snow cover for March-April. All changes are relative to corresponding averages for the period 1961-1990. Smoothed curves represent decadal averaged values while circles show yearly values. The shaded areas are the uncertainty intervals estimated from a comprehensive analysis of known uncertainties (A and B) and from the time series (C). SOURCE: Figure SPM.3 in IPCC (2007a).

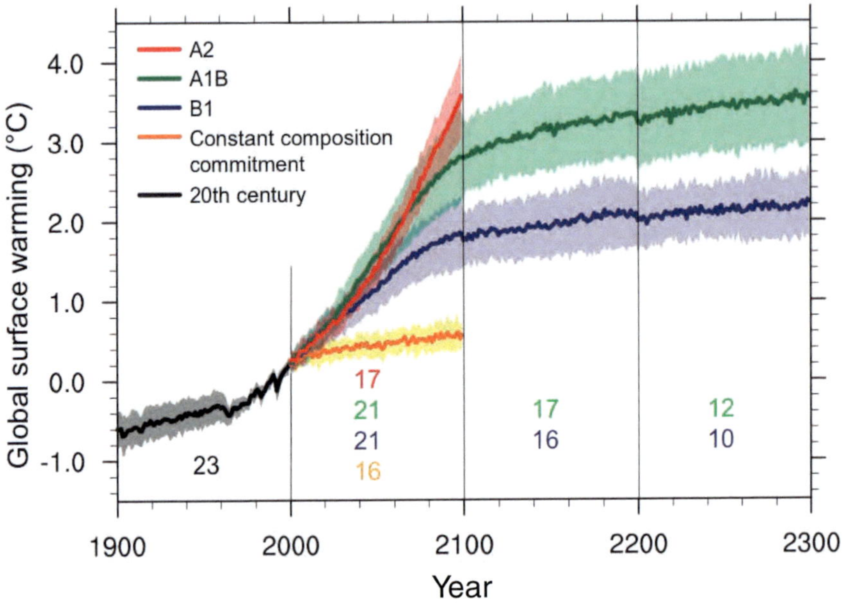

FIGURE 1-2 Multimodel averages and assessed ranges for surface warming (compared to the 1980-1999 base period) for the SRES scenarios A2 (red), A1B (green), and B1 (blue), shown as continuations of the twentieth-century simulation. The latter two scenarios are continued beyond the year 2100 with forcing kept constant (committed climate change as it is defined in Box TS.9 [of IPCC, 2007a]). An additional experiment, in which the forcing is kept at the year 2000 level is also shown (orange). Linear trends from the corresponding control runs have been removed from these time series. Lines show the multimodel means, shading denotes the ±1 standard deviation range. Discontinuities between different periods have no physical meaning and are caused by the fact that the number of models that have run a given scenario is different for each period and scenario (numbers indicated in figure). For the same reason, uncertainty across scenarios should not be interpreted from this figure.
SOURCE: Figure TS.32 in IPCC (2007a).

The study of potential associations between climate change and health poses a number of methodological challenges including the need to consider confounding factors as possible explanations of apparent associations between climatic variables and health outcomes. Such confounding factors may include changes in resistance to insecticides (in the case of vector-borne diseases); changes in resistance to commonly used drugs for treatment (e.g., in the case of malaria); migration of populations, which may result in the exposure of nonimmune populations to infectious diseases; and changes in the performance of disease surveillance systems over time. Some diseases, such as malaria, exhibit differences in

BOX 1-1
Regional-Scale Changes

Changes include the following:

- Warming greatest over land and at most high northern latitudes and least over the Southern Ocean and parts of the North Atlantic Ocean, continuing recent observed trends
- Contraction of snow cover area, increases in thaw depth over most permafrost regions, and decrease in sea ice extent; in some projections, Arctic late-summer sea ice disappears almost entirely by the latter part of the twenty-first century
- Very likely increase in frequency of hot extremes, heat waves, and heavy precipitation
- Likely increase in tropical cyclone intensity; less confidence in global decrease of tropical cyclone numbers
- Poleward shift of extratropical storm tracks with consequent changes in wind, precipitation, and temperature patterns
- Very likely precipitation increases in high latitudes and likely decreases in most subtropical land regions, continuing observed recent trends

SOURCE: IPCC (2007a).

local transmission dynamics that complicate attempts to model the likely effect of climate change. Improvements in public health infrastructure leading to improved adaptation to climate change could, in the future, attenuate relationships between the changing climate and health outcomes. Climate change is likely to be a long-term process that will evolve over decades and centuries while our understanding of the linkages between climate and health is based largely on studies of short-term variability.

There are likely to be interactions between climate change and other environmental changes, such as deforestation, growth in global travel, increased local population mobility, and depletion of aquifers in some regions. For example, deforestation may change the distribution of disease vectors as well as contributing to climate change, and migration of populations into formerly forested areas may result in increased exposure to a number of diseases. A study in the Peruvian Amazon suggested that the abundance of the malaria vector *Anopheles darlingi* was two hundredfold higher in deforested areas than in pristine forest and this could not be attributed solely to increased population density (Vittor et al., 2006). Deforestation may increase malaria risk in the Americas and Africa, but reduce it in Southeast Asia (Guerra et al., 2006).

This paper focuses on potential relationships between infectious disease and climate change, but there are a range of other health outcomes such as the effects of heat waves and potential reductions in cold-related deaths, particularly in temperate countries; the direct effect of wind, storms, and floods causing deaths and injuries; and the effect of droughts on malnutrition and food security. Floods and droughts, together with less extreme changes in precipitation, can also have important implications for a range of water-related diseases, as well as impacting health through the effects of increases in malnutrition and consequent reductions in immunity to disease. A summary of the pathways through which climate change may affect human health including infectious diseases is shown in Figure 1-3.

Rainfall, Temperature, and Disease

Water-related diseases include water-borne diseases due to ingestion of pathogens in contaminated water and water-washed diseases as a result of poor hygiene. A recent global overview of cross-sectional studies based on 36 published reports from low- and middle-income countries between 1954 and 2000 suggests an average of 4 percent increase in diarrhea incidence in children less than 5 years of age per 10 mm decrease in rainfall per month (Lloyd et al., 2007). Currently, more than 2 billion people live in relatively dry regions of the world and are likely to suffer disproportionately from lack of access to clean water. While there have been substantial improvements in the management of diarrheal disease, particularly because of the widespread use of oral rehydration therapy, child mortality remains unacceptably high, particularly in sub-Saharan Africa. Diarrheal diseases claim almost 2 million lives a year in children under 5. However, it is important to recognize that most freshwater is used for irrigation rather than personal consumption; therefore, the relationship between reduced freshwater availability and diarrheal diseases may be indirect. Hand washing with soap has a protective effect against diarrheal disease (Curtis and Cairncross, 2003), and reduced availability and/or increased costs of freshwater may lead to reduced hand washing where rainfall is low.

Heavy precipitation may be associated with outbreaks of water-borne diseases, such as cryptosporidiosis (Curriero et al., 2001). Although the global overview referred to above did not find a relationship with temperature, other studies have found associations between higher temperatures and increased episodes of diarrheal diseases in Peru, the Pacific Islands, and Australia (Checkley et al., 2004; McMichael et al., 2003; Singh et al., 2001). The association between sea surface temperatures and cholera transmission has been most convincingly shown in the Bay of Bengal (Colwell, 1996).

Time-series analysis of weekly cases of salmonellosis using data from 16 sites in industrialized countries suggested an approximately linear increase in reported cases above the threshold of about 6°C. The study focused on the analysis of sporadic cases only. The association appeared to be particularly evident in the case of *Salmonella enteritidis*, with a lag of around 1 week between the

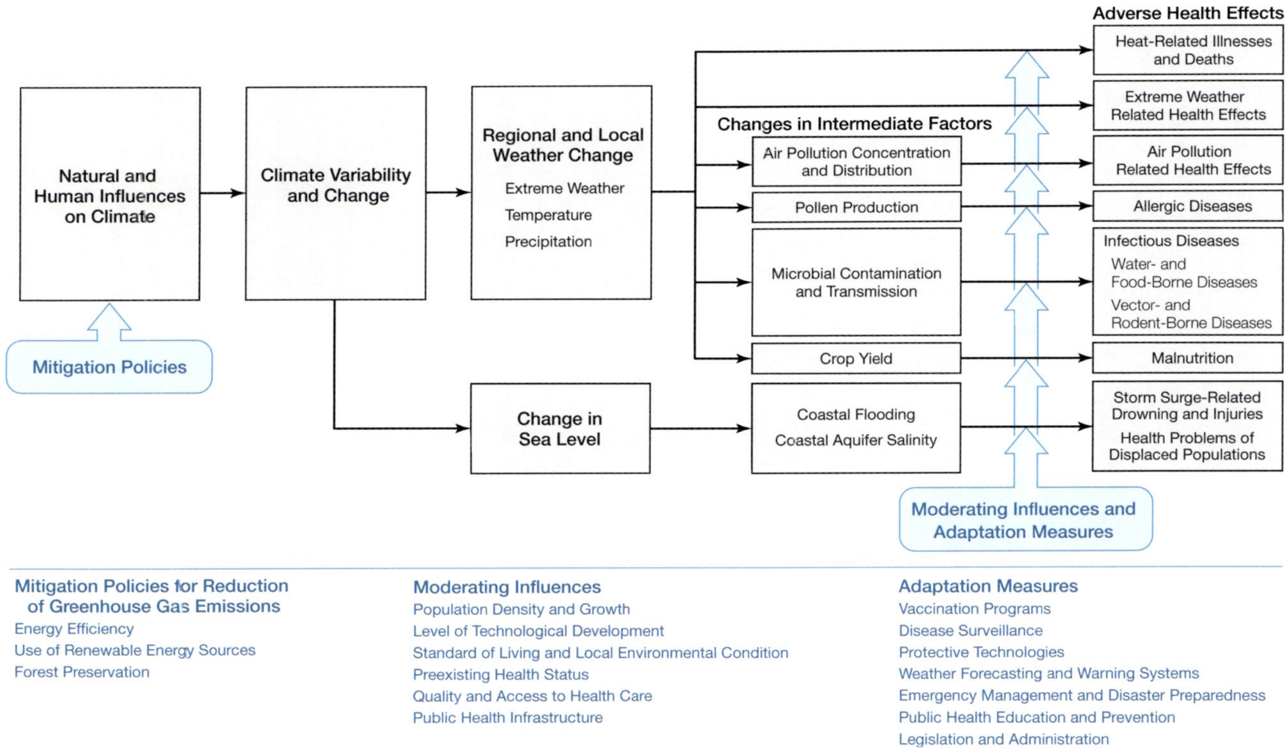

FIGURE 1-3 Pathways by which climate change may affect human health, including infectious diseases.
SOURCE: Reprinted from Haines and Patz (2004) with permission from the American Medical Association. Copyright 2004. All rights reserved; adapted from Patz et al. (2000).

increased temperature and the increase in cases, suggesting that the effect may be on the replication of salmonella after food has been prepared (Kovats et al., 2004).

One of the best documented examples of the way in which climate can drive the onset of disease is the relationship between the dry northern winds (called the Harmattan) and meningococcal meningitis epidemics in West Africa. The mean weeks of the winter maximum wind speed and of the onset of the epidemic are identical and usually occur between February 7 and 15 (Sultan et al., 2005). Although the causal mechanism is not fully understood, the disease may result from the effects of mucosal drying and abrasion as a result of the strong dust-laden winds.

There is some evidence that the geographical distribution of meningococcal meningitis in West Africa has expanded in the recent past, possibly as a result of changes in land use and climate (Molesworth et al., 2003).

The El Niño/Southern Oscillation and Health

El Niño events are large-scale ocean-atmospheric climate phenomena emanating from the central and east-central equatorial Pacific Ocean that have occurred for thousands of years. A major feature is the upwelling of warm water off the coast of Peru and Ecuador. Associated by distant connections (teleconnections) with climatic changes in Australia, Indonesia, the Pacific highlands, and East Africa, as well as parts of Latin and North America, the El Niño cycle is usually between 3 and 7 years. The El Niño (warm event) is frequently followed by a La Niña (opposite, cold) event. The Southern Oscillation is the name given to the seesaw of air pressure differences between the east and west Pacific, which is associated with the El Niño phenomenon and leads to the full name El Niño/Southern Oscillation (ENSO).

Figure 1-4 demonstrates some of the teleconnections associated with El Niño and indicates where health outcomes, such as increased risk of epidemic malaria, are experienced after the onset of an El Niño event (Kovats et al., 2003). ENSO is the most important climatic cycle that contributes to natural disasters. Drought is twice as frequent worldwide in the year after the onset of the El Niño than in other years, particularly in southern Africa and South Asia. In an average El Niño year, around 35 people per 1,000 worldwide are affected by a natural disaster, more that four times that of non-El Niño years based on analysis of data for three decades between 1963 and 1993 (Bouma et al., 1997).

The relationship between El Niño and intense rainfall is also strong in many areas, although unlike drought it is not seen on a global scale because flood-related disasters are relatively localized and the risk is increased during both El Niño and La Niña phases in different parts of the world. During an El Niño event, storm activity in parts of the Pacific is increased and decreased in the Atlantic so that hurricanes in the Caribbean and the Gulf of Mexico tend to be less com-

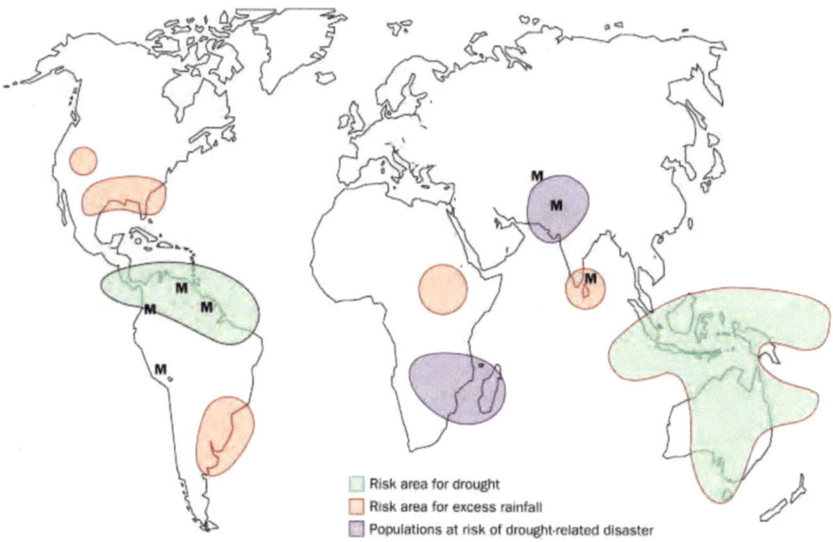

FIGURE 1-4 ENSO teleconnections and risk map for malaria.
SOURCE: Reprinted from Kovats et al. (2003) with permission from Elsevier.

mon during an El Niño but more common during a La Niña, while typhoons are more likely to occur near the Marshall Islands in the Pacific Ocean during an El Niño event compared to other times because of the western shift in storm tracks (Spennemann and Marschner, 1995).

The study of the relationship between ENSO events and health outcomes provides useful evidence for causal relationships between climatic variability and human health, which can help to improve understanding of the potential impacts of climate change. It is not known precisely how climate change will affect the El Niño phenomenon, but such events will occur in the future against a background of climate change, which is likely to increase the health and social impacts.

For a causal relationship between the ENSO cycle and a health outcome to be inferred, it is necessary to have appropriate climatic data for the region to be studied. There also needs to be a plausible biological relationship between the health outcome in question and climate variability, and appropriate statistical analysis is required, taking into account autocorrelation and using, where available, long time series of data over decadal periods. There have also been numerous case reports of associations of disease outbreaks with single ENSO events, although these provide less compelling evidence of association.

The analysis of linkages between health outcomes and El Niño events is complicated by the fact that they are not identical but vary in intensity and regional impacts. For example, in 1997 and 1998 the anticipated drought did not occur in the southern African region and some areas received above average rainfall.

Despite the limitations of available data, a number of robust associations have been described at various sites throughout the world, particularly with epidemic malaria in countries such as Colombia, Venezuela, and Guyana (Kovats et al., 2003). In highland areas, such as northern Pakistan, higher temperatures associated with El Niño may increase the transmission of malaria. In many desert fringe regions, such as the Punjab and the Thar Desert, increased rainfall during the post-El Niño and La Niña years has historically increased epidemic risk. Conversely, El Niño-related droughts have been associated with malaria outbreaks in Sri Lanka, Colombia, and Venezuela, possibly as a result of reduced river flows, which in turn permitted increased breeding of mosquitoes.

The yearly incidence of dengue has also been associated with the El Niño year in Indonesia. Monthly cases of dengue on some but not all Pacific Islands have been associated with the Southern Oscillation index. In Australia, epidemic years for the Ross River virus disease over a 70-year period up to 1998 showed some association with the Southern Oscillation index, but two other studies did not show a relationship between yearly notifications of this disease and the index. Associations of health outcomes with climatic factors may be localized and may not be detectable with aggregated data.

The association between cholera and ENSO has been established in Dhaka, Bangladesh, and the relationship has become apparent in recent decades (Rodo et al., 2002).

The annual incidence of visceral leishmaniasis in Bahia State, northeastern Brazil, was significantly related to sea surface temperature in the Pacific Ocean (Franke et al., 2002) probably as a result of the association between the ENSO cycle and drought in northeast Brazil.

The El Niño event of 1993 resulted in increased rainfall in the southern United States, which was then followed by a drought that facilitated the emergence of hantavirus pulmonary syndrome. There was also a fivefold increase in reported cases associated with the 1997-1998 El Niño. Although there is a plausible mechanism for ways in which climatic factors can lead to substantial increases in the rodent population and thus hantavirus transmission, a consistent association with ENSO has not been established (Engelthaler et al., 1999). Likewise in the case of Rift Valley fever in East Africa, there is good evidence that epidemics in the dry grasslands are triggered by heavy rainfall events, but there is no association between this disease in Kenya and the ENSO index (Linthicum et al., 1999).

Extreme Climate Events and Infectious Diseases

Case reports suggest that extreme climatic events such as floods, droughts, and storms can be associated with outbreaks of infectious disease, but the association is not always seen and is likely to depend on socioeconomic factors, damage to infrastructure, and the extent of population displacement.

There are a number of mechanisms by which the risk of flooding is likely to increase as a result of climate change. These include the melting of glaciers, increased frequency of episodes of heavy precipitation, and sea level rise. However, an extensive review of the effects of floods in Europe (Hajat et al., 2003) showed that infectious disease outbreaks were rarely a major public health problem. A more consistent finding seems to be increased prevalence of common mental disorders (i.e., anxiety, depression) following exposure to floods. This is likely to be due to loss of familiar possessions, forced evacuation, loss of livelihood, and increased poverty.

However, following floods, increases in diarrheal and respiratory diseases have been reported in both high- and low-income countries. Crowding of displaced populations may contribute to transmission. In low-income countries in particular, there may be outbreaks of leptospirosis, hepatitis, and vector-borne diseases, including malaria and Rift Valley fever (Ahern et al., 2005). On occasions, such as in Tanzania in 1997-1998, flooding had the opposite effect and washed away vector breeding sites.

Populations in low-lying areas are vulnerable to the effects of sea level rise, particularly in low-income countries where flood defenses may not be upgraded. For example, up to 57 percent of people in unprotected dry land areas in Bangladesh could be inundated if there was a 4°C temperature increase and, as a consequence, a 100 cm increase in sea level together with increases in monsoon precipitation and discharge into major rivers (BCAS/RA/Approtech, 1994). Deltaic regions are particularly vulnerable to the effects of sea level rise as illustrated in Figure 1-5. Salinization of freshwater aquifers may occur as a result of incursion of seawater. Even in industrialized countries, densely populated urban areas are at risk from sea level rise, as shown by the impact of Hurricane Katrina on New Orleans. After Hurricane Katrina there were increases in diarrheal disease incidence due to fecal contamination of drinking water (Manuel, 2006).

The potential health effects of drought in developing countries are likely to be wide ranging (see Figure 1-6). It is important to note, however, that domestic water consumption represents only 2 percent of the global total, and a flow sufficient to meet the domestic water requirements of around 1,000 people would be sufficient to irrigate only 1 hectare of land, which is capable of feeding a couple of families (Shiklomanov, 2000). It has also been pointed out that the weight of evidence from studies in low-income countries suggests that ready access to water that results in increased quantities used for hygiene is likely to be more important than water quality improvements by themselves in determining benefits for health (Cairncross, 2003). This is likely to be due to the fact that most endemic diarrheal disease is transmitted person-to-person by hands, food, and other fomites because of poor hygiene.

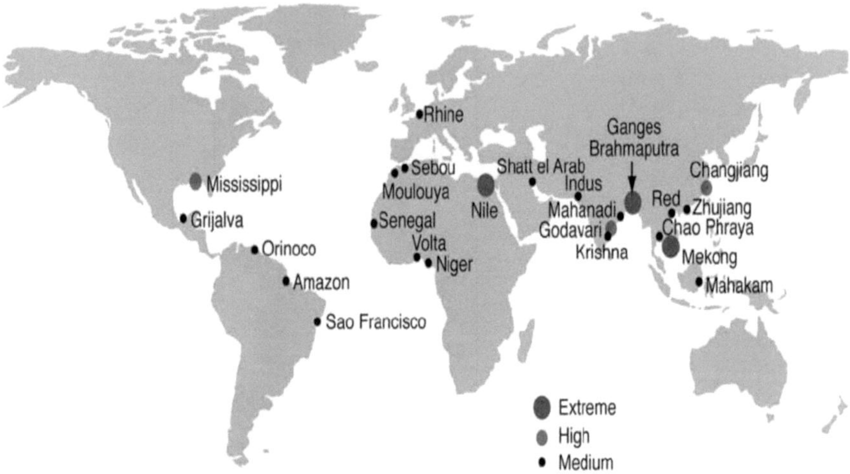

FIGURE 1-5 Relative vulnerability of coastal deltas as indicated by estimates of the population potentially displaced by current sea-level trends to 2050 (extreme >1 million; high 1 million to 50,000; medium 50,000 to 5,000). Climate change would exacerbate these impacts.
SOURCE: Figure TS.8 in IPCC (2007b).

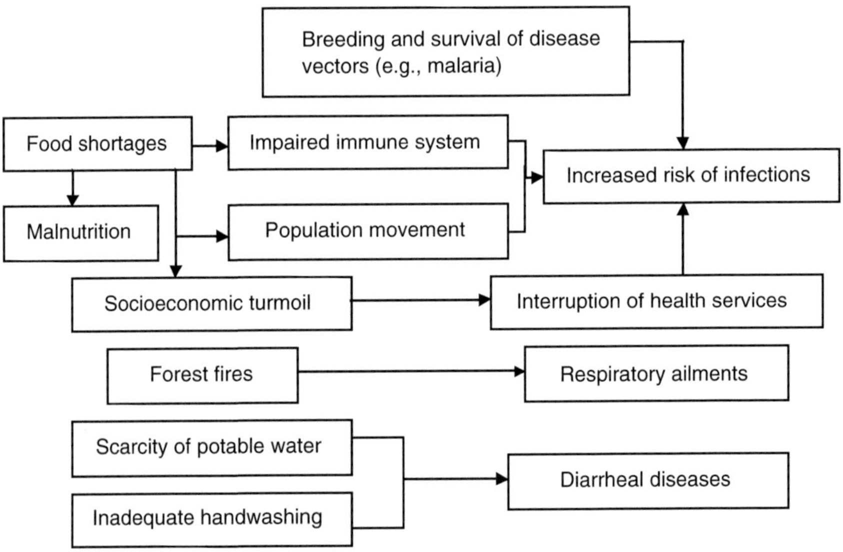

FIGURE 1-6 Potential health effects of drought in developing countries.
SOURCE: Adapted from Kovats et al. (2003) with permission from Elsevier.

A study in six sub-Saharan African countries showed the potential for HIV/AIDS to amplify the effects of drought on childhood malnutrition, particularly in periurban and urban populations where HIV prevalence was high (Mason et al., 2005). Malnutrition associated with drought increases susceptibility to diseases such as measles, particularly in countries such as Somalia where immunization rates are low (Shepherd-Johnson, 2006).

Drought has variable effects on the incidence and distribution of vector-borne diseases such that, for example, reductions in mosquito activity during droughts may be followed by increases in disease transmission once the drought ends because of the increased number of susceptible hosts (Woodruff et al., 2002). In other cases, stagnation of water in residual pools may cause short-term increases in the transmission of malaria. Long-term drought may result in the contraction of areas suitable for malaria transmission.

Assessment of the likely effect of climate change on malnutrition is complex because the impacts on food production and consumption depend on a range of factors, including agricultural practices, the potential role of carbon dioxide fertilization in improving some crop yields (the effect of increasing concentrations of carbon dioxide in improving the yields of some crops may be reduced when crops are stressed as a result of high temperatures or changes in precipitation), patterns of land ownership, and the ability of disadvantaged populations to purchase food. Concern has also been expressed about the potential competition for land between biofuels and agricultural production for food, although this is not an inevitable consequence and will depend on the policy choices that are made (Haines et al., 2007). Vulnerability to increased malnutrition as a result of climate change is likely to be greatest in regions currently most vulnerable to food insecurity, particularly sub-Saharan Africa (FAO, 2005).

Both fatalities and direct economic losses of national per capita income from natural disasters are higher by orders of magnitude in low- and middle-income countries compared to high-income countries (Linnerooth-Bayer et al., 2005). For example, a study of the impact of Hurricane Mitch on the livelihood of rural poor in Honduras showed that one of every two households surveyed incurred medical, housing, or other costs due to the hurricane. Relief amounted to less than one-tenth of the losses incurred by households (Morris et al., 2002). Such economic losses accentuate poverty and contribute substantially to the adverse effects of climate-related disasters on public health. For example, an estimate of mortality due to floods in Mozambique in 2000 suggested that the increase in infant mortality associated with around a 14 percent drop in gross domestic product in the flooded provinces made a substantial contribution to the overall deaths due to flooding (Cairncross and Alvarinho, 2006).

Is There Evidence That Climate Change Has Begun to Affect Human Health?

There is good evidence that climate change has caused an earlier onset of the spring pollen season in the Northern Hemisphere, with resulting changes in the seasonality of allergic rhinitis (Emberlin et al., 2002). The summer of 2003 was probably the hottest in Europe since 1500, and climate change is thought to have at least doubled the risk of a heat wave such as that experienced in 2003 (Stott et al., 2004).

However, in the case of infectious diseases there is still considerable controversy about the degree to which climate change has been responsible for changes in the incidence and distribution of disease. This is due to the potential contribution of other factors, such as changing land use patterns, human behavior, and methodological issues, including the use and analysis of appropriate climate data. Northern shifts in tick distribution have been observed in Sweden (Lindgren et al., 2000), and the incidence of tick-borne encephalitis (TBE) in Sweden has increased considerably since the mid-1980s. A study of cases of TBE in Stockholm County between 1960 and 1998 showed that increases in disease incidence were significantly related to a combination of two consecutive mild winters, as well as temperatures favoring spring development and extended autumn activity in the prior year, and temperatures allowing activity early in the incidence year (Lindgren and Gustafson, 2001). This suggests that milder winters and the early arrival of spring may have contributed to the increased incidence of TBE, but other factors may be implicated, such as changes in land use and land cover leading to increases in the wildlife hosts of ticks, together with the presence of more people in endemic locations (Randolph, 2001).

There has been considerable interest in the possible role of climate change as a factor in the increases in malaria incidence in the East African highlands over recent decades. Using an analysis of data for four sites, Hay and colleagues (2002) asserted that there were no significant trends in climate variables and therefore concluded that climate change had played no role in malaria resurgence in the region. However, the use of low-resolution data to investigate the relationship was criticized, and it was suggested that the apparent lack of an association could not be interpreted as convincing evidence that climate change had not played a role (Patz, 2002). Subsequently, an updated time series (Pascual et al., 2006) using an additional 5 years of data demonstrated that, using both nonparametric and parametric statistical analyses, there was evidence of significant warming trends of around 0.5°C at all sites. It was suggested that the "observed temperature changes would be significantly amplified by the mosquito population dynamics with a difference in the biological response at least 1 order of magnitude larger than that in the environmental variable" (Pascual et al., 2006).

Other authors have shown that climatic factors play a role in malaria epidemics in the East African highlands. The study of climate variability, seasonality, and the number of monthly malaria outpatients over 10- to 15-year periods in seven highland sites in East Africa showed substantial spatial variation in the sensitivity of malaria outpatient numbers to climate variability, with between 12 and 63 percent of variance attributed to climate variability (Zhou et al., 2004). The study of malaria epidemic risk in Ethiopia showed that epidemics were significantly more often preceded by a month of abnormally high minimum temperature during the previous 3 months than would be expected by chance (Abeku et al., 2003).

The recent United Nations Development Programme (UNDP) *Human Development Report* (UNDP, 2007) has documented the growing burden of climate disasters, which is greater than can be explained by population growth alone. Weather-related insurance losses are increasing faster than the population, inflation, and coverage, but the greatest impacts are in developing countries where the majority lack insurance. Between 2000 and 2004, more than 250 million people per year were affected by hydrometeorological disaster (with increases in floods, droughts, lightning strikes, and the intensity of tropical cyclones), compared to less than 50 million per year between 1975 and 1979. A recent report has documented the contribution of increased sea surface temperature to increased Atlantic hurricane activity in recent years (Saunders and Lea, 2008).

Estimating the Impact of Climate Change on Infectious Diseases

A growing number of studies that have modeled projected impacts of climate change on health have been reviewed by the IPCC (IPCC, 2007b). A study of the potential effect of climate change on malaria transmission in Africa that assessed the impacts of 3 climate scenarios suggests a modest (5 to 7 percent) increase in the population at risk, largely due to expansion into higher altitudes. It also suggested a prolongation of the transmission season in some areas, leading to a 16 to 28 percent increase in the total number of person-months exposure (Tanser et al., 2003). This analysis, although based on a very large database of historical malaria surveillance data, has been criticized for oversimplifying the situation (1) by underestimating the variability in response of local vector species to climatic change and (2) because extension of the transmission season does not necessarily translate into a proportional increase in mortality or clinical disease (Reiter et al., 2004). Inevitably, however, all models involve some simplification of assumptions, and for example, few studies take adaptive capacity into account.

Exercises to estimate the global burden of disease due to climate change have been undertaken under the auspices of the World Health Organization (WHO). Epidemiological models were used to estimate the impact of climate change on a number of health outcomes (malaria, diarrheal disease, malnutrition, flood deaths, and direct effects from the heat and cold). The analysis suggests that although there are likely to be some benefits, particularly lower cold-related mortality in temperate zones, these benefits will be greatly exceeded by negative impacts on health, particularly in terms of infectious diseases and malnutrition in low-income countries. The methodological approach has been outlined elsewhere (McMichael et al., 2004), and on aggregate, the estimates suggest that compared to baseline, climate change had caused around 150,000 deaths annually by 2000, an equivalent to 0.3 percent of global deaths per year or 0.4 percent of global disability-adjusted life-years (DALYs) lost annually. The estimate may well be conservative because the baseline used was the average climate for 1960-1991 when climate change was probably already under way and the range of health outcomes was limited. Although increasing wealth and some level of adaptation could blunt the adverse effects, it is likely that the disease burden as a result of climate change will increase substantially over time and will be particularly concentrated in the poorer populations. Nevertheless, populations in all regions of the world are likely to experience some adverse effects, particularly if temperature increases exceed 2°C, at which temperature the probability of major adverse events—such as melting of ice caps and disruption of ecosystems—appears to be unacceptably high, as judged for example by policy makers of the European Union (EU) who have undertaken to pursue negotiations with the aim of keeping temperature increases below that level (European Commission, 2007).

Adaptation Strategies

Adaptation to climate change may take a number of forms; physiological and behavioral adaptation may take place without policy changes, but properly designed adaptation strategies can result in near-term benefits to public health, as well as improving the resilience of populations to future climate change. Policies to improve access to clean water and sanitation, promote hygiene behaviors, promote uptake of immunization, and strengthen health systems are needed in any event in order to improve the chances of attaining the UN Millennium Development Goals (Haines and Cassels, 2004). Increased use of insecticide-impregnated nets and appropriate antimalarial drug combinations that take into account prevailing patterns of drug resistance, as well as effective vector control strategies such as indoor residual spraying, can all help to reduce the burden of malaria.

The threat of climate change has resulted in increased interest in climate-based early warning health systems for heat waves and climate-sensitive diseases. Early warning systems must be integrated into local health systems if they are to have an impact. One example is the highland malaria project (HIMAL), which aims to create and test functional systems for malaria early warning and early detection including district-level surveillance and predictive modeling (Abeku et al., 2004). Most routine disease surveillance systems lack the ability to provide accurate and timely indications of increases in the number of cases of malaria. There is a need to improve the routine collection of data on parasitologically confirmed cases of malaria because febrile illnesses other than malaria have to be considered as possible causes of outbreaks (Cox et al., 2007). Seasonal forecasts can also be used to increase preparedness for climate variability and extreme events associated with phenomena such as El Niño; these approaches were used to warn specific governments when a strong El Niño was developing in 1997-1998 (Hamnett, 1998).

Meeting the Energy Needs of the Poor While Reducing Greenhouse Gas Emissions

Meeting the energy needs of the poor will also help to reduce vulnerability to climate change. Currently there are around 1.6 billion people without electricity (Modi et al., 2006), and 2.4 billion use solid fuels (wood, dung, coal) in their households. Meeting the essential energy needs of the poor will take around 1 percent of current world energy use and in addition reduce exposure to high levels of indoor air pollution (with an attributable annual mortality of about 1.6 million) (WHO, 2002). Concerted action is needed to improve access to more efficient cook stoves and to assist poor populations to move to cleaner fuels, such as kerosene, liquefied petroleum gas, or biogas. Electrification, using renewable energy where possible, can improve adaptation by supplying electricity to maintain the cold chain for vaccines, to provide a reliable power source for health facilities, and to make possible the use of information and communication technologies. The need for policies that prevent dangerous anthropogenic interference with the climate while addressing the energy needs of disadvantaged people is an essential challenge for the current era (Haines et al., 2007).

Although most renewable energy technologies can provide near-term benefits for health, for example by reducing exposure to air pollution, as well as mitigating greenhouse gas emissions, it is important that health impact assessments are undertaken. For example, a range of public health problems related to dams (which can be used to generate hydroelectricity and to promote adaptation to climate change through improved irrigation) have been documented, including increases in the prevalence of schistosomiasis, the introduction of Rift Valley fever, and increases in the burden of malaria. A recent systematic review concluded that although it was not possible to quantify the attributable fraction of the malaria burden due to dam building and irrigation, future water resource development projects should include in-depth assessment of potential effects (Keiser et al., 2005).

Conclusions

Climate change is likely to have far-reaching implications for human health and development. *The Stern Review* (Stern, 2006) has extensively reviewed the economic rationale for mitigation and adaptation polices, and also suggested that there is a strong economic case for action in the near term because the effects of climate change could result in losing around 5 percent of the gross world product (GWP) by the middle of the twenty-first century, perhaps even reaching 20 percent or more if the full range of effects is considered.

Infectious diseases are one of a number of categories of health outcomes that are likely to be affected adversely by climate change. Public health policies should take into account the need to adapt to a changing climate, as well as the potential for near-term benefits to health from a range of policies to mitigate climate change. Research funders should increase resources available to improve our understanding of the linkages between climate change, other environmental changes, and human health.

CLIMATE CHANGE AND HUMAN HEALTH[1]

Paul R. Epstein, M.D., M.P.H.

In 1998, Hurricane Mitch dropped six feet of rain on Central America in three days. In its wake, the incidence of malaria, dengue fever, cholera, and leptospirosis soared. In 2000, rain and three cyclones inundated Mozambique for six weeks, and the incidence of malaria rose fivefold. In 2003, a summer heat wave in Europe killed tens of thousands of people, wilted crops, set forests ablaze, and melted 10 percent of the Alpine glacial mass.

This summer's blistering heat wave was unprecedented with regard to intensity, duration, and geographic extent. More than 200 U.S. cities registered new record high temperatures. In Phoenix, Arizona, sustained temperatures above 100°F (38°C) for 39 consecutive days, including a week above 110°F (43°C), took a harsh toll on the homeless. Then came Hurricane Katrina, gathering steam from the heated Gulf of Mexico and causing devastation in coastal communities.

These sorts of extreme weather events reflect massive and ongoing changes in our climate to which biologic systems on all continents are reacting. So concluded the United Nations Intergovernmental Panel on Climate Change[2], a collaboration of more than 2000 scientists from 100 countries. In 2001, the panel concluded that humans are playing a major role in causing these changes, largely through deforestation and the combustion of fossil fuels that produce heat-trapping gases such as carbon dioxide.

---

[1]This article was originally published in the *New England Journal of Medicine* 353(14):1433-1436 (2005). Reprinted with permission from the Massachusetts Medical Society. Copyright 2005. All rights reserved. An interview with Dr. Epstein can be heard at www.nejm.org. Dr. Epstein is the associate director of the Center for Health and the Global Environment, Harvard Medical School, Boston.

[2]Houghton JT, Ding Y, Griggs DJ, et al., eds. Climate change 2001: the scientific basis: contribution of the Working Group I to the third assessment report of the Intergovernmental Panel on Climate Change. Cambridge, England: Cambridge University Press, 2001.

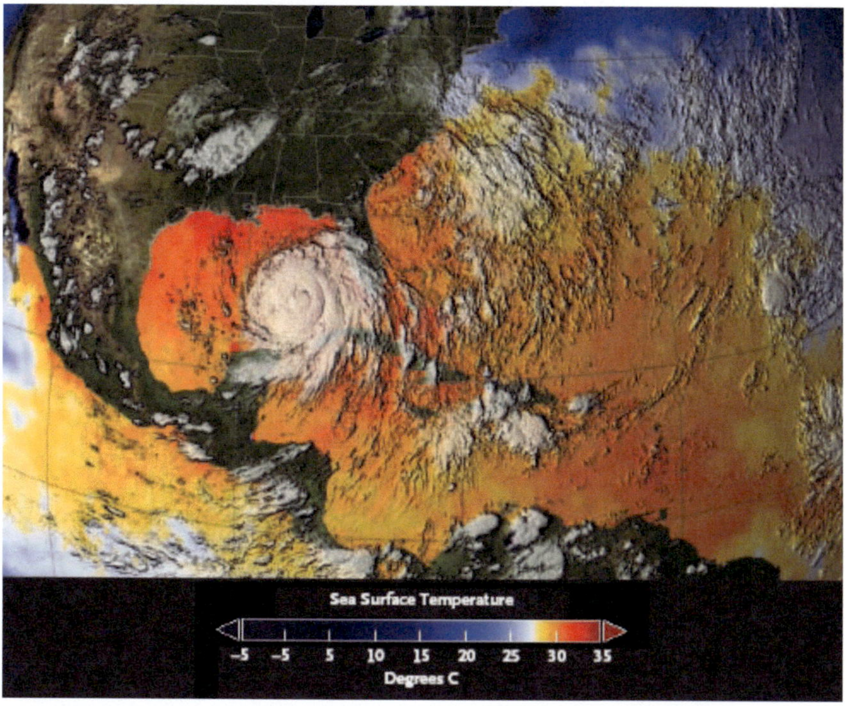

FIGURE 1-7 Hurricane Katrina passing over the Gulf of Mexico. The map shows the three-day average of sea-surface temperatures from August 25, 2005, through August 27, 2005, and Hurricane Katrina growing in strength and breadth as it passes over the unusually warm Gulf of Mexico. Yellow, orange, and red areas are at or above 82°F (27.8°C, the temperature required for hurricanes to strengthen). Since the 1970s, the number of category 4 and 5 hurricanes has increased as sea temperatures have risen.
SOURCE: Scientific Visualization Studio at NASA.

Since 2001, we've learned substantially more. The pace of atmospheric warming and the accumulation of carbon dioxide are quickening; polar and alpine ice is melting at rates not thought possible several years ago[3]; the deep ocean is heating up, and circumpolar winds are accelerating; and warming in the lower atmosphere is retarding the repair of the protective "ozone shield" in the stratosphere. Moreover, ice cores that are drilled in Greenland indicate that the climate can change abruptly. Given the current rate of carbon dioxide buildup and the projected degree of global warming, we are entering uncharted seas.

[3]Hassol SJ. ACIA, Impacts of a warming Arctic: arctic climate impact assessment. Cambridge, England: Cambridge University Press, 2004.

As we survey these seas, we can see some of the health effects that may lie ahead if the increase in very extreme weather events continues[4]. Heat waves like the one that hit Chicago in 1995, killing some 750 people and hospitalizing thousands, have become more common[2]. Hot, humid nights, which have become more frequent with global warming, magnify the effects. The 2003 European heat wave—involving temperatures that were 18°F (10°C) above the 30-year average, with no relief at night—killed 21,000 to 35,000 people in five countries.

But even more subtle, gradual climatic changes can damage human health. During the past two decades, the prevalence of asthma in the United States has quadrupled, in part because of climate-related factors. For Caribbean islanders, respiratory irritants come in dust clouds that emanate from Africa's expanding deserts and are then swept across the Atlantic by trade winds accelerated by the widening pressure gradients over warming oceans. Increased levels of plant pollen and soil fungi may also be involved. When ragweed is grown in conditions with twice the ambient level of carbon dioxide, the stalks sprout 10 percent taller than controls but produce 60 percent more pollen. Elevated carbon dioxide levels also promote the growth and sporulation of some soil fungi, and diesel particles help to deliver these aeroallergens deep into our alveoli and present them to immune cells along the way.

The melting of the earth's ice cover has already become a source of physical trauma. In Alaska, Inuits report an increase in accidents caused by walking on thin ice[3]. Ocean warming and Arctic thawing are also spawning severe winter storms and hazardous travel conditions in the continental United States. Although tropical sea surfaces are warming and becoming saltier, parts of the North Atlantic are freshening from melting polar ice and increased amounts of rain falling at high latitudes. Contrasting barometric pressures over changing oceans increase winds and propel storms.

Meanwhile, in the past three decades, widening social inequities and changes in biodiversity—which alter the balance among predators, competitors, and prey that help keep pests and pathogens in check—have apparently contributed to the resurgence of infectious diseases. Global warming and wider fluctuations in weather help to spread these diseases: temperature constrains the range of microbes and vectors, and weather affects the timing and intensity of disease outbreaks[5]. Disease-bearing ticks in Sweden are moving northward as winters become warmer, and models project a similar shift in the United States and Canada. The encroachment of human housing on wilderness and reductions in the populations of predators of deer and competitors of mice are largely responsible for the current spread of Lyme disease.

---

[4]Leaf A. Potential health effects of global climatic and environmental changes. N Engl J Med 1989; 321:1577-83.

[5]McMichael AJ, Campbell-Lendrum DH, Corvalán CF, et al., eds. Climate change and human health: risks and responses. Geneva: World Health Organization, 2003:250.

Mosquitoes, which can carry many diseases, are very sensitive to temperature changes. Warming of their environment—within their viable range—boosts their rates of reproduction and the number of blood meals they take, prolongs their breeding season, and shortens the maturation period for the microbes they disperse. In highland regions, as permafrost thaws and glaciers retreat, mosquitoes and plant communities are migrating to higher ground[6].

The increased weather variability that accompanies climate instability contributed to the emergence of both the hantavirus pulmonary syndrome and West Nile virus in the United States. Six years of drought in the Southwest apparently reduced the populations of predators, and early heavy rainfall in 1993 produced a bounty of piñon nuts and grasshoppers for rodents to eat. The resulting legions of white-footed mice heralded the appearance of hantavirus in the Americas. The origin of the 1999 outbreak of West Nile virus in New York City remains a mystery, but city-dwelling, bird-biting *Culex pipiens* mosquitoes thrive in shallow pools of foul water that remain in drains during droughts. When dry springs yield to sweltering summers, viral development accelerates and, with it, the cycle of mosquito-to-bird transmission. During the hot, arid summer of 2002, West Nile virus traveled across the country, infecting 230 species of animals, including 138 species of birds, along the way. Many of the affected birds of prey normally help to rein in rodent populations that can spread hantaviruses, arenaviruses, and yersinia and leptospira bacteria, as well as ticks infected with *Borrelia burgdorferi*.

Extremely wet weather may bring its own share of ills. Floods are frequently followed by disease clusters: downpours can drive rodents from burrows, deposit mosquito-breeding sites, foster fungus growth in houses, and flush pathogens, nutrients, and chemicals into waterways. Milwaukee's cryptosporidium outbreak, for instance, accompanied the 1993 floods of the Mississippi River, and norovirus and toxins spread in Katrina's wake. Major coastal storms can also trigger harmful algal blooms ("red tides"), which can be toxic, help to create hypoxic "dead zones" in gulfs and bays, and harbor pathogens.

Prolonged droughts, for their part, can weaken trees' defenses against infestations and promote wildfires, which can cause injuries, burns, respiratory illness, and deaths. Shifting weather patterns are jeopardizing water quality and quantity in many countries, where groundwater systems are already being overdrawn and underfed. Most montane ice fields are predicted to disappear during this century—removing a primary source of water for humans, livestock, and agriculture in some parts of the world.

A still greater threat to human health comes from illnesses affecting wildlife, livestock, crops, forests, and marine organisms. The Millennium Ecosystem Assessment of 2005 revealed that 60 percent of the resources and life-support

---

[6]Epstein PR, Diaz HF, Elias S, et al. Biological and physical signs of climate change: focus on mosquito-borne diseases. Bull Am Meteorol Soc 1998;78:409-17.

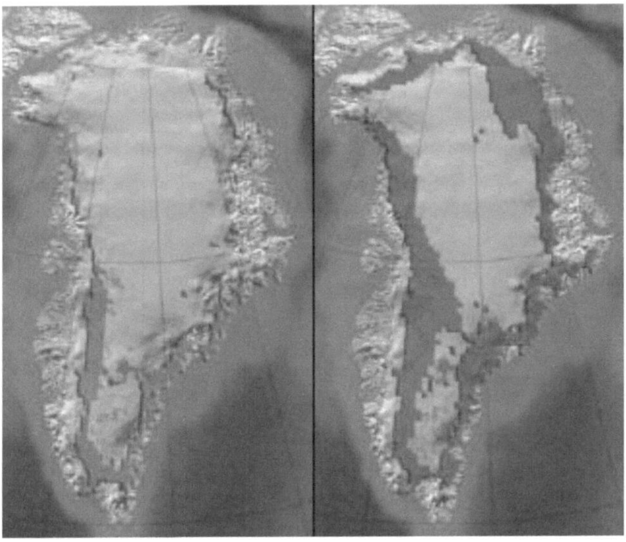

FIGURE 1-8 Increase from 1992 (left) to 2002 (right) in the amount of the Greenland ice sheet melted in the summer. The extent of seasonal melting on the Greenland ice sheet has been observed by satellite since 1979. The melt zone (dark gray), where summer warmth turns snow and ice around the edges of the ice sheet into slush and water, has been expanding inland and to record-high elevations in recent years. When the meltwater seeps through cracks in the ice sheet, it may accelerate melting and allow ice to slide more easily over bedrock, speeding its movement to the sea. In addition to contributing to a rising sea level, this process adds freshwater to the ocean, with potential effects on ocean circulation and regional climate.
SOURCE: Map by C. Grabhorn. Reprinted from the Arctic Climate Impact Assessment (2004) with permission from Cambridge University Press and C. Grabhorn.

systems examined—from fisheries to fresh water—are already in decline or are being used in unsustainable ways. The resulting biologic impoverishment may have important consequences for our air, food, and water.

Crops are being confronted with more volatile weather, vanishing pollinators, and the proliferation of pests and pathogens. One fungal disease, soybean rust, is thought to have been ushered into the United States by Hurricane Ivan last fall. Warmth and moisture will favor its propagation.

And many habitats are not faring well. Our coastal zones, for example, are in trouble: coral reefs are suffering from warming-induced "bleaching," excess waste, physical damage, overfishing, and fungal and bacterial diseases. Reefs provide a buffer against storms and groundwater salinization and offer protection for fish, the primary protein source for many inhabitants of island nations. One reef resident, the cone snail, produces a peptide that is 1000 times as potent as morphine and that is not addictive. We may never know what other potential treatments will be lost as reefs deteriorate.

All in all, it would appear that we may be underestimating the breadth of biologic responses to changes in climate. Treating climate-related ills will require preparation, and early-warning systems forecasting extreme weather can help to reduce casualties and curtail the spread of disease. But primary prevention would require halting the extraction, mining, transport, refining, and combustion of fossil fuels—a transformation that many experts believe would have innumerable health and environmental benefits and would help to stabilize the climate.

The good news is that we may also be underestimating the economic benefits of the clean-energy transition. When the financial incentives are adequate, renewable energy, energy-efficient and hybrid technologies, "green buildings," and expanded public transportation systems can constitute an engine of growth for the 21st century.

CLIMATE CHANGE FUTURES: HEALTH, ECOLOGICAL, AND ECONOMIC DIMENSIONS[7]

The Center for Health and the Global Environment
Harvard Medical School

Preamble

"Imagining the unmanageable" was to be the subtitle for the Climate Change Futures report. But the devastating series of intense, immense fall hurricanes besetting the United States displaced it. What were once extreme scenarios for the US have occurred, and the consequences have cascaded across the physical landscape, overwhelming the capacities of health, ecological and economic systems to absorb, adapt to and manage the change.

Hurricane Katrina killed over 1,000 people, displaced over a million people, and spread oil, toxins, microorganisms and deep losses throughout the US Gulf Coast. It revealed deep-seated inequities and vulnerabilities, and the shock waves have reverberated through all sectors of society. The need for prevention has become embedded into our future political landscape.

While no one event is diagnostic of climate change, the relentless pace of unusually severe weather since 2001—prolonged droughts, heat waves of extraordinary intensity, violent windstorms and more frequent "100-year" floods—is descriptive of a changing climate.

The reasons for the changed weather patterns are well understood. Five

[7]Reprinted with permission from the Center for Health and the Global Environment. 2005. *Climate change futures: health, ecological and economic dimensions.* Cambridge, MA: Harvard Medical School. Sponsored by Swiss Re and the United Nations Development Programme.

years ago, Levitus and colleagues at the US Department of Commerce's National Oceanic and Atmospheric Administration reported that the world's oceans had warmed to a depth of two miles in five decades. This year Barnett and colleagues at the Scripps Institution of Oceanography reported that the oceans had absorbed 84% of the globe's warming and that the warming pattern is unmistakably attributable to human activities.

Because of the natural cycles on which global warming is superimposed, the overall frequency of hurricanes ebbs and flows. But, since the 1970s, tropical storm destructiveness (peak winds and duration) (Emanuel 2005) and the frequency of category 4 and 5 storms (Webster et al. 2005) have essentially doubled. These observations are correlated with warming tropical seas, and the scientists project that continued warming will likely enhance the frequency of large storms still further.

Warm sea surfaces evaporate quickly and, with the deep ocean warming, the water replenishing that which evaporates is also warm and fuels subsequent storms. A warmer atmosphere also holds more water vapor, and the accelerated water cycle generates more droughts and more floods (see Figure 1-9; Trenberth, 2005).

This fall's succession of megastorms is, at the very least, a harbinger of what we can expect more of in a changing climate (Kerr 2005). The series may also mark a turning point in our understanding of how an energized climate system is exaggerating natural phenomena and of just how rapidly climate has changed.

This multidimensional assessment of climate change includes trends, case studies and scenarios—with a focus on health, ecological and economic dimensions. One surprise is the vulnerability of the energy sector—the primary source of increased heat in Earth systems. The risks to oil production compound the threats to the electricity grid from heat waves and the instabilities of pipelines grounded in thawing tundra.

At the same time, recovery, adaptation and prevention open the door to enormous opportunities. Developing a diversified portfolio of safe, well-distributed and nonpolluting energy sources, with hybrids and complementing technologies, can fortify energy security, bolster public health, promote economic activity and help stabilize the climate. Bold initiatives and innovative measures spearheading a well-funded, well-insured clean energy transition may be just the components needed to build a sustainable engine of growth for the 21st century.

Paul Epstein and Evan Mills

Executive Summary

Climate is the context for life on earth. Global climate change and the ripples of that change will affect every aspect of life, from municipal budgets for snowplowing to the spread of disease. Climate is already changing, and quite rapidly.

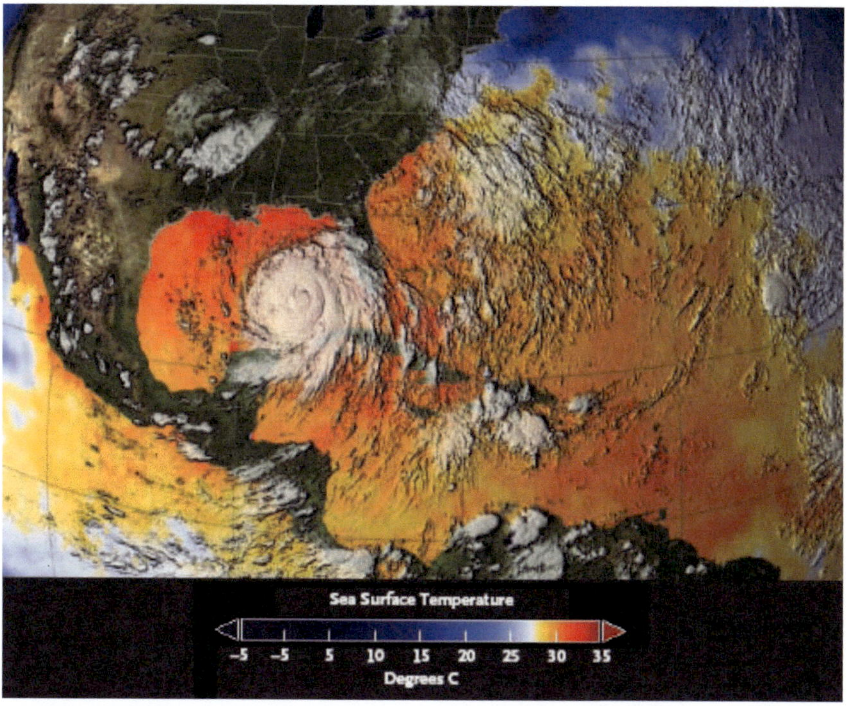

FIGURE 1-9 Warm ocean waters fuel hurricanes. This image depicts the three-day average of sea surface temperatures (SSTs) from August 25-27, 2005, and the growing breadth of Hurricane Katrina as it passed over the warm Gulf of Mexico. Yellow, orange and red areas are at or above 82°F (27.7°C), the temperature needed for hurricanes to strengthen. By late August SSTs in the Gulf were well over 90°F (32.2°C).
SOURCE: NASA/Goddard Space Flight Center/Scientific Visualization Studio.

With rare unanimity, the scientific community warns of more abrupt and greater change in the future.

Many in the business community have begun to understand the risks that lie ahead. Insurers and reinsurers find themselves on the front lines of this challenge since the very viability of their industry rests on the proper appreciation of risk. In the case of climate, however, the bewildering complexity of the changes and feedbacks set in motion by a changing climate defy a narrow focus on sectors. For example, the effects of hurricanes can extend far beyond coastal properties to the heartland through their impact on offshore drilling and oil prices. Imagining the cascade of effects of climate change calls for a new approach to assessing risk.

The worst-case scenarios would portray events so disruptive to human enterprise as to be meaningless if viewed in simple economic terms. On the other hand, some scenarios are far more positive (depending on how society reacts to the threat of change). In addition to examining current trends in events and costs, and exploring case studies of some of the crucial health problems facing society and the natural systems around us, "Climate Change Futures: Health, Ecological and Economic Dimensions" uses scenarios to organize the vast, fluid possibilities of a planetary-scale threat in a manner intended to be useful to policymakers, business leaders and individuals.

Most discussions of climate impacts and scenarios stay close to the natural sciences, with scant notice of the potential economic consequences. In addition, the technical literature often "stovepipes" issues, zeroing in on specific types of events in isolation from the real-world mosaic of interrelated vulnerabilities, events and impacts. The impacts of climate change cross national borders and disciplinary lines, and can cascade through many sectors. For this reason we all have a stake in adapting to and slowing the rate of climate change. Thus, sound policymaking demands the attention and commitment of all.

While stipulating the ubiquity of the threat of climate change, understanding the problem still requires a lens through which the problem might be approached. "Climate Change Futures" focuses on health. The underlying premise of this report is that climate change will affect the health of humans as well as the ecosystems and species on which we depend, and that these health impacts will have economic consequences. The insurance industry will be at the center of this nexus, both absorbing risk and, through its pricing and recommendations, helping business and society adapt to and reduce these new risks. Our hope is that Climate Change Futures (CCF) will not only help businesses avoid risks, but also identify opportunities and solutions. An integrated assessment of how climate change is now adversely affecting and will continue to affect health and economies can help mobilize the attention of ordinary citizens around the world and help generate the development of climate-friendly products, projects and policies. With early action and innovative policies, business can enhance the world's ability to adapt to change and help restabilize the climate.

Why Scenarios?

CCF is not the first report on climate change to use scenarios. The Intergovernmental Panel on Climate Change (IPCC) employs six of the very long-term and very broad scenarios representative of the many scenarios considered. Other organizations have explored scenarios of climate trajectories, impacts for some sectors and the mix of energy sources, to explore the potential consequences of trends and actions taken today. Scenarios are not simple projections, but are stories that present alternative images of how the future might unfold. Handled carefully, scenarios can help explore potential consequences of the interplay of multiple variables and thereby help us to make considered and comprehensive decisions.

The IPCC scenarios, contained in The Special Report on Emissions Scenarios (SRES), make projections into the next century and beyond and assume that climate change will be linear and involve gradual warming. But events of the last five years have overtaken the initial SRES scenarios. Climate has changed faster and more unpredictably than the scenarios outlined. Many of the phenomena assumed to lie decades in the future are already well underway. This faster pace of change on many fronts indicates that more sector-specific, short-term scenarios are needed.

With this in mind, the CCF scenarios are designed to complement the far-reaching IPCC framework. Drawing upon the full-blown, long-term scenarios offered by the IPCC, we have developed two scenarios that highlight possibilities inadequately considered in past assessments of climate change impacts.

Both CCF scenarios envision a climate context of gradual warming with growing variability and more weather extremes. Both scenarios are based on "business-as usual," which, if unabated, would lead to doubling of atmospheric CO_2 from pre-industrial values by midcentury. Both are based on the current climate trends for steady warming along with an increase in extremes, with greater and costlier impacts. The compilation of extreme weather events of all types shows a clear increase over the past decade in the number of extremes occurring in both hemispheres (see Figure 1-10, below).

Overall costs from catastrophic weather-related events rose from an average of US $4 billion per year during the 1950s, to US $46 billion per year in the 1990s, and almost double that in 2004. In 2004, the combined weather-related losses from catastrophic and small events were US $107 billion, setting a new record. (Total losses in 2004, including non-weather-related losses, were US $123 billion; Swiss Re 2005a). With Hurricanes Katrina and Rita, 2005 had, by September, broken all-time records yet again.

Meanwhile, the insured percentage of catastrophic losses nearly tripled from 11% in the 1960s to 26% in the 1990s and reached 42% (US $44.6 billion) in 2004 (all values inflation-corrected to 2004 dollars, Munich Re NatCatSERVICE).

As an insurer of last resort, the US Federal Emergency Management Agency has experienced escalating costs for natural disasters since 1990. Moreover, in the past decade, an increasing proportion of extreme weather events have been occurring in developed nations (Europe, Japan and the US) (see chart below).

The first impact scenario, or CCF-I, portrays a world with an increased correlation and geographical simultaneity of extreme events, generating an overwhelming strain for some stakeholders. CCF-I envisions a growing frequency and intensity of weather extremes accompanied by disease outbreaks and infestations that harm humans, wildlife, forests, crops and coastal marine systems. The events and their aftermaths would strain coping capacities in developing and developed nations and threaten resources and industries, such as timber, tourism, travel and the energy sector. The ripples from the damage to the energy sector would be felt throughout the economy.

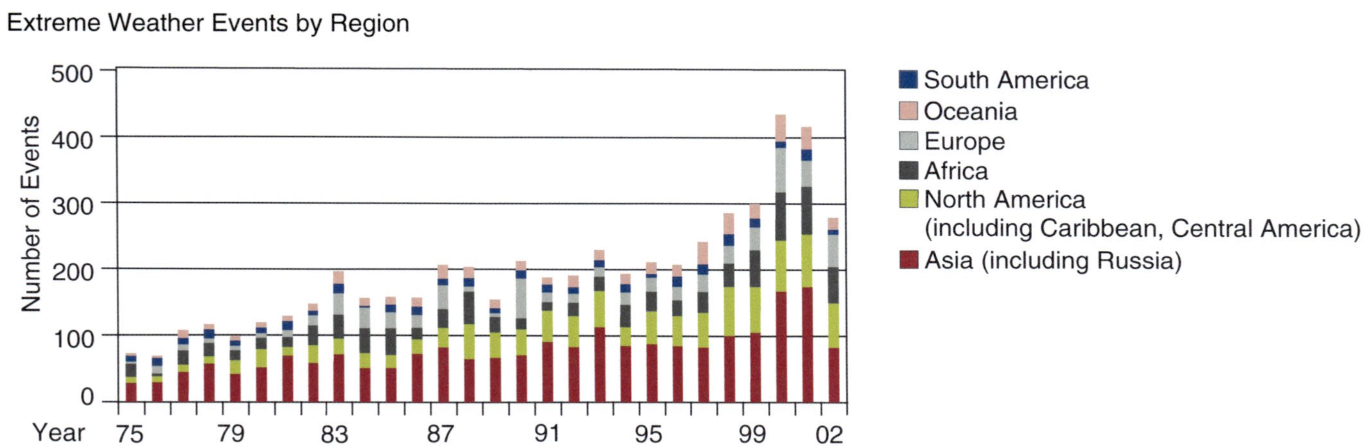

FIGURE 1-10 These data are taken from EMDAT (Emergency Events Database) from 1975 to 2002. Compiled by the Center for Research on the Epidemiology of Disasters (CRED) at the Universite-Catholique de Louvain in Brussels, Belgium, this data set draws from multiple disaster relief organizations (such as OFDA and USAID), and is therefore skewed toward events that have large human impacts. The data taken from EMDAT were created with the initial support of WHO and the Belgian government.

BOX 1-2
Key Points

1. Warming favors the spread of disease.
2. Extreme weather events create conditions conducive to disease outbreaks.
3. Climate change and infectious diseases threaten wildlife, livestock, agriculture, forests and marine life, which provide us with essential resources and constitute our life-support systems.
4. Climate instability and the spread of diseases are not good for business.
5. The impacts of climate change could increase incrementally over decades.
6. Some impacts of warming and greater weather volatility could occur suddenly and become widespread.
7. Coastal human communities, coral reefs and forests are particularly vulnerable to warming and disease, especially as the return time between extremes shortens.
8. An increasingly unstable climate could shift and settle into a new equilibrium, allowing a measure of adaptation and the opportunity to rapidly reduce the global environmental influence of human activities, namely fossil fuel combustion and deforestation.
9. A well-funded, well-insured program to develop and distribute a diverse suite of means to generate energy cleanly, efficiently and safely offers enormous business opportunities and may present the most secure means of restabilizing the climate.
10. Solutions to the emerging energy crisis must be thoroughly scrutinized as to their life cycle impacts on health and safety, environmental integrity, global security and the international economy.

In CCF-I, an accelerated water cycle and retreat of most glaciers undermine water supplies in some regions and land integrity in others. Melting of permafrost (permanently frozen land) in the Arctic becomes more pronounced, threatening native peoples and northern ecosystems. And gradually rising seas, compounded by more destructive storms cascading over deteriorating barrier reefs, threaten all low-lying regions.

Taken in aggregate, these and other effects of a warming and more variable climate could threaten economies worldwide. In CCF-I, some parts of the developed world may be capable of responding to the disruptions, but the events would be particularly punishing for developing countries. For the world over, historical weather patterns would diminish in value as guides to forecasting the future.

The second impact scenario, CCF-II, envisions a world in which the warming and enhanced variability produce surprisingly destructive consequences. It explores a future rife with the potential for sudden, wide-scale health, environmental and economic impacts as climate change pushes ecosystems past tipping points. As such, it is a future inherently more chaotic and unpredictable than CCF-I.

BOX 1-3
Vulnerabilities in the Energy Sector

- Heat waves generate blackouts.
- Sequential storms disrupt offshore oil rigs, pipelines, refineries and distribution systems.
- Diminished river flows reduce hydroelectric capacity and impede barge transport.
- Melting tundra undermines pipelines and power transmission lines.
- Warmed inland waters shut down power plant cooling systems.
- Lightning claims rise with warming.

Each stage in the life cycle of oil, including exploration, extraction, transport, refining and combustion, carries hazards for human health and the environment. More intense storms, thawing permafrost and dried riverbeds, make every stage more precarious.

Some of the impacts envisioned by the second scenario are very severe and would involve catastrophic, widespread damages, with a world economy beset by increased costs and chronic, unmanageable risks. Climate-related disruptions would no longer be contained or confined.

Threshold-crossing events in both terrestrial and marine systems would severely compromise resources and ecological functions, with multiple consequences for the species that depend upon them. For example:

- Repeated heat waves on the order of the 2003 and 2005 summers could severely harm populations, kill livestock, wilt crops, melt glaciers and spread wildfires.
- The probability of such extreme heat has already increased between two and four times over the past century and, based on an IPCC climate scenario, more than half the years by the 2040s will have summers warmer than that of 2003.
- Chronic water shortages would become more prevalent, especially in semi-arid regions, such as the US West.
- With current usage levels, more environmentally displaced persons and a changing water cycle, the number of people suffering water stress and scarcity today will triple in two decades.

Other non-linear impact scenarios include:

- Widespread diebacks of temperate and northern forests from drought and pests.

- Coral reefs, already multiply stressed, collapse from the effects of warming and diseases.
- Large spikes occur in property damages from a rise of major rivers. (A 10% increase in flood peak would produce 100 times the damage of previous floods, as waters breach dams and levees.)
- Severe storms and extreme events occurring sequentially and concurrently across the globe overwhelm the adaptive capacities of even developed nations; large areas and sectors become uninsurable; major investments collapse; and markets crash.

CCF-II would involve blows to the world economy sufficiently severe to cripple the resilience that enables affluent countries to respond to catastrophes. In effect, parts of developed countries would experience developing nation conditions for prolonged periods as a result of natural catastrophes and increasing vulnerability due to the abbreviated return times of extreme events.

Still, CCF-II is not a worst-case scenario.

A worst-case scenario would include large-scale, nonlinear disruptions in the climate system itself—slippage of ice sheets from Antarctica or Greenland, raising sea levels inches to feet; accelerated thawing of permafrost, with release of large quantities of methane; and shifts in ocean thermohaline circulation (the stabilizing ocean "conveyor belt").

Finally, there are scenarios of climate stabilization. Restabilizing the climate will depend on the global-scale implementation of measures to reduce greenhouse gas emissions. Aggressively embarking on the path of non-fossil fuel energy systems will take planning and substantive financial incentives—not merely a handful of temporizing, corrective measures.

This assessment examines signs and symptoms suggesting growing climate instability and explores some of the expanding opportunities presented by this historic challenge.

Applying the Scenarios

In choosing how to apply the two impact scenarios, we have focused on case studies of specific health and ecological consequences that extend beyond the more widely studied issue of property damages stemming from warming and natural catastrophes. In each case study, we identify current trends under way and envision the future consequences for economies, social stability and public health.

Infectious diseases have resurged in humans and in many other species in the past three decades. Many factors, including land-use changes and growing poverty, have contributed to the increase. Our examination of malaria, West Nile virus and Lyme disease explores the role of warming and weather extremes in expanding the range and intensity of these diseases and both linear and non-linear projections for humans and wildlife.

The rising rate of asthma (two to threefold increase in the past two decades; fourfold in the US) receives special attention, as air quality is affected by many aspects of a changing climate (wildfires, transported dust and heat waves), and by the inexorable rise of atmospheric CO_2 in and of itself, which boosts ragweed pollen and some soil molds.

We also examine the public health consequences of natural catastrophes themselves, including heat waves and floods. An integrated approach exploring linkages is particularly useful in these instances, since the stovepipe perspective tends to play down the very real health consequences and the manifold social and economic ripples stemming from catastrophic events.

Another broad approach of the CCF scenarios is to study climate change impacts on ecological systems, both managed and natural. We examine projections for agricultural productivity that, to date, largely omit the potentially devastating effects of more weather extremes and the spread of pests and pathogens. Crop losses from pests, pathogens and weeds could rise from the current 42% to 50% within the coming decade.

BOX 1-4
Case Studies in Brief

Infectious and Respiratory Diseases

- **Malaria:** Malaria is the deadliest, most disabling and most economically damaging mosquito-borne disease worldwide. Warming affects its range, and extreme weather events can precipitate large outbreaks. This study documents the fivefold increase in illness following a 6-week flood in Mozambique, explores the surprising role of drought in northeast Brazil, and projects changes for malaria in the highlands of Zimbabwe.

- **West Nile Virus:** West Nile virus (WNV) is an urban-based, mosquito-borne infection, afflicting humans, horses and more than 138 species of birds. Present in the United States, Europe, the Middle East and Africa, warm winters and spring droughts play roles in amplifying this disease. To date, there have been over 17,000 human cases and over 650 deaths from WNV in North America.

- **Lyme Disease:** Lyme disease is the most widespread vector-borne disease in the US and can cause long-term disability. Lyme disease is spreading in North America and Europe as winters warm, and models project that warming will continue to shift the suitable range for the deer ticks that carry this infection.

continued

BOX 1-4 Continued

- **Asthma:** Asthma prevalence has quadrupled in the US since 1980, and this condition is increasing in developed and underdeveloped nations. New drivers include rising CO_2, which increases the allergenic plant pollens and some soil fungi, and dust clouds containing particles and microbes coming from expanding deserts, compounding the effects of air pollutants and smog from the burning of fossil fuels.

Extreme Weather Events

- **Temperature:** Heat waves are becoming more common and more intense throughout the world. This study explores the multiple impacts of the highly anomalous 2003 summer heat wave in Europe and the potential impact of such "outlier" events elsewhere for human health, forests, agricultural yields, mountain glaciers and utility grids.

- **Flooding:** Floods inundated large parts of Central Europe in 2002 and had consequences for human health and infrastructure. Serious floods occurred again in Central Europe in 2005. The return times for such inundations are projected to decrease in developed and developing nations, and climate change is expected to result in more heavy rainfall events.

Natural and Managed Systems

- **Drought:** Forests are experiencing numerous pest infestations. Warming increases the range, reproductive rates and activity of pests, such as spruce bark beetles, while drought makes trees more susceptible to the pests. This study examines the synergies of drought and pests, and the dangers of wildfire. Large-scale forest diebacks are possible, and they would have severe consequences for human health, property, wildlife, timber and Earth's carbon cycle.

- **Agriculture:** Agriculture faces warming, more extremes and more diseases. More drought and flooding under the new climate, and accompanying outbreaks of crop pests and diseases, can affect yields, nutrition, food prices and political stability. Chemical measures to limit infestations are costly and unhealthy.

- **Marine Environments:** Marine ecosystems are under increasing pressure from overfishing, excess wastes, loss of wetlands, and diseases of bivalves that normally filter and clean bays and estuaries. Even slightly elevated ocean temperatures can destroy the symbiotic relationship between algae and animal polyps that make up coral reefs, which buffer shores, harbor fish and contain organisms with powerful chemicals useful to medicine. Warming seas and diseases may cause coral reefs to collapse.

- **Fresh Water Availability:** Water, life's essential ingredient, faces enormous threats. Underground stores are being overdrawn and underfed. As weather patterns shift and mountain ice fields disappear, changes in water quality and availability will pose growth limitations on human settlements, agriculture and hydropower. Flooding can lead to water contamination with toxic chemicals and microbes, and natural disasters routinely damage water-delivery infrastructure.

IMPACTS OF GLOBAL CLIMATE CHANGE ON INFECTIOUS DISEASES

Donald S. Burke, M.D.[8]
University of Pittsburgh

Relationships between weather and health have been appreciated since the dawn of civilization. Hippocrates, in 400 B.C.E. (*On Airs, Waters, and Places*), observed that

> [w]hoever would study medicine aright must learn of the following subjects. First he must consider the effect of the *seasons* of the year and the differences between them. Secondly he must study the *warm and cold winds*, both those which are in common to every country and those peculiar to a particular locality. Lastly, the *effect of water* on the health must not be forgotten.

Can measurements of climate and weather be used to predict epidemics? The purpose of this paper is to review exactly what is and what is not known about the causal relationships between changes in climate and weather and infectious diseases. I do not attempt to present an exhaustive review, but I touch on key issues and guide the interested reader to more comprehensive sources. I focus especially on some computational approaches that can be used to strengthen the state of the art.

Recent Expert Summary Reports

In 2001, I chaired a National Research Council (NRC) committee that issued a report entitled *Under the Weather: Climate, Ecosystems, and Infectious Diseases* (NRC, 2001). This expert committee concluded that although it is clear that weather fluctuations and seasonal-to-interannual climate variability influence many infectious diseases, published observational and modeling studies must be interpreted cautiously, and consequently the exact future impacts of global climate change remain uncertain. On a brighter note, the *Under the Weather* committee did go on to say that recent technological and modeling advances are improving our abilities to identify patterns in infectious disease epidemiology.

More recently, two Nobel Prize-winning documents have appeared that illuminate the current knowledge about the relationships of climate change and human health. Last year the IPCC released its Fourth Assessment Report including the "Impacts, Adaptation and Vulnerability" Working Group section (Chapter 8) on human health (IPCC, 2007). The IPCC meticulously explained the potential range of direct, indirect, and socially mediated effects of climate change on human health, including careful caveats about how these effects may

[8]Dean, Graduate School of Public Health.

be modulated by future environmental, societal, and health system changes. The IPCC then ranked its findings under the headings very high confidence, high confidence, and medium confidence. High confidence negative impacts of climate change on human health included more malnutrition; more deaths, injuries, and diseases from extreme weather events; and more cardiorespiratory diseases from the deterioration of air quality. However, findings related to infectious diseases were careful and circumscribed. The IPCC expressed very high confidence that there would be changes in the distribution of malaria, but noted that in various locales the direction of change could be either an increase *or* a decrease in malaria, depending on predicted redistributions of rainfall, without a clear conclusion as to whether the net global impact would be positive or negative. Similarly, the group predicted with high confidence that there would be changes in the distribution of many insect vectors of other diseases, in most instances with a negative impact on human health but in some locales a favorable impact. The group expressed medium confidence that there would be a net global increase in the burden of diarrheal diseases. Overall, I interpret the IPCC conclusions to be remarkably measured and quite consistent with our *Under the Weather* findings.

Former U.S. Vice President Al Gore's documentary *An Inconvenient Truth* also treads carefully around possible relationships between global warming and epidemic diseases (Gore, 2006). Here is what he says about infectious diseases in the Oscar-winning feature film:

> There are cities that were founded because they were just above the mosquito line. Nairobi is one. Harare is another. There are plenty of others. Now the mosquitoes with warming are climbing to the higher altitudes. There are a lot of vectors for infectious diseases that are worrisome to us that are also expanding their ranges, not only mosquitoes but all of these others as well. And we've had 30 so-called new diseases that have emerged just in the last quarter-century. And a lot of them like SARS [severe acute respiratory syndrome] have caused tremendous problems. The resistant forms of tuberculosis. There are others. And there's been a reemergence of some diseases that were once under control. The avian flu, of course, [is] quite a serious matter as you know. West Nile virus. It came to the eastern shore of Maryland in 1999. Two years later it was across the Mississippi. And two years after that it had spread across the continent. But these are very troubling signs.

Although Vice President Gore cites changes in vector distributions and the threat of emerging diseases, he stops short of claiming a strong causal linkage of particular diseases to climate change.

In sum, assessments by respected scientific bodies have stayed clear of apocalyptic pronouncements. Conclusions about the probable impact of global climate change on epidemic infectious diseases have been cautious and measured.

Uncertainties and Complexities

Why so cautious? My own view is that this caution honestly reflects the uncertainties involved, which in turn reflect the difficulty of the underlying scientific problems. While the idea that infectious diseases are driven by weather and climate is inescapable, our predictive powers remain limited. Foremost among the underlying scientific problems is the fact that epidemics are inherently dynamic, nonlinear processes, composed of many interacting subsystems, only one of which is weather and climate. Other subsystems of the dynamic epidemic system include human social interactions, waxing and waning of immunity at the population level, various health system interventions, and other factors. Ability to detect a strong cause-and-effect relationship between any single weather factor—say ambient temperature—and disease incidence requires a good understanding of the entire system and its dynamics. It is now painfully clear that any robust assessment of the impact of climate change on epidemic infectious diseases must be evaluated using dynamical models; the word arguments and static arrow diagrams ("St. Sebastian diagrams") so often used in the past are simply not persuasive.

Various infectious diseases are affected differently by environmental conditions (NRC, 2001). As may be seen in Table 1-1, for example, warm temperatures favor insect vector-borne diseases such as malaria and dengue, as evidenced by their tropical distributions and warm season peak incidences; cold temperatures favor influenza, as evidenced by its typical winter appearance; dryness favors meningocci (meningitis) and coccidioidomycosis, as evidenced by their association with arid conditions and dust storms; and wet conditions favor cryptosporidiosis and Rift Valley fever, as evidenced by their association with flooding. Obviously, while climate change in any given direction may increase the incidence of one disease, it may reduce the incidence of other diseases. For any given geographic locale, the net effects on human health may be negative, neutral, or positive, depending on local infectious disease circumstances. Predic-

TABLE 1-1 Examples of Environmental Factors Known to Be Strongly Associated with Certain Specific Infectious Diseases

Environmental Condition	Disease Favored	Evidence
Warm	Malaria, dengue	Primarily tropical distribution, seasonal transmission pattern
Cold	Influenza	Seasonal transmission pattern
Dry	Meningococcal meningitis, coccidioidomycosis	Associated with arid conditions, dust storms
Wet	Cryptosporidiosis, Rift Valley fever	Associated with flooding

SOURCE: NRC (2001).

tions about the net effects of global warming will require a better understanding of its effects on each particular infectious disease.

Influenza and dengue are both globally important infectious diseases, capable of serious epidemic spread. Although both are caused by small enveloped RNA viruses, one (influenza) is a cool season disease, and one (dengue) is a warm season disease. A comparison is instructive.

Seasonality of Influenza

A comparison of the incidence curves in the United States, Mexico, Colombia, Brazil, and Argentina shows that the timing of annual influenza incidence varies closely with latitude. Epidemic peaks occur in the winter season in high-latitude countries (alternating with the Northern and Southern Hemisphere winters), but no clear annual peak is seen in countries closer to the equator. In addition to seasonal changes in ambient temperature, a variety of explanations have been proposed to account for the strong wintertime seasonality of influenza (for some recent reviews, see Dowell, 2001; Finkelman et al., 2007; Fisman, 2007; Lofgren et al., 2007). One set of explanations focuses on altered host immunity, possibly due to changes in the photoperiod with shortened days in winter or to low vitamin D caused by decreased exposure to sunshine. Another set of explanations focuses on increased human-to-human exposure in winter—for example, increased social contact among children during the typical school season or breathing shared air in confined spaces. Indeed it is possible that more than one factor contributes to the seasonality of influenza.

Experimental studies in laboratory animals have been valuable in understanding the effects of altered environmental conditions on influenza transmission from animal to animal. More than 40 years ago, Kilbourne and colleagues, studying influenza transmission in caged mice, showed that low temperature and low relative humidity can increase influenza transmission (Schulman and Kilbourne, 1963). Recent studies by Palase and colleagues have provided even more compelling evidence and are summarized here (Lowen et al., 2007). They studied guinea pigs in a climate-controlled setup in which air flow direction, temperature, and relative humidity could be carefully controlled. Pairs of guinea pigs were placed in adjacent cages, one intentionally inoculated with the H3N2 A/Panama/99 strain influenza virus (a typical strain) and placed in the cage upwind of the other noninfected animal. Four pairs of animals were studied at a time under controlled conditions of temperature (5 to 30°C) and relative humidity (20 to 80 percent). Results are shown in Figure 1-11. Under conditions of high temperature or high relative humidity, none of the four downwind guinea pigs breathing infectious air became infected. In contrast, under conditions of low temperature and/or low relative humidity, four of four downwind animals invariably became infected. Increased viral shedding into respiratory secretions was found in animals held at lower temperatures. The authors concluded that both low temperature and

	20%RH	35%RH	50%RH	65%RH	80%RH
5 deg		4	4	2	2
5 deg		4	3	2	2
20 deg	4	4	1	3	0
20 deg	3	4	1	3	0
30 deg		0			
30 deg		0			

FIGURE 1-11 Transmission of influenza from infected guinea pigs to uninfected exposed guinea pigs under different experimental conditions in which ambient temperature and relative humidity were varied. Each box shows the results of one experimental study in which four guinea pigs were exposed. Results are shown as the number (of these four) that became infected. Boxes are colored, with red indicating that all (4/4) exposed animals became infected, green indicating that no (0/4) exposed animals became infected, and yellow indicating that some but not all (1, 2, or 3 out of 4) became infected. Black indicates that no experiments were conducted for those specific conditions.
SOURCE: Based on data provided in Lowen et al. (2007).

low relative humidity increased transmission, the temperature effect mediated by increased virus shedding and the relative humidity effect probably by bioaerosol formation of droplet nuclei. Although these fine experiments do provide strong evidence of the mechanism relating winter to flu, they provide little help in formulating predictive models on which to base future influenza transmission scenarios. How much of a seasonal change in transmissibility (e.g., shedding, droplet nuclei formation) is needed to see the typical seasonal oscillations of influenza? The answer is that it depends on how the magnitude and timing of these seasonal forces interact with other oscillations inherent in the influenza epidemic processes.

Dushoff and colleagues recently demonstrated that epidemic systems such as influenza can display intrinsic oscillations without *requiring* any seasonal effects on transmissibility (Dushoff et al., 2004). Instead they showed that seasonal effects on transmissibility can either resonate with or dampen the intrinsic oscillations of the system, depending on their respective relative frequencies. They formulated a relatively simple model system in which individuals progress from susceptible to infected to recovered or immune and back to susceptible again (S-I-R-S). The incidence of new cases in the system = $\beta \times I \times S$ or beta (a measure of transmission) × the number of infected × the number of susceptibles. Even when the transmission variable beta is held constant, the system shows oscillations, and the intrinsic period of the oscillation of incidence is a function of several standard epidemic parameters, including the basic reproductive ratio (R_0), the length of infection (L), and the duration of immunity (D). Decreasing R_0, prolonging the duration of infection, or lengthening the duration of immunity can all serve to slow the intrinsic periodicity. Importantly, this intrinsic periodicity need not be annual.

Dushoff et al. (2004) then asked what happens if seasonality is imposed. That is, instead of being a fixed constant, β is varied sinusoidally (seasonally). How does this exogenous *forcing* oscillation of transmissibility interact with the intrinsic S-I-R-S oscillation? They found that a very small seasonal oscillation in β (the transmission parameter) could markedly increase the peak-to-trough amplitude of the system. Figure 1-12 illustrates the effects of a small 4 percent change in the amplitude of the seasonal transmission parameter. Random combinations of R_0, D, and L were seeded to generate intrinsically oscillating systems. The graph shows the oscillation amplitudes without and with a tiny 4 percent seasonally varying change in transmission (β). There is a strong nonlinear resonance of the effects of the seasonal forcing on oscillation amplitude when the intrinsic oscillation period is also seasonal (i.e., equal to 1). If natural influenza systems follow the same behaviors as this model system, then it may prove difficult indeed to measure seasonal changes in disease incidence as a function of seasonal changes in the transmission parameter β.

Regrettably, real-world influenza dynamics are more complex than the simple model of Dushoff et al. One aspect of this complexity is that "seasonal influenza" is in fact not one disease process but three: influenza A/H3, influenza A/H1, and influenza B. All three viruses co-circulate worldwide and contribute to the reported total cases of influenza. Data on influenza isolates in the United States were obtained from the WHO "FluNet" website[9] and graphed as a continuous 11-year time series, from 1997 through 2007, as shown in Figure 1-13. It can be seen that A/H1 and A/H3 influenza strains dominate in different years; it is possible that they may reciprocally interfere with each other. Another factor complicating any analysis of the relationships between weather and influenza epidemiology is the fact that the three influenza viruses may well have somewhat different values of transmission parameters (βs); however, this has not been directly measured.

[9]See http://gamapserver.who.int/GlobalAtlas/home.asp.

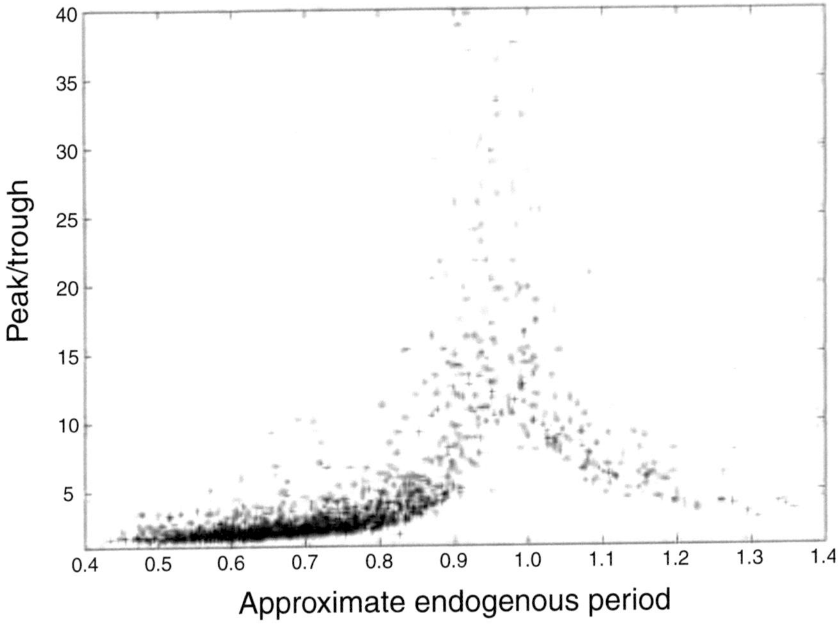

FIGURE 1-12 Graph showing the amplitude of oscillations (*y* axis, peak-trough ratio) as a function of the endogenous oscillation period (*x* axis) in a stochastic forced S-I-R-S epidemic model for 2,000 sets of randomly chosen parameters. The imposed 1-year seasonal variation in transmission for all instances is set as a sinusoidal curve with only ±4 percent variation. Strong resonance between the endogenous and imposed oscillations occurs when the approximate endogenous period is near 1 year.
SOURCE: Dushoff et al. (2004).

Epidemics as Partially Decomposable Systems

Because dynamical systems, such as epidemics, are often composed of two or more semi-independent but partially interacting dynamical subsets (e.g., environmental conditions, immunity, health systems), it is essential to isolate and analyze these component dynamical subsystems so as to be able to understand the effects of any particular forcing factor (e.g., ambient temperature). This field is still in its infancy, but there are a number of techniques available. My colleague Derek Cummings and I have experimented with various time-series decomposition techniques borrowed from physics to tease apart the component subsystems from long-term records of the waxing and waning of epidemic diseases (Cummings et al., 2004). Working with our colleagues at the Thailand Ministry of Public Health, we reviewed epidemiological records over many years, entered them into digital format, and applied decomposition methods. Some of these

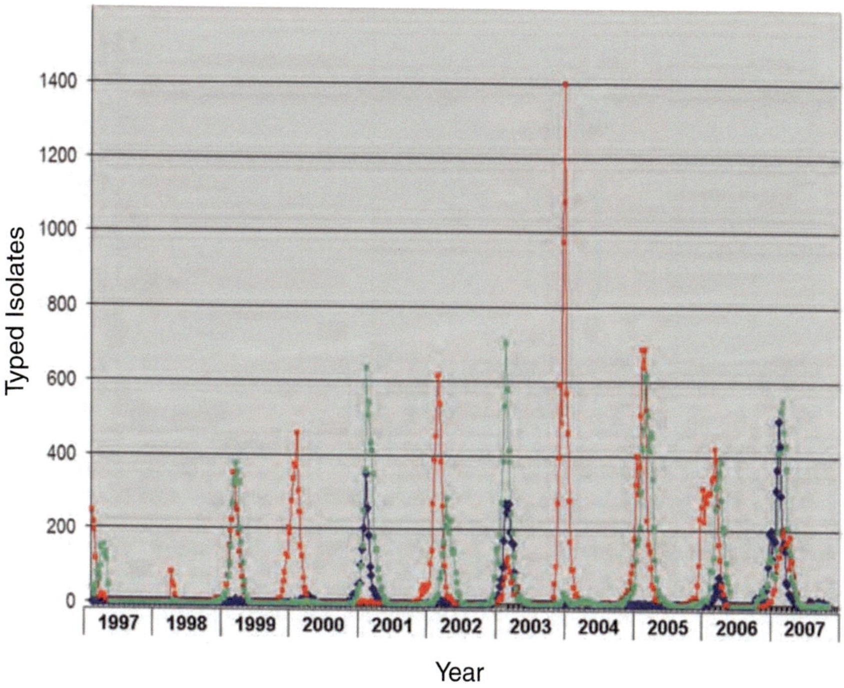

FIGURE 1-13 Influenza virus types isolated in the United States between 1997 and 2007. Influenza H3N2 is shown in red, H1N1 in blue, and B in green. Numbers shown are total isolates typed and reported.
SOURCE: WHO (2008).

time-series decomposition methods include the well-known Fourier decomposition methods, but we also examined various wavelet decomposition methods, and the Empirical Mode Decomposition. Figure 1-14 is just one example of an analysis of longitudinal epidemic time-series data, showing that this is a partially decomposable system. We applied the Empirical Mode Decomposition (Huang et al., 1998) to analyze 15 consecutive years of the incidence of dengue hemorrhagic fever in Bangkok. A major feature of the Empirical Mode Decomposition is that the method identifies component "modes" of differing frequency, from fast to slow, that together contribute to the full tracing of the epidemic time series. Note that in the Empirical Mode Decomposition, the identified modes are not single standing frequencies, but instead are patterns whose frequencies may vary. For dengue in Bangkok, we identified several major component frequency modes, including a slow 3- to 4-year oscillation that we believe is due to changes in host immunity, a clear 1-year annual oscillation that is probably driven by seasonal changes in weather, and a spiky, irregular, rapid (faster-than-annual) oscillation

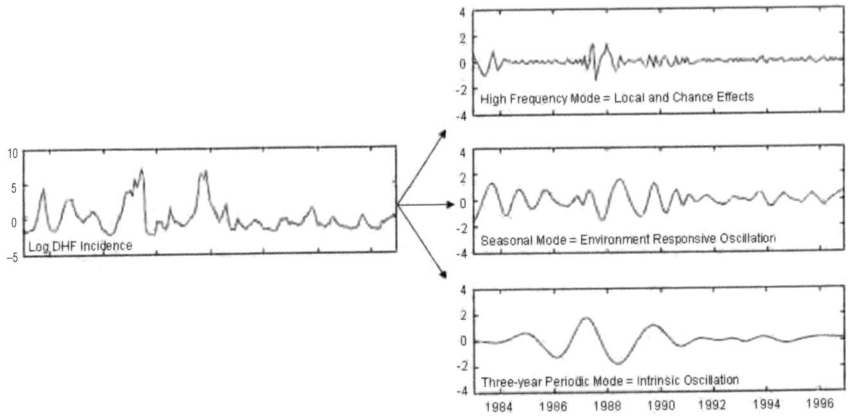

FIGURE 1-14 Multiyear time series of incidence of dengue hemorrhagic fever cases in Bangkok decomposed using the Empirical Mode Decomposition method into three modes of different approximate frequencies. When summed, the three modes add up to the original series. We hypothesize that the modes represent the isolation of different dynamic processes.
SOURCE: Adapted with permission from Macmillan Publishers Ltd: *Nature*, Cummings et al. (2004), copyright 2004.

that may represent local and chance events. Our expectation is that decomposition methods such as these will make it possible to eliminate—as "noise"—the changes in incidence contributed by other modes and to focus on a single mode to understand its intrinsic behaviors and responsiveness to external forces.

Conclusions

In this paper, I have argued that our current understanding of the relationships between climate and weather and epidemic infectious diseases is insufficient to make quantitative predictions about future threats posed by infectious diseases under various global climate change scenarios. Nonetheless, I am confident that new state-of-the-art methods, including computational tools, are now available to apply to these difficult scientific problems. I am hopeful that within a few years it may be possible to robustly predict such risks and take steps to intervene to avert potential crises.

It is safe to say that the main "Recommendations for Future Research and Surveillance" of our *Under the Weather* group continue to be relevant:

- Research on the linkages between climate and infectious diseases must be strengthened.

- Further development of disease transmission models is needed to assess the risks posed by climatic and ecological changes.
- Epidemiological surveillance programs should be strengthened.
- Observational, experimental, and modeling activities are all highly interdependent and must progress in a coordinated fashion.
- Research on climate and infectious disease linkages inherently requires interdisciplinary collaborations.

REFERENCES

Overview References

Center for Health and the Global Environment. 2005. *Climate change futures: health, ecological and economic dimensions*. Cambridge, MA: Harvard Medical School.

Dushoff, J., J. B. Plotkin, S. A. Levin, and D. J. D. Earn. 2004. Dynamical resonance can account for seasonality of influenza epidemics. *Proceedings of the National Academy of Sciences* 101(48):16915-16916.

Hilborn, R., and M. Mangel. 1997. *The ecological detective*. Princeton, NJ: Princeton University Press.

NRC (National Research Council). 2001. *Under the weather: climate, ecosystems, and infectious diseases*. Washington, DC: National Academy Press.

Stott, P. A., D. A. Stone, and M. R. Allen. 2004. Human contribution to the European heatwave of 2003. *Nature* 432(7017):610-614.

Haines References

Abeku, T. A., G. J. van Oortmarssen, G. Borsboom, S. J. de Vlas, and J. D. Habbema. 2003. Spatial and temporal variations of malaria epidemic risk in Ethiopia: factors involved and implications. *Acta Tropica* 87(3):331-340.

Abeku, T. A., S. I. Hay, S. Ochola, P. Langi, B. Beard, S. J. de Vlas, and J. Cox. 2004. Malaria epidemic early warning and detection in African highlands. *Trends in Parasitology* 20(9):400-405.

Ahern, M., R. S. Kovats, P. Wilkinson, R. Few, and F. Matthies. 2005. Global health impacts of floods: epidemiologic evidence. *Epidemiologic Reviews* 27(1):36-46.

BCAS/RA/Approtech. 1994. *Vulnerability of Bangladesh to climate change and sea level rise: concepts and tools for calculating risk in integrated coastal zone management*. Dhaka, Bangladesh: Bangladesh Centre for Advanced Studies (BCAS).

Bouma, M. J., R. S. Kovats, S. A. Goubet, J. S. Cox, and A. Haines. 1997. Global assessment of El Niño's disaster burden. *Lancet* 350(9089):1435-1438.

Cairncross, S. 2003. Water supply and sanitation: some misconceptions. *Tropical Medicine and International Health* 8(3):193-195.

Cairncross, S., and M. Alvarinho. 2006. *Floods, hazards and health: responding to present and future risks*, edited by R. Few and F. Matthies. London, UK: Earthscan Publications Ltd.

Checkley, W., R. H. Gilman, R. E. Black, L. D. Epstein, L. Cabrera, C. R. Sterling, and L. H. Moulton. 2004. Effect of water and sanitation on childhood health in a poor Peruvian peri-urban community. *Lancet* 363(9403):112-118.

Colwell, R. R. 1996. Global climate and infectious disease: the cholera paradigm. *Science* 274(5295): 2025-2031.

Cox, J., T. Abeku, J. Beard, J. Turyeimuka, E. Tumwesigye, M. Okia, and J. Rwakimari. 2007. Detecting epidemic malaria, Uganda. *Emerging Infectious Diseases* 13(5):779-780.

Curriero, F. C., J. A. Patz, J. B. Rose, and S. Lele. 2001. The association between extreme precipitation and waterborne disease outbreaks in the United States, 1948-1994. *American Journal of Public Health* 91(8):1194-1199.

Curtis, V., and S. Cairncross. 2003. Effect of washing hands with soap on diarrhoea risk in the community: a systematic review. *Lancet Infectious Diseases* 3(5):275-281.

Emberlin, J., M. Detandt, R. Gehrig, S. Jaeger, N. Nolard, and A. Rantio-Lehtimaki. 2002. Responses in the start of betula (birch) pollen seasons to recent changes in spring temperatures across Europe. *International Journal of Biometeorology* 46(4):159-170.

Engelthaler, D. M., D. G. Mosley, J. E. Cheek, C. E. Levy, K. K. Komatsu, P. Ettestad, T. Davis, D. T. Tanda, L. Miller, J. W. Frampton, R. Porter, and R. T. Bryan. 1999. Climatic and environmental patterns associated with hantavirus pulmonary syndrome, Four Corners region, United States. *Emerging Infectious Diseases* 5(1):87-94.

European Commission. 2007. *EU action against climate change: leading global action to 2020 and beyond.* Luxembourg: European Communities.

FAO (Food and Agriculture Organization). 2005. *The state of food insecurity around the world: eradicating hunger—key to achieving the millennium development goals.* Rome: FAO.

Franke, C. R., M. Ziller, C. Staubach, and M. Latif. 2002. Impact of the El Niño/Southern Oscillation on visceral leishmaniasis, Brazil. *Emerging Infectious Diseases* 8(9):914-917.

Guerra, C. A., R. W. Snow, and S. I. Hay. 2006. A global assessment of closed forests, deforestation and malaria risk. *Annals of Tropical Medicine and Parasitology* 100(3):189-204.

Haines, A., and A. Cassels. 2004. Can the millennium development goals be attained? *British Medical Journal* 329(7462):394-397.

Haines, A., and J. Patz. 2004. Health effects of climate change. *Journal of the American Medical Association* 291(1):99-103.

Haines, A., K. R. Smith, D. Anderson, P. R. Epstein, A. J. McMichael, I. Roberts, P. Wilkinson, J. Woodcock, and J. Woods. 2007. Policies for accelerating access to clean energy, improving health, advancing development, and mitigating climate change. *Lancet* 370(9594):1264-1281.

Hajat, S., K. L. Ebi, S. Kovats, B. Menne, S. Edwards, and A. Haines. 2003. The human health consequences of flooding in Europe and the implications for public health: a review of the evidence. *Applied Environmental Science and Public Health* 1:13-21.

Hamnett, M. P. 1998. *The Pacific ENSO Applications Centre and the 1997-98 El Niño.* Online. Pacific ENSO update 4(4), http://www.soest.hawaii.edu/MET/Enso/peu/update.html (accessed May 28, 2008).

Hay, S. I., J. Cox, D. J. Rogers, S. E. Randolph, D. I. Stern, G. D. Shanks, M. F. Myers, and R. W. Snow. 2002. Climate change and the resurgence of malaria in the east African highlands. *Nature* 415(6874):905-909.

IPCC (Intergovernmental Panel on Climate Change). 2007a. *Climate change 2007: the physical science basis. Summary for policymakers.* Contribution of Working Group I to the fourth assessment report of the Intergovernmental Panel on Climate Change, edited by S. Solomon, D. Qin, M. Manning, Z. Chen, M. Marquis, K. B. Averyt, M. Tignor, and H. L. Miller. Cambridge, UK: Cambridge University Press.

———. 2007b. *Climate change 2007: impacts, adaptation and vulnerability. Technical summary.* Contribution of Working Group II to the fourth assessment report of the Intergovernmental Panel on Climate Change, edited by M. Parry, O. Canziani, J. Palutikof, P. van der Linden, and C. Hanson. Cambridge, UK: Cambridge University Press.

Keiser, J., M. C. De Castro, M. F. Maltese, R. Bos, M. Tanner, B. H. Singer, and J. Utzinger. 2005. Effect of irrigation and large dams on the burden of malaria on a global and regional scale. *American Journal of Tropical Medicine and Hygiene* 72(4):392-406.

Kovats, R. S., M. J. Bouma, S. Hajat, E. Worrall, and A. Haines. 2003. El Niño and health. *Lancet* 362(9394):1481-1489.

Kovats, R. S., S. J. Edwards, S. Hajat, B. G. Armstrong, K. L. Ebi, and B. Menne. 2004. The effect of temperature on food poisoning: a time-series analysis of salmonellosis in ten European countries. *Epidemiology and Infection* 132(3):443-453.

Lindgren, E., and R. Gustafson. 2001. Tick-borne encephalitis in Sweden and climate change. *Lancet* 358(9275):16-18.

Lindgren, E., L. Talleklint, and T. Polfeldt. 2000. Impact of climatic change on the northern latitude limit and population density of the disease-transmitting European tick *Ixodes ricinus*. *Environmental Health Perspectives* 108(2):119-123.

Linnerooth-Bayer, J., R. Mechler, and G. Pflug. 2005. Refocusing disaster aid. *Science* 309(5737): 1044-1046.

Linthicum, K. J., A. Anyamba, C. J. Tucker, P. W. Kelley, M. F. Myers, and C. J. Peters. 1999. Climate and satellite indicators to forecast Rift Valley fever epidemics in Kenya. *Science* 285(5426):397-400.

Lloyd, S. J., R. S. Kovats, and B. G. Armstrong. 2007. Global diarrhoea morbidity, weather and climate. *Climate Research* 34(2):119-127.

Manuel, J. 2006. In Katrina's wake. *Environmental Health Perspectives* 114(1):A32-A39.

Mason, J. B., A. Bailes, K. E. Mason, O. Yambi, U. Jonsson, C. Hudspeth, P. Hailey, A. Kendle, D. Brunet, and P. Martel. 2005. AIDS, drought, and child malnutrition in southern Africa. *Public Health Nutrition* 8(6):551-563.

McMichael, A. J., R. Woodruff, P. Whetton, K. Hennessy, N. Nicholls, S. Hales, A. Woodward, and T. Kjellstrom. 2003. *Human health and climate change in Oceania: risk assessment 2002.* Canberra, Australia: Department of Health and Ageing.

McMichael, A., D. Campbell-Lendrum, S. Kovats, S. Edwards, P. Wilkinson, T. Wilson, R. Nicholls, S. Hales, F. Tanser, D. Le Sueur, M. Schlesinger, and N. Andronova. 2004. *Global climate change.* Geneva, Switzerland: World Health Organization.

Modi, V., S. McDade, D. Lallement, and J. Saghir. 2006. *Energy services for millennium development goals.* New York: Energy Sector Management Assistance Programme, United Nations Development Programme, UN Millennium Project, and World Bank.

Molesworth, A. M., L. E. Cuevas, S. J. Connor, A. P. Morse, and M. C. Thomson. 2003. Environmental risk and meningitis epidemics in Africa. *Emerging Infectious Diseases* 9(10):1287-1293.

Morris, S. S., O. Neidecker-Gonzales, C. Carletto, M. Munguia, J. M. Medina, and Q. Wodon. 2002. Hurricane Mitch and the livelihoods of the rural poor in Honduras. *World Development* 30(1):49-60.

NOAA (National Oceanic and Atmospheric Administration) Climate Prediction Center. 2008. *El Niño/Southern Oscillation (ENSO) diagnostic discussion*, http://www.cpc.ncep.noaa.gov/products/analysis_monitoring/enso_advisory/ensodisc.html (accessed June 16, 2008).

Pascual, M., J. A. Ahumada, L. F. Chaves, X. Rodo, and M. Bouma. 2006. Malaria resurgence in the east African highlands: temperature trends revisited. *Proceedings of the National Academy of Sciences* 103(15):5829-5834.

Patz, J. A. 2002. A human disease indicator for the effects of recent global climate change. *Proceedings of the National Academy of Sciences* 99(20):12506-12508.

Patz, J. A., M. A. McGeehin, S. M. Bernard, K. L. Ebi, P. R. Epstein, A. Grambsch, D. J. Gubler, P. Reither, I. Romieu, J. B. Rose, J. M. Samet, J. Trtanj. 2000. The potential health impacts of climate variability and change for the United States: executive summary of the report of the health sector of the U.S. National Assessment. *Environmental Health Perspectives* 108(4):367-376.

Randolph, S. E. 2001. The shifting landscape of tick-borne zoonoses: tick-borne encephalitis and Lyme borreliosis in Europe. *Philosophical Transactions of the Royal Society of London. Series B Biological Sciences* 356(1411):1045-1056.

Reiter, P., C. J. Thomas, P. M. Atkinson, S. I. Hay, S. E. Randolph, D. J. Rogers, G. D. Shanks, R. W. Snow, and A. Spielman. 2004. Global warming and malaria: a call for accuracy. *Lancet Infectious Diseases* 4(6):323-324.

Rodo, X., M. Pascual, G. Fuchs, and A. S. Faruque. 2002. ENSO and cholera: a nonstationary link related to climate change? *Proceedings of the National Academy of Sciences* 99(20):12901-12906.

Saunders, M. A., and A. S. Lea. 2008. Large contribution of sea surface warming to recent increase in Atlantic hurricane activity. *Nature* 451(7178):557-560.

Shepherd-Johnson, D. 2006. *Drought 2006—boy's long journey to treatment*, http://www.unicef.org/somalia/reallives_2557.html (accessed May 12, 2008).

Shiklomanov, I. A. 2000. Appraisal and assessment of world water resources. *Water International* 25:11-32.

Singh, R. B., S. Hales, N. de Wet, R. Raj, M. Hearnden, and P. Weinstein. 2001. The influence of climate variation and change on diarrheal disease in the Pacific islands. *Environmental Health Perspectives* 109(2):155-159.

Spennemann, D. H., and I. C. Marschner. 1995. The association between El Niño/Southern Oscillation events and typhoons in the Marshall Islands. *Disasters* 19(3):194-197.

Stern, N. 2006. *The Stern review: the economics of climate change*. Cambridge, UK: Cambridge University Press.

Stott, P. A., D. A. Stone, and M. R. Allen. 2004. Human contribution to the European heatwave of 2003. *Nature* 432(7017):610-614.

Sultan, B., K. Labadi, J. F. Guegan, and S. Janicot. 2005. Climate drives the meningitis epidemics onset in west Africa. *PLoS Medicine* 2(1):e6.

Tanser, F. C., B. Sharp, and D. le Sueur. 2003. Potential effect of climate change on malaria transmission in Africa. *Lancet* 362(9398):1792-1798.

UNDP (United Nations Development Programme). 2007. *Human development report 2007/2008. Fighting climate change: human solidarity in a divided world*. New York: United Nations Development Programme.

Vittor, A. Y., R. H. Gilman, J. Tielsch, G. Glass, T. Shields, W. S. Lozano, V. Pinedo-Cancino, and J. A. Patz. 2006. The effect of deforestation on the human-biting rate of *Anopheles darlingi*, the primary vector of falciparum malaria in the Peruvian Amazon. *American Journal of Tropical Medicine and Hygiene* 74(1):3-11.

WHO (World Health Organization). 2002. *The world health report 2002: reducing risks, promoting healthy life*. Geneva, Switzerland: World Health Organization.

Woodruff, R. E., C. S. Guest, M. G. Garner, N. Becker, J. Lindesay, T. Carvan, and K. Ebi. 2002. Predicting Ross River virus epidemics from regional weather data. *Epidemiology* 13(4):384-393.

Zhou, G., N. Minakawa, A. K. Githeko, and G. Yan. 2004. Association between climate variability and malaria epidemics in the east African highlands. *Proceedings of the National Academy of Sciences* 101(8):2375-2380.

The Center for Health and the Global Environment References

Emanuel, K. Increasing destructiveness of tropical cyclones over the past 30 years. Nature 436, 686-688 (2005).

Kerr, Richard A. Is Katrina a harbinger of still more powerful hurricanes? Science 309: 1807 (2005).

Swiss Re. Natural Catastrophes and Man-Made Disasters in 2004. In: sigma 1/2005. (Swiss Reinsurance Company, Zurich, 2005a).

Trenberth, K. Uncertainty in hurricanes and global warming. Science 308, 1753-1754 (2005).

Webster, P. J., Holland, G. J., Curry, J. A., and Chang, H.-R. Changes in tropical cyclone number, duration, and intensity in a warming environment. Science 2005 309: 1844-1846.

Burke References

Cummings, D. A. T., R. A. Irizarry, N. E. Huang, T. P. Endy, A. Nisalak, K. Ungchusak, and D. S. Burke. 2004. Traveling waves in the occurrence of dengue haemorrhagic fever in Thailand. *Nature* 427(6972):344-347.

Dowell, S. F. 2001. Seasonal variation in host susceptibility and cycles of certain infectious diseases. *Emerging Infectious Diseases* 7(3):369-374.

Dushoff, J., J. B. Plotkin, S. A. Levin, and D. J. D. Earn. 2004. Dynamical resonance can account for seasonality of influenza epidemics. *Proceedings of the National Academy of Sciences* 101(48):16915-16916.

Finkelman, B. S., C. Viboud, K. Koelle, M. J. Ferrari, N. Bharti, and B. T. Grenfell. 2007. Global patterns in seasonal activity of influenza A/H3N2, A/H1N1, and B from 1997 to 2005: viral coexistence and latitudinal gradients. *PLoS One* 2(12):e1296.

Fisman, D. N. 2007. Seasonality of infectious diseases. *Annual Review of Public Health* 28: 127-143.

Gore, A. A., Jr. 2006. *An inconvenient truth*. Directed by D. Guggenheim. Hollywood, CA: Paramount.

Huang, N. E., Z. Shen, S. R. Long, M. C. Wu, H. H. Shih, Q. Zheng, N.-C. Yen, C. C. Tung, and H. H. Liu. 1998. The empirical mode decomposition and the Hilbert spectrum for nonlinear and non-stationary time series analysis. *Proceedings of the Royal Society of London A* 454(1971):903-995.

IPCC (Intergovernmental Panel on Climate Change). 2007. *Climate change 2007: impacts, adaptation, and vulnerability*. Contribution of Working Group II to the fourth assessment report of the Intergovernmental Panel on Climate Change. Cambridge, UK: Cambridge University Press. Chapter 8.

Lofgren, E., N. H. Fefferman, Y. N. Naumov, J. Gorski, and E. N. Naumova. 2007. Influenza seasonality: underlying causes and modeling theories. *Journal of Virology* 81(11):5429-5436.

Lowen, A. C., S. Mubareka, and P. Palese. 2007. Influenza virus transmission is dependent on relative humidity and temperature. *PLoS Pathogens* 3(10):1470-1476.

NRC (National Research Council). 2001. *Under the weather: climate, ecosystems, and infectious diseases*. Washington, DC: National Academy Press.

Schulman, J. L., and E. D. Kilbourne. 1963. Transmission of influenza virus infection in mice. II. Some factors affecting the incidence of transmitted infection. *Journal of Experimental Medicine* 118:267-275.

WHO (World Health Organization). 2008. *Global atlas of infectious diseases*, http://gamapserver.who.int/GlobalAtlas/home.asp (accessed March 24, 2008).

2

Climate, Ecology, and Infectious Disease

OVERVIEW

As depicted in the convergence model of infectious disease emergence, illustrated in Figure SA-4, climate interacts with, and can alter, the complex ecological relationships underlying infectious disease transmission patterns. This chapter examines such interactions from several perspectives:

- Their consequences throughout the aquatic-marine food web, which defines ecological relationships for water-dwelling animals
- In patterns of distribution and transmission dynamics of individual infectious diseases (cholera, Rift Valley fever, chikungunya, and plague)
- Their effects on the dynamics of plant diseases, and their effects on agriculture and natural ecosystems
- As manifested in the public health challenges posed by climate change to human populations in the Arctic

Research on the effects of climate variation on infectious disease incidence and geographic range in these diverse contexts is providing the basis for developing climate-based early warning systems for disease risk. Such studies also represent a necessary first step toward anticipating how climate change may alter infectious disease dynamics in various ecological frameworks.

In her workshop presentation, Leslie Dierauf, director of the U.S. Geological Survey's National Wildlife Health Center in Madison, Wisconsin, described the apparent and predicted effects of climate on a broad cross-section of animal species that inhabit fresh- and saltwater ecosystems, as well as the intertidal zones that unite aquatic and marine environments. Ecological connections among these environments are illustrated in Figure SA-8, which depicts the marine food web.

Dierauf also emphasized the physical connectedness of aquatic and marine environments, which makes it possible for infectious diseases of fish and wildlife to move from freshwater sources to intertidal zones to marine environments, affecting species that may not have encountered these disease agents before. Salmon, for example, hatch in small freshwater streams, travel hundreds of kilometers downstream to the ocean where they live for several years, only to return to the same streams where they hatched to spawn and die shortly thereafter. Thus, she observed, "if the temperature of the streams changes or the fish themselves pick up novel disease agents, because a vector, or an intermediate host, or a disease agent thrives in the new warmer environment, infectious disease may result."

Evidence-based studies of the effects of climate change on the health of aquatic and marine wildlife are few, Dierauf reported; therefore, current understanding of this topic derives from such sources as historical comparisons (of climatic conditions and of animal health and behaviors), long-term ecological research, correlation studies, and recognition of the physical, chemical, and biological processes governing climate change. Following the flow of water from inland streams to estuaries and into the open ocean, Dierauf considered the possible impacts of climate change in each of the three main elements of the aquatic continuum and how these changes may affect the health of their animal inhabitants.

In freshwater ecosystems, extreme weather events that produce flooding can trigger overwhelming influxes of nutrients into ecosystems. Storms can cause a range of environmental disturbances; Dierauf described the release of Nile tilapia into Mississippi streams from aquaculture facilities damaged by Hurricane Katrina. Several emerging diseases of inland aquatic animals, described and depicted in Box SA-2 in the Summary and Assessment, may also be influenced by climate change.

Intertidal areas, such as salt marshes and estuaries, are essential for maintaining a delicate balance among many complex and interactive variables (such as temperature, light, salinity, wave action, sea level rise, erosion, and sediment deposition) that characterize the transition from freshwater to saltwater environments, Dierauf explained. Storms, such as hurricanes, greatly affect intertidal zones. Heavy inland rainfall increases the speed and volume of the run-off that reaches estuaries, while marine storms drive saltwater and its contents past the intertidal buffer, affecting inland ecosystem health.

Climate change is expected to produce a range of important effects on oceans (as well as on large, deep-water lakes such as the Great Lakes), according to Dierauf. These include increased wave intensity, increased nutrient turnover, changes in nutrients, and changes in the food web. In addition, she noted, higher concentrations of atmospheric carbon dioxide are dramatically increasing the acidity of ocean waters, which in turn is weakening the carbonate shells and skeletons of many marine species that comprise coral reef systems. She also noted the effects of harmful algal blooms (HABs), which are thought to result from nutrient influxes to the ocean (see Summary and Assessment). HABs appear to be increasing in both frequency and size as the climate warms, she said; this could result from increased upwelling of nutrients within the ocean or changes in ocean currents, as well as from the effects of extreme weather events inland. "What we do know is that HABs are affecting and often killing living things in the food web, like zooplankton, shellfish, fish, birds, and marine mammals, like manatees," she said.

Ocean warming, which is reducing the availability of food and sea ice for marine mammals, may also be compromising their resistance to infectious disease, Dierauf said. "Already, climate change and thinning of sea ice has reduced the time mother polar bears have to build the fat stores they need to sustain themselves over winter and to feed their young come spring when they emerge from their dens," she noted. Faced with shortages of food in their native waters, some marine mammals move to new territories where they both encounter and introduce novel disease agents (see Summary and Assessment).

"Climate change and climate variability will affect aquatic and marine species worldwide," Dierauf concluded. "We must act now at personal, professional, local, and global levels to protect vulnerable ecosystems and the aquatic and marine species that depend on these habitats for survival."

In contrast to the broad perspective on the effect of climate change on aquatic ecosystems offered by Dierauf, this chapter's first paper, by Rita Colwell of the University of Maryland, focuses on the specific and well-characterized effects of climate on cholera, a water-borne disease that affects an estimated 100,000 people per year, resulting in 10,000 deaths. The incidence and distribution of cholera are controlled by water temperature, precipitation patterns, and water salinity—all of which are influenced by global climate—and conducted through a complex web of ecological relationships. Sanitation and infrastructure also play a role in the incidence and distribution of cholera. Colwell noted, however, that "by simply educating women to filter drinking water through several layers of 'sari cloth,' we were able to reduce cholera incidence by 50 percent." Colwell described how, over the course of decades, she and coworkers deduced the circumstances under which the causal agent of cholera, the bacterium *Vibrio cholerae*, is transmitted to humans by the plankton species with which the bacterium associates. This knowledge led to the development of remote sensing systems capable of predicting the onset of cholera epidemics in the Ganges delta, known as the "home of cholera," because of its long history of epidemic disease.

This chapter's second paper also describes the use of remote sensing to monitor the effects of climate variation on specific infectious diseases. Speaker Jean-Paul Chretien, of the Department of Defense Global Emerging Infections

Surveillance and Response System (DOD-GEIS), and coauthors describe the use of satellite and epidemiological data to examine connections between the El Niño/Southern Oscillation (ENSO) and recent epidemics of two mosquito-borne viral diseases: Rift Valley fever (RVF) and chikungunya fever. In the first case, the association of RVF outbreaks in East Africa with periods of heavy rainfall, which occur during the El Niño phase of ENSO, led researchers to develop a model to forecast RVF risk in that region based on vegetation density (a marker for rainfall), as measured by satellite (Linthicum et al., 1999). The authors describe the operation of this model in the El Niño season of 2006-2007, when its prediction of elevated risk of disease prompted intensified surveillance for RVF in Kenya and, ultimately, to an international effort to stem a pending epidemic.

Chikungunya fever caused a series of outbreaks along the Kenyan coast in 2004, from which it apparently spread to several western Indian Ocean islands and India, resulting in the largest chikungunya fever epidemic on record (Chretien et al., 2007). At the time of the initial outbreaks in Kenya, a regional drought—corresponding to the La Niña phase of ENSO—had gripped the region. Chretien and coauthors discuss several possible, nonexclusive mechanisms connecting the epidemic with the drought, some of which may have also have influenced the first appearance of chikungunya fever in Europe in 2007.

In the chapter's third paper, speaker Nils Stenseth of the University of Oslo provides a much longer view of climate variation and its effects on infectious disease dynamics. Throughout recorded history, the various forms of plague, caused by the bacterium *Yersinia pestis* and transmitted by fleas among a wide range of hosts, are known to have caused both endemic and epidemic disease. Stenseth examines the dynamic ecology and epidemiology of plague in its ancient reservoir in Central Asia, and compares these patterns with local climate variation over the course of decades (as recorded in regular measurements of temperature and rainfall) and centuries (as reflected in tree-ring data for the past 1,000 years).

Using data collected twice annually between 1949 and 1995 in Kazakhstan, a focal region for plague where human cases are regularly reported, Stenseth and colleagues determined that *Y. pestis* prevalence increases dramatically in its primary host, the great gerbil (*Rhombomys opimus*), during warmer springs and wetter summers (Stenseth et al., 2006). Rodent populations also tend to increase under these conditions and, along with them, the possibility that plague will be transmitted to humans. Analyses of historical climate variation, as reflected in tree-ring patterns, suggest that similar warm, wet conditions existed in Central Asia during the onset of the Black Death in the fourteenth century, as well as in the years preceding a mid-nineteenth-century plague pandemic. As Earth's climate warms, warmer springs and wetter summers are expected to become more common in Central Asia (as well as in North America) therefore raising the possibility that plague activity—and therefore the potential for epidemic disease—will increase.

"Although the number of human cases of plague is relatively low, it would be a mistake to overlook its threat to humanity, because of the disease's inherent communicability, rapid spread, rapid clinical course, and high mortality if left untreated," Stenseth notes. Moreover, he adds, even a minor plague outbreak can result in panic, with severe economic repercussions; a 1994 plague outbreak in India that caused 50 deaths also led to a nationwide collapse in tourism and trade, costing the nation an estimated $600 million (Fritz et al., 1996). "Plague remains a fairly poorly understood threat that we cannot afford to ignore," Stenseth concludes. "Only by knowing more about how the eco-epidemiological plague systems in the different parts of the world will respond to given climate scenarios can we take the necessary precautionary measures to reduce the risks of human infections."

While climate-based early warning systems for human disease are in an early stage of development, plant disease forecasting systems based on variables such as temperature and precipitation have been used for many years, according to speaker Karen Garrett of Kansas State University. However, she adds, these well-established models will need to be adapted (based on sound science) to account for climate change, as will plant disease management policies that flow from climate-based forecasts. In her contribution to this chapter, Garrett establishes a framework for this critical effort. She describes standard methods for managing plant disease, reviews observed effects of climate variation on plant diseases and their implications given projected future climatic conditions, and discusses research and policy needs for plant disease management in response to climate change. In considering the consequences of climate change for plant health, Garrett emphasizes threshold effects: environmental perturbations that produce disproportionate ecological upheaval. Examples of such thresholds include longer growing seasons; pathogen introductions and range shifts; pathogen overwintering; and the removal of constraints on pathogen reproduction at a critical population size.

Much as it has been argued that the most effective available protective measures against the adverse human health effects of climate change are basic public health interventions (see Campbell-Lendrum in Chapter 4), Garrett observes that "the good news for formulation of strategies for plant disease management under changing climate conditions is that much of what needs to be done is the same with or without climate change." Thus, she advocates research to advance our understanding of plants' adaptive capacities and mechanisms, and policies to encourage the development of "diverse, flexible, and resilient agricultural systems that can adapt more readily to new climatic conditions."

The chapter's final paper, by Alan Parkinson of the Centers for Disease Control and Prevention's (CDC's) Arctic Investigations Program in Anchorage, Alaska, presents a panoramic view of the public health challenges faced by people living in the Arctic, where the physical effects of climate change are dramatically apparent. Temperatures in this region have increased at nearly twice the global average over the past century, causing widespread melting of land and sea ice (see Figure SA-13; Borgerson, 2008; IPCC, 2007). These conditions are exposing the Arctic's human inhabitants, many of whom have limited access to public health and/or sanitation services, to an increasingly broad range of infectious disease threats (among other health challenges). Parkinson describes the observed and projected effects of climate change in the Arctic environment, discusses the direct effects of higher ambient temperatures on the health of Arctic inhabitants, and catalogs the many ways in which climate change may increase the risk of infectious disease for Arctic residents.

Indeed, Parkinson observes, infectious disease risks are already increasing in the Arctic through the indirect influence of climate change on the populations and ranges of disease vector species (e.g., mosquitoes, ticks) and the population density and range of reservoir hosts that can transmit disease (e.g., rodents, foxes). Flooding and the loss of permafrost are also damaging the sanitation infrastructure of Arctic communities, thereby increasing the risk of water-borne infectious diseases, respiratory diseases, and skin infections. Meanwhile, increasing mean ambient temperatures raise the risk of food-borne diseases, particularly for Arctic residents who rely on traditional methods of subsistence and food preservation (e.g., fermentation, air-drying, burying).

In the face of these public health challenges, Parkinson recommends a range of public health responses, including monitoring of high-risk, climate-sensitive infectious diseases with potentially large public health impacts (e.g., water-borne diseases such as giardiasis), prompt investigation of infectious disease outbreaks that may be related to climate change, and research on the relationship between climate and infectious disease emergence to guide early detection and public health interventions. He also encourages the creation of infectious disease monitoring networks to connect typically small, isolated Arctic communities and link them to regional, national, and international health organizations. Such networks would encourage the standardization of monitoring methods, the sharing of data, and the detection of infectious disease trends over a larger geographic area.

THE MARINE ENVIRONMENT AND HUMAN HEALTH: THE CHOLERA MODEL

Rita Colwell, Ph.D.[1]
University of Maryland

Cholera, a disease I have studied for more than 30 years, is a model of the complex interactions between climate, ecology, environment, and weather related to epidemics of infectious diseases. Revealing cholera's secrets has required inter-

[1]Chairman, Canon U.S. Life Sciences, Inc., and Distinguished University Professor at both the University of Maryland at College Park and at the Johns Hopkins University Bloomberg School of Public Health.

disciplinary research examining all of these influences, as well as a point of view that I call biocomplexity: recognizing that infectious diseases operate on a wide range of time and space scales. Thus, we employ gene probes, environmental measurements (ground truth), and other precise techniques for pathogen detection, but at the same time, we take a holistic approach that integrates information from the atomic to the atmospheric—and perhaps, in some cases, even the cosmic—in order to build a predictive model for cholera outbreaks.

Cholera is a significant, global public health problem, as shown in Table 2-1. Annually, it results in approximately 100,000 hospitalizations and approximately 10,000 deaths, varying from year to year. A few cases of cholera appear each year in the United States, usually associated with seafood harvested from closed beds near sewage outfalls in the Gulf of Mexico.

Most of my group's research on cholera has focused on the Ganges delta, which feeds into the Bay of Bengal. This area is known as the home of cholera due to spring and fall epidemics, of varying but predictable intensity, that have recurred there for hundreds of years (see Figure 2-1). During the monsoon season, flooding rains wash nutrients down from the Himalayas, while winds drive water from the Bay of Bengal up into the Ganges and its tributaries, creating ideal conditions (discussed later) for cholera outbreaks. The fall 2007 epidemic, which followed massive flooding, was catastrophic. The Center for Diarrheal Disease

TABLE 2-1 Cholera Cases Officially Reported to WHO, 2004—Selected Countries

Country	Number of Cases	Imported	Deaths	Mortality Rate (%)
Benin	642		9	1.40
Burundi	819		14	1.71
Cameroon	8,005		137	1.71
Comoros	1		0	0.00
Côte d'Ivoire	105		9	8.57
DROC (Congo)	7,665		228	2.97
Niger	2,178		57	2.62
Nigeria	3,186		185	5.81
Somalia	4,490		26	0.58
Uganda	3,380		91	2.69
Tanzania	10,319		272	2.64
Zambia	12,149		373	3.07
Zimbabwe	119		9	7.56
India	4,695		7	0.15
Japan	66	55	0	0.00
Singapore	11	1	1	9.09
Total	57,830	56	1,418	2.45

SOURCE: WHO (2005).

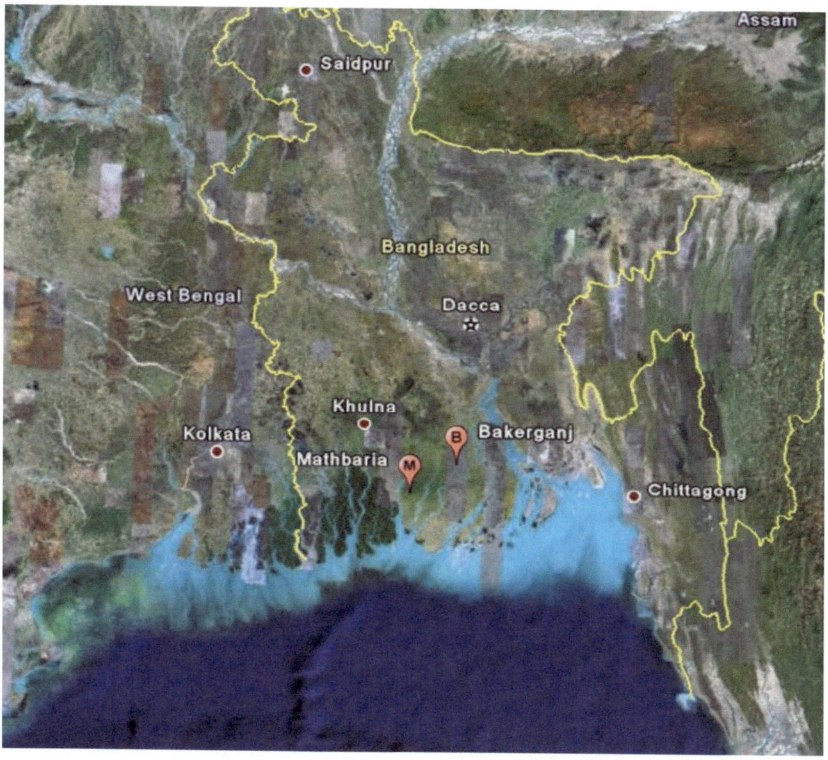

FIGURE 2-1 Bangladesh border, barrier islands, and location of Dacca, Matlab, Mathbaria, and Bakerganj.
SOURCE: Printed with permission from Google.

Research in Dacca admitted about a thousand new cases per day for almost 30 days and had to use temporary space to house cholera victims. We are working to create predictive models to provide advance warning of conditions that produce severe epidemics in this region of the world.

However, *V. cholerae,* the bacterium, is a natural inhabitant of rivers, estuaries, and coastal waters throughout the world. Currently, we are sequencing approximately 50 different strains of *Vibrio cholerae*, the causative agent of cholera collected from many geographic locations to examine their genetic relationships. Preliminary sequencing studies of *V. cholerae* collected at a depth of 2,000 m at a site located off the coast of Oregon indicate that those isolates may represent ancestral strains; interestingly, one strain studied in detail has genes in common with other *Vibrio* pathogens, as well, including *Vibrio vulnificus* and *Vibrio parahaemolyticus*, the latter being the most common food-borne pathogen in Asian countries, where raw seafood is consumed.

The Ecology of Cholera

My laboratory accomplished the first isolation of *Vibrio cholerae* from the Chesapeake Bay more than two decades ago, and we now know that this bacterium is found in estuaries of similar salinity, (*ca.* 15 parts per thousand), where the water temperature rises seasonally to 15°C or higher and where an influx of nutrients encourages plankton blooms (Colwell, 1996). Other species of *Vibrio*, including *V. parahaemolyticus* and *V. vulnificus*, also thrive under these conditions. One of my current graduate students, Brad Haley, has just returned from Iceland, where he was able to isolate *V. cholerae* at locations where geothermal effluent flows into bays. Clearly, water temperature is critical to the growth of this pathogen.

Vibrio cholerae also has a dormant state, which it assumes between epidemics and during which it cannot be cultured but can be detected with probes (fluorescent antibodies and gene signature sequences). Only during the peak of the zooplankton bloom, in the spring and the fall, is *V. cholerae* easily culturable. We were able to show that by adding nalidixic acid and nutrient (yeast extract) to water containing the quiescent bacterium, we can stimulate cell elongation and metabolism.

Another important discovery was that cholera is transmitted by plankton. Thus, it is not enough to say that its growth correlates with sea surface temperature and salinity; it is important to recognize the ecological interactions that produce these correlations. There is a commensal relationship—which may prove to be symbiosis—between *Vibrio* bacteria and zooplankton. Vibrios are chitinolytic (i.e., capable of breaking down chitin, the material that forms the carapaces of zooplankton and crustaceans (e.g., crabs, shrimp). *V. cholerae* also produces a powerful proteolytic enzyme that the bacterium apparently uses to perform an additional function for zooplankton: breaking down its egg sac, enabling the eggs to disperse into the water column. We are discovering that interactions between *V. cholerae* and various zooplankton species are quite intricate; for example, certain strains of the bacterium attach preferentially to certain species of zooplankton (Rawlings et al., 2007). All of this leads to the conclusion that *V. cholerae* is integral to marine ecosystems, and therefore cannot be eradicated.

The Epidemiology of Cholera

We have determined in earlier studies that between 10,000 and 50,000 *Vibrio cholerae* bacteria may be attached to an individual copepod (the zooplankton favored by *V. cholerae*). A liter of water drawn by a villager from a pond in Bangladesh between epidemics may contain 10 copepods. However, during a zooplankton bloom, that concentration can increase a hundredfold or more per liter, carrying a dose of cholera bacteria sufficient to cause cholera. The severity of the disease is dose dependent: a low concentration of bacterial cells will produce mild diarrhea; hospitalized cases—which represent about 25 percent of all infections—require more since one million bacteria per milliliter has been shown to be required to produce the disease. Thus, it has been estimated that only 25 percent of those with cholera end up in hospitals and many more may have been infected (Colwell and Huq, 2004).

Cholera is a disease with rapid onset. Within 24 to 48 hours, the typical patient can lose up to 18 liters of fluid. If that fluid can be replenished quickly, either intravenously or through oral rehydration (using a simple mixture of bicarbonate of soda, table salt, and sugar), recovery is fairly rapid.

From years of study in Bangladesh, we have determined several factors that interact and are associated with the massive annual biennial (spring and fall) cholera epidemics, so that we can predict the onset and severity of epidemics. Our recent research focuses on the communities of Mathbaria and Bakerganj, which are located in the barrier islands region of the Ganges delta (see Figure 2-1). Mangrove-based ecosystems are abundant in copepods. Thus, the *Vibrio* population is also abundant, and during the zooplankton/*Vibrio* bloom, cholera results from drinking untreated water.

In Bakerganj and Mathbaria, copepods comprise the majority of zooplankton species. We now have evidence that the severity of a given local cholera epidemic can be determined by copepod population dynamics, with intense epidemics occurring during times of abundance of those copepod species to which epidemic strains of *V. cholera* preferentially attach. We are currently conducting a seasonal study of zooplankton species in an attempt to determine which species carry *V. cholera* and to identify factors that influence population size; we will use this information, with other environmental data, to build a predictive capacity for cholera epidemics.

We are also using our knowledge of cholera epidemiology to help the people of Bangladesh to avoid contracting cholera. In one study, for example, we found that by simply educating women to filter drinking water through several layers of sari cloth, we were able to reduce cholera incidence by 50 percent. This result supported our hypothesis that plankton and particulates—to which the bacteria are attracted—transmit cholera and when removed by simple filtration, the incidence of the disease is significantly reduced.

Predictive Models of Cholera

Currently, the spring bloom of phytoplankton in the Bay of Bengal can be measured by satellite sensors that measure chlorophyll intensity and, therefore, the phytoplankton population. Phytoplankton blooms are followed by zooplankton blooms, but the latter cannot yet be measured directly by satellite sensors. However, the zooplankton peak can be inferred using a series of calculations from measurements of the phytoplankton populations that precede the zooplankton population peak. This information taken together with salinity, temperature, and other environmental factors, provides a more complete picture.

We have also gathered ground truth data over the past 10 years in the Bakerganj area, including conductivity of the water, presence of inorganic nutrients, temperature, and salinity. With these data, we are able to improve our prediction of the timing and, possibly, the severity of cholera epidemics.

In our original work, we were able to use satellite imagery to measure sea surface temperature and sea surface height in the Bay of Bengal. As shown in Figure 2-2, the correlation of chlorophyll and temperature data, measured by satellite sensors, provides a predictive capacity for conditions conducive to cholera outbreaks. We are currently working on a predictive model that takes into account ocean currents to monitor the movement of plankton into the Bay of Bengal estuaries from the southern tip of India. This could provide as much as a 3-month warning prior to an impending cholera outbreak.

In Latin America, the 1991-1992 El Niño event corresponded with a cholera epidemic that was initially attributed to the dumping of ballast water by a ship in the harbor of Lima, Peru (Gil et al., 2004). We were able to disprove this hypothesis by demonstrating that cholera outbreaks had occurred in three different cities along the coast of Peru, starting before the peak of the 1991-1992 El Niño event. The epidemic more likely resulted from the effect of increased sea surface temperatures on existing plankton and *V. cholerae* populations.

Our most sophisticated predictive model for cholera incorporates chlorophyll, sea surface height, temperature, and extensive ground truth data. Within a few years, the National Oceanic and Atmospheric Administration (NOAA) will launch a satellite that may provide salinity data. We are also refining our model, based on the 40 years of data accumulated on cholera in Bangladesh and in India, which we are presently analyzing. Nevertheless, with the analyses we have performed to date—sea surface temperature and sea surface height from satellite sensors; measurements of chlorophyll intensity (corrected for the time lag from chlorophyll-phytoplankton bloom to the zooplankton bloom that feeds on the phytoplankton); and measurements of vibrio dispersion in the water—we are able to determine significant correlations and, thus, a foundation from which to predict cholera epidemics.

Conclusion

Climate change is likely to increase the burden of cholera in Bangladesh, but even greater suffering will occur if sea levels rise to predicted levels, displacing millions of people. However, our interdisciplinary, international (as demonstrated by our large number of collaborators from many countries), and biocomplexity approach to studying cholera extends well beyond Bangladesh and even beyond the disease itself. By gaining an understanding of the complex interactions between infectious disease, ecology, and the physical environment, we can

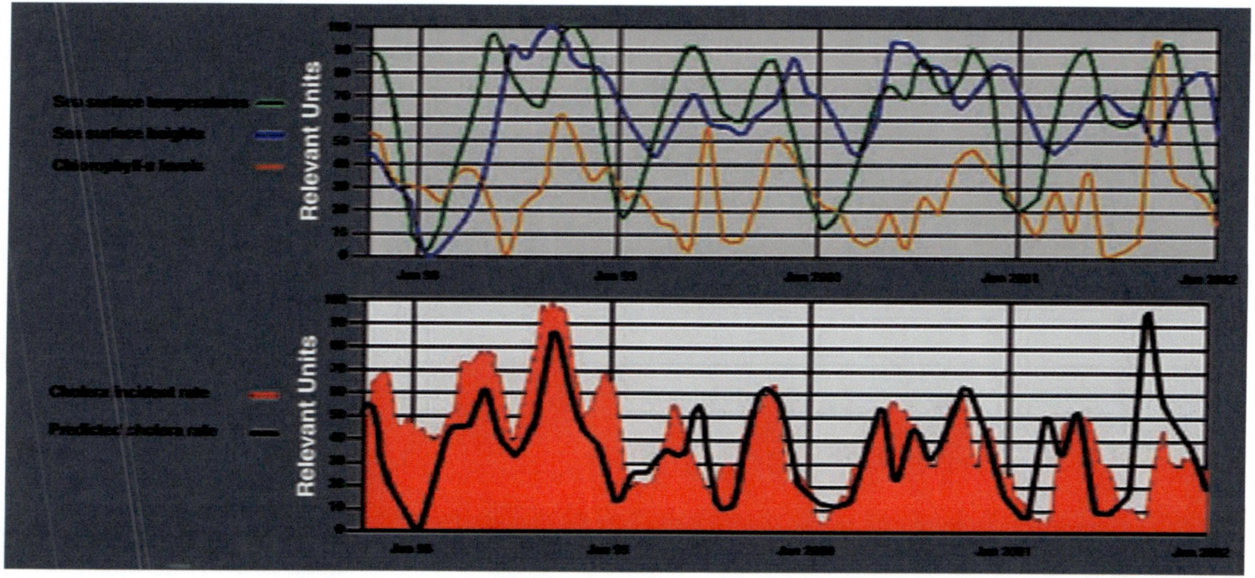

FIGURE 2-2 Environmental parameters (top) and predicted versus actual cholera incidence rate (bottom).
SOURCE: Printed with permission from John Calkins, ESRI User Conference (2004).

develop predictive models of infectious diseases that in turn will allow us to develop a preemptive medicine: that is, to mitigate the impact of infectious disease, if not to prevent it by having an early warning system to initiate appropriate and responsive public health measures.

EXTREME WEATHER AND EPIDEMICS: RIFT VALLEY FEVER AND CHIKUNGUNYA FEVER[2]

Jean-Paul Chretien, M.D., Ph.D.[3]
Department of Defense

Assaf Anyamba, Ph.D.[4]
NASA Goddard Space Flight Center

Jennifer Small, M.A.[4]
NASA Goddard Space Flight Center

Compton J. Tucker, Ph.D.[4]
NASA Goddard Space Flight Center

Seth C. Britch, Ph.D.[5]
U.S. Department of Agriculture

Kenneth J. Linthicum, Ph.D.[5]
U.S. Department of Agriculture

As Earth's climate changes, the frequency and intensity of heat waves, droughts, floods, and other extreme weather events are expected to increase over large regions (IPCC, 2007b). Trends already are apparent, with regions affected by drought and the frequency of heavy precipitation that leads to flooding increasing since the 1950s (IPCC, 2007a). Besides obvious, direct effects on human health, extreme events can facilitate infectious disease epidemics—for example, through effects on disease vector ecology, infrastructure, and human behavior.

Satellite observations and modeling allow prediction of some extreme weather events and consequent infectious disease activity. In this paper, we use

[2]The views expressed in this paper are the private views of the authors and are not to be construed as official or representing the true views of the Department of Defense.

[3]Coordinator, Overseas Laboratories, Global Emerging Infections Surveillance & Response System, 2900 Linden Lane, Silver Spring, MD 20910; Phone: 301-319-9418; Fax: 301-319-9213; E-mail: Jean-Paul.Chretien@us.army.mil.

[4]Biospheric Sciences Branch, Greenbelt, MD.

[5]Agricultural Research Service, Center for Medical, Agricultural, and Veterinary Entomology, Gainesville, FL.

satellite and epidemiological data to examine connections between the El Niño/Southern Oscillation (ENSO) phenomenon and two recent mosquito-borne epidemics in Africa: Rift Valley fever (RVF) and chikungunya fever, which followed heavy rains and drought, respectively. These case studies suggest considerations in developing early warning systems for extreme weather-associated epidemics.

El Niño/Southern Oscillation and Rift Valley Fever Prediction

The ENSO is an irregular, but natural, feature of the global climate system. It results from interactions between the oceans and the atmosphere across the Indo-Pacific region and affects the weather around the world. In the warm, or El Niño, phase of the cycle, sea surface temperatures are warmer than usual in the eastern-central equatorial Pacific Ocean. El Niño sometimes is followed by a cool, or La Niña, phase with colder-than-usual temperatures in the eastern-central equatorial Pacific. The warm and cool phases cycle over irregular intervals of several years but have characteristic effects on precipitation and temperature throughout much of the tropics.

In areas where it influences climate, El Niño is associated with increased risk of some infectious diseases (Kovats et al., 2003). For example, in East Africa, El Niño is associated with flooding and RVF activity (Linthicum et al., 1999)—epizootics among economically important livestock, with humans infected incidentally by the mosquito vectors or by handling or consuming infected animal products. Outbreaks begin near natural depressions ("dambos") that harbor *Aedes* mosquito eggs infected directly by the parent during development. The eggs hatch with dambo flooding, producing an initial wave of RVF vectors; other species that transmit the virus emerge over subsequent weeks (Linthicum et al., 1984) and propagate the outbreak. The largest recorded RVF outbreak, in 1997-1998, coincided with a strong El Niño. There were an estimated 89,000 human infections and hundreds of deaths in northeastern Kenya and southern Somalia (CDC, 1998).

Following the 1997-1998 outbreak, scientists at the U.S. National Aeronautics and Space Administration Goddard Space Flight Center (NASA-GSFC) and the Department of Defense Global Emerging Infections Surveillance and Response System (DOD-GEIS) initiated a partnership to forecast conditions favorable for RVF activity in Africa by monitoring ENSO and other climatic phenomena. The program uses satellite data from ongoing NASA and NOAA climate and environmental observation programs to provide predictions of areas at elevated RVF risk. The primary data sets are sea surface temperature (SST), rainfall, outgoing longwave radiation (OLR; which is correlated with cloud cover and rainfall), and Normalized Difference Vegetation Index (NDVI; a key measure for identifying risk areas). NDVI is correlated with rainfall but integrates effects of other climatic parameters, responds most to sustained rather than intermittent rains, and is available globally since 1981, while ground-based rain gauge coverage is limited in Africa.

Updated forecasts are available monthly, or more frequently if conditions warrant, on the DOD-GEIS public website.[6] Forecasts and alerts also are communicated to public health agencies that can act on them to enhance surveillance or community preparedness in at-risk areas. Important partners in responding to forecasts and alerts include the World Health Organization (WHO), Food and Agriculture Organization of the United Nations (FAO), the U.S. Centers for Disease Control and Prevention's (CDC's) International Emerging Infections Program in Kenya, and two members of the DOD-GEIS network: the U.S. Army Medical Research Unit-Kenya (USAMRU-K) in Nairobi and the U.S. Naval Medical Research Unit-3 (NAMRU-3) in Cairo.

Rift Valley Fever Outbreaks in East Africa, 2006-2007

In September 2006, the NASA-GSFC/DOD-GEIS monitoring program identified indications of an impending El Niño episode, with SSTs anomalously elevated in the central-eastern Pacific Ocean (+2°C) and the western Indian Ocean (+1°C) (see Figure 2-3). These conditions enhanced precipitation over these areas and the Horn of Africa through November (see Figure 2-4). Rainfall increased through December, with vegetation response (see Figure 2-5A) and conditions favorable for RVF activity in large areas of northeastern Kenya and nearby areas in Somalia and Ethiopia, as well as in southern Kenya and northern Tanzania (see Figure 2-5B).

The NASA-GSFC/DOD-GEIS program released a series of epidemic warnings based on these observations. In September 2006, it issued a global, regional-scale forecast covering late 2006-early 2007 for possible El Niño-linked outbreaks, including RVF in East Africa, to the DOD-GEIS network (these forecasts were published online in the *International Journal of Health Geographics,* an open access journal, in December; Anyamba et al., 2006). As rainfall increased in the Horn of Africa, the FAO Emergency Prevention System for Transboundary Animal Diseases issued an RVF alert for the Horn in November, identifying areas flagged as conducive to RVF activity (FAO, 2006). NASA-GSFC/DOD-GEIS also communicated with the WHO, which transmitted alerts to the countries at risk for RVF activity and called for enhanced surveillance and community awareness.

USAMRU-K, in coordination with Kenya Medical Research Institute (KEMRI) and CDC's International Emerging Infections Program (IEIP), deployed a field team in early December to assess high-risk areas in the Garissa district of northeastern Kenya (which was experiencing severe flooding). USAMRU-K tested mosquitoes collected by the team in Garissa and from established collection sites in other areas (see Figure 2-6), identifying RVF virus-infected mosquitoes from Garissa. The field team also investigated local reports of possible

[6] See http://www.geis.fhp.osd.mil.

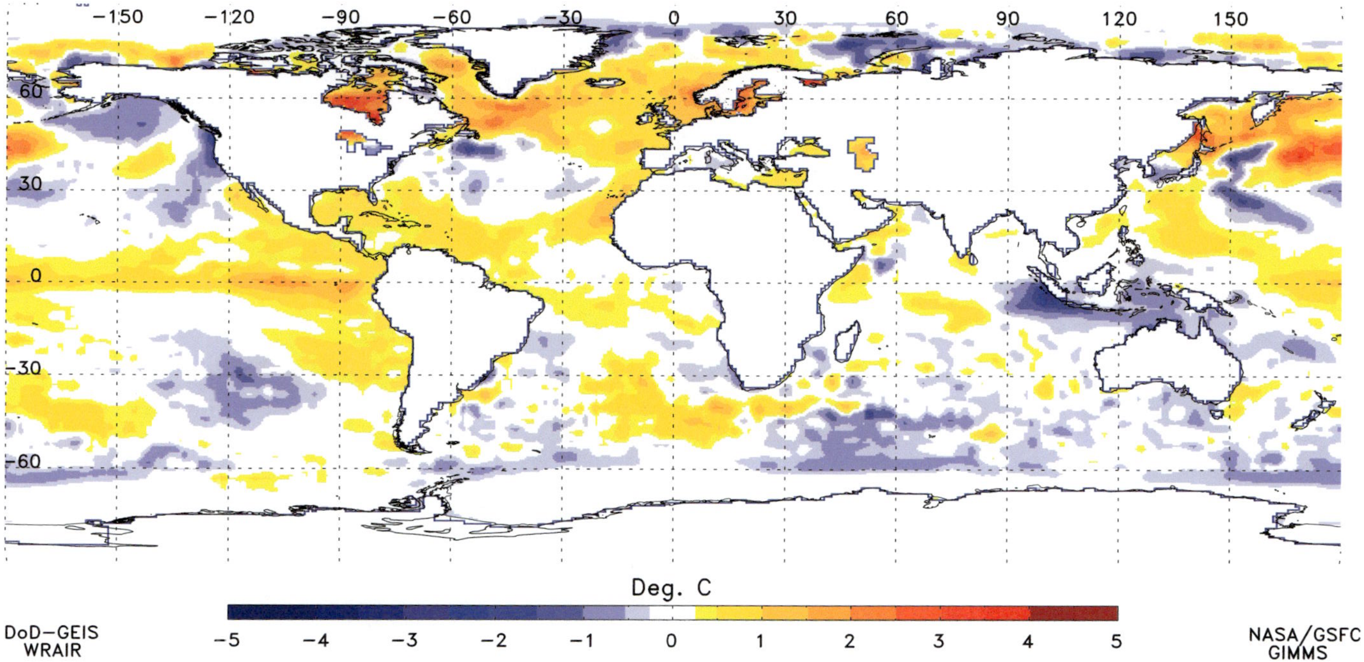

FIGURE 2-3 Global SST anomalies, September 2006.

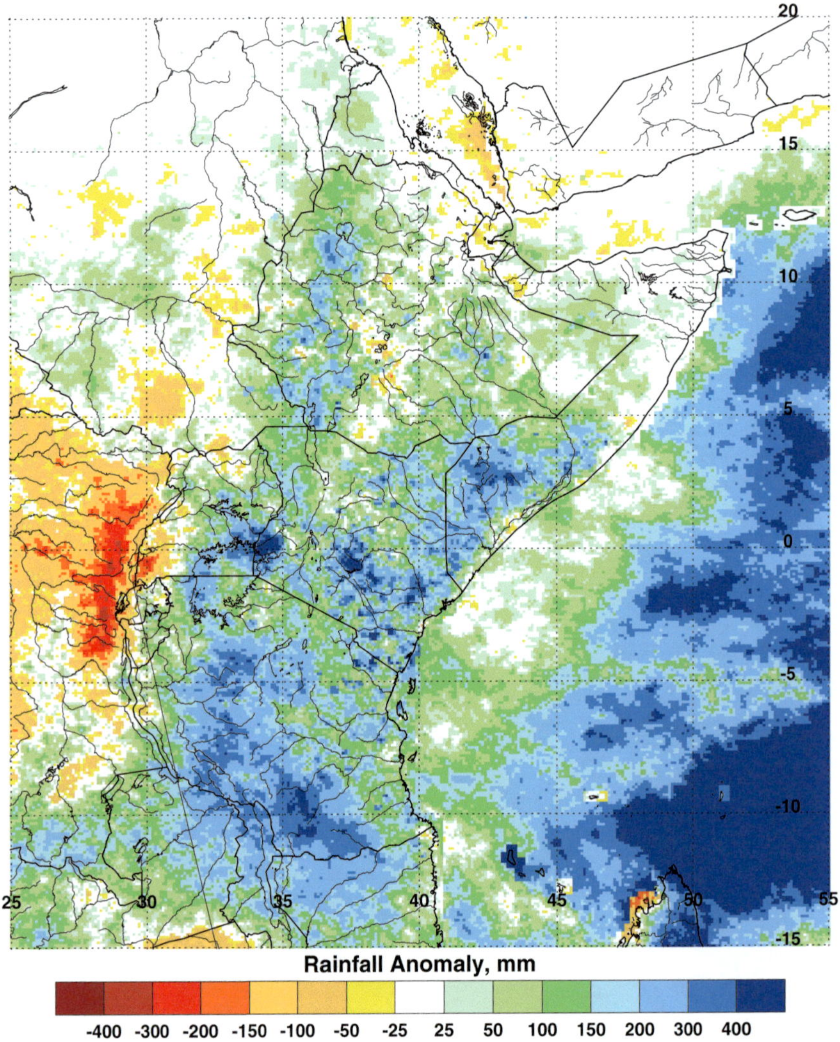

FIGURE 2-4 Seasonal rainfall anomalies in the Horn of Africa, September 2006–January 2007.

animal RVF cases and traveled with Ministry of Health staff to hospitals that recently had admitted patients with suspected RVF, obtaining specimens for testing at KEMRI.

On December 21, KEMRI confirmed RVF virus infection in specimens taken from several patients in the Garissa district (WHO, 2007a). The Kenya Ministry

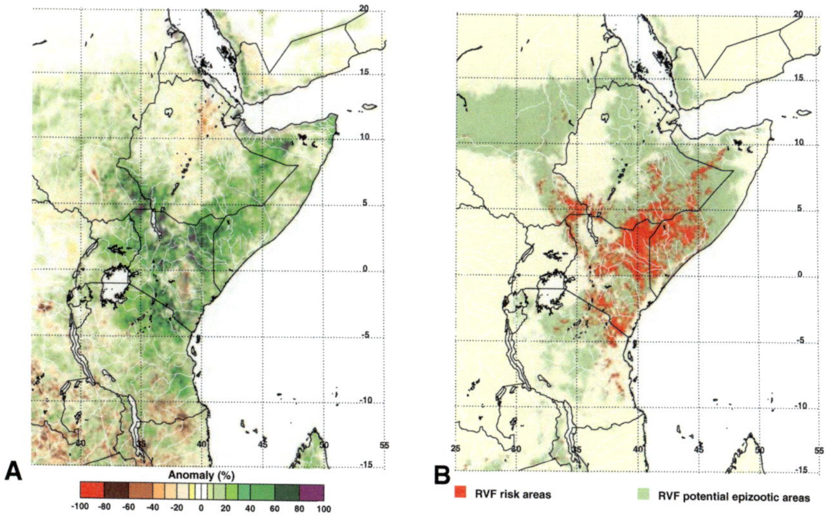

FIGURE 2-5 NDVI anomalies (A) and RVF calculated risk (B) in the Horn of Africa, January 2007. In (B), green identifies areas included in the NDVI-based RVF risk assessment (based on permissive permanent environmental features) and red indicates areas currently at elevated risk, based on persistence of positive NDVI anomalies over at least 3 months.

of Health initiated a response with international partners, including WHO, CDC, USAMRU-K, NAMRU-3, and the U.S. Department of Agriculture. An intensive social mobilization campaign began in northeastern Kenya in late December, along with a locally enforced ban on animal slaughtering over most of Eastern and North Eastern Provinces (animal vaccination began in January, but by then the epidemic was waning). NASA-GSFC/DOD-GEIS provided frequent, high-spatial-resolution risk assessment updates to facilitate targeted surveillance during the epidemic response.

Between November 30, 2006, retrospectively identified as the date of onset for the index case, and March 9, 2007, when the last case was identified, 684 cases with 155 deaths were reported in Kenya. North Eastern province, which includes the Garissa district, reported the most cases of affected provinces (N = 333). Smaller RVF epidemics in Somalia and Tanzania followed the Kenya outbreaks: in Somalia, 114 cases with 51 deaths were reported between late December 2006 and February 2007; in Tanzania, 264 cases with 109 deaths were reported between mid-January and early May.

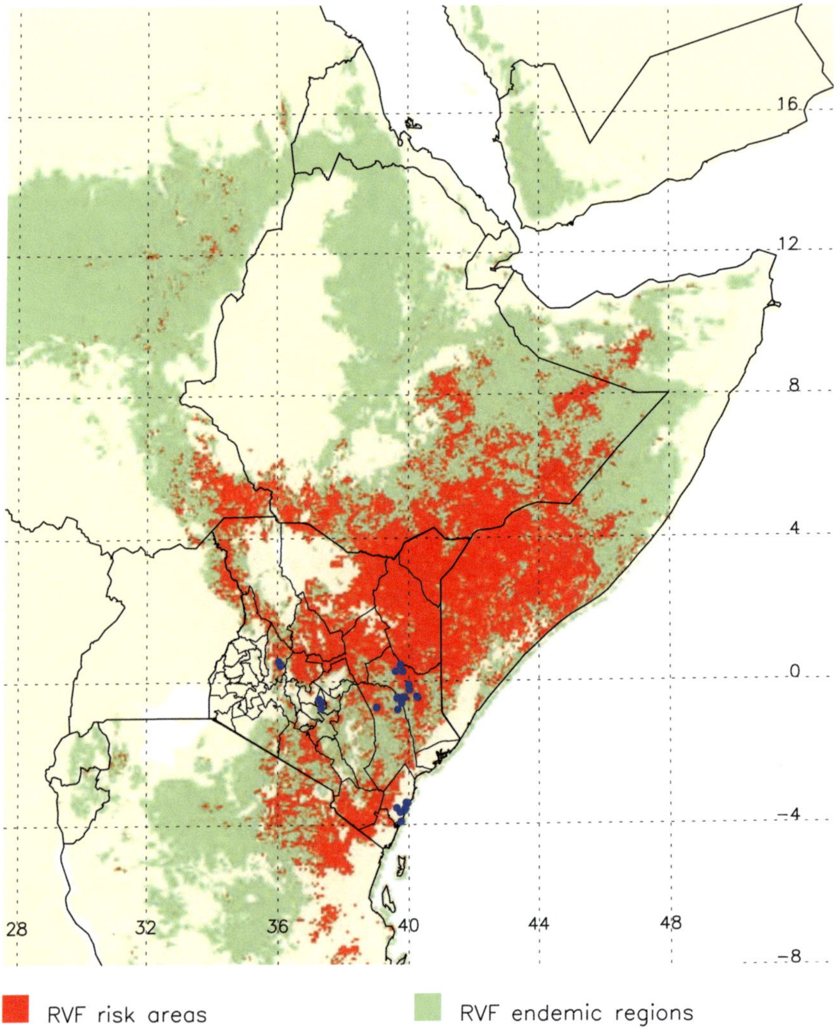

FIGURE 2-6 USAMRU-K mosquito collection sites (blue dots) and RVF risk assessment, December 2006.

Chikungunya Fever Outbreaks in Kenya and Other Regions, 2004-2008

In July 2004, while East Africa experienced a severe drought, a public hospital in Lamu, a coastal island city of Kenya, noted a sharp increase in cases of acute febrile illness. Many patients reported joint pain and had negative malaria blood smears (Bedno et al., 2006). The Ministry of Health launched an outbreak investigation, which was supported by USAMRU-K and the CDC's IEIP. Laboratory testing of outbreak specimens identified chikungunya virus as the cause. After the outbreak, a population-based serological study led by the Kenya Field Epidemiology Training Program estimated that 13,500 people, or 75 percent of the Lamu population, were infected (Sergon et al., 2008). In November, a chikungunya outbreak was reported in Mombasa, around 200 miles south of Lamu on the Kenya coast.

Though rarely fatal, chikungunya virus infection may cause prolonged and debilitating joint pain. The disease is endemic throughout much of tropical Africa, maintained by transmission cycles involving forest-dwelling *Aedes* mosquitoes and wild primates in which humans are infected incidentally. Urban *Aedes aegypti* and *Aedes albopictus* cause epidemics in tropical Asia without nonhuman hosts. The vectors in urban Lamu and Mombasa were thought to be peridomestic *Aedes aegypti*, which were found in unprotected domestic water sources that were not changed frequently because of water shortages during the drought. The outbreaks marked the first confirmation of chikungunya fever transmission in coastal Kenya.

Retrospective analysis of climate data preceding the Lamu outbreak (assumed to have begun in June 2004) showed anomalously warm, dry conditions over much of East Africa, but especially coastal Kenya, during May 2004 (Chretien et al., 2007). NDVI anomalies in Lamu were the most negative in the available record (1998-2003), reflecting substantially reduced rainfall. When the outbreaks occurred in Lamu and Mombasa, each city had experienced a cumulative rainfall deficit of approximately 100 mm compared to the average (see Figure 2-7).

The warm, dry conditions may have enabled the epidemic in two ways: unsafe domestic water storage practices, along with infrequent changes of water stores because of the drought, may have increased peridomestic *Aedes* vector abundance; and the warm, dry conditions may have enhanced *Aedes* vectorial capacity by decreasing the extrinsic incubation period (Watts et al., 1987).

Following the Kenya chikungunya fever outbreaks, the epidemics spread to other areas with susceptible human populations and competent vectors: to western Indian Ocean islands, including Reunion, where viral mutation may have facilitated adaptation to the highly efficient *Aedes albopictus* vector (Tsetsarkin et al., 2007) and more than 200,000 people likely were infected (WHO, 2006), and to India, which reported well over 1 million cases (Mavalankar et al., 2007).

Also, for the first time ever, chikungunya fever reached Europe. In a northeastern Italian province, public health authorities identified 205 cases during July-September 2007 (Rezza et al., 2007). The presumed index case developed symptoms after visiting relatives in an affected area of India. Local *Aedes albopictus* mosquitoes, an invasive species introduced into Italy around 1990 (tire importation is suspected as the mechanism), then propagated the epidemic. While the role of climatic conditions in the Italian outbreak is unclear, much of southern Europe had experienced an anomalously warm, dry summer (see Figure 2-8)

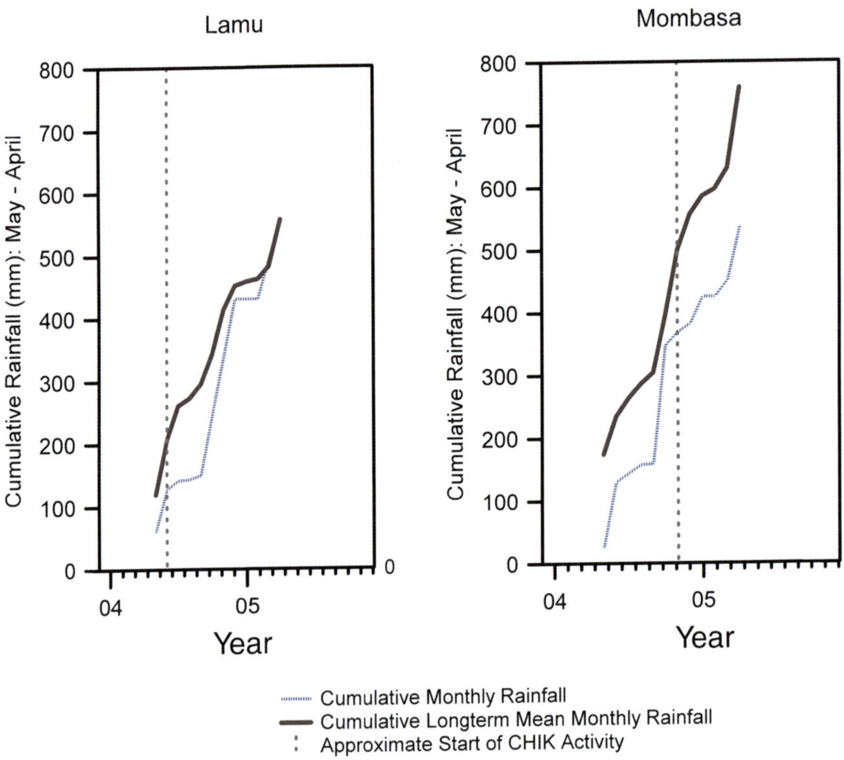

FIGURE 2-7 Cumulative monthly rainfall (dotted line) and long-term mean cumulative monthly rainfall (solid line) in Lamu and Mombasa. Vertical dashed line indicates approximate starting dates for the outbreaks (Lamu, June 2004; Mombasa, November 2004).

that, along with historically poor vector control, may have contributed to the abundance of mosquitoes in the affected area at the time of the outbreak (reported anecdotally; Rezza et al., 2007).

Developing Early Warning Systems for Extreme Weather-Linked Infections

In both the RVF and the chikungunya fever examples, climate appears to have interacted with other factors to facilitate the outbreaks (see Table 2-2), consistent with the "Convergence Model" of infectious disease emergence proposed by the Institute of Medicine's (IOM's) Committee on Microbial Threats to Health in the Twenty-First Century (Figure 2-9; IOM, 2003). For example, besides flooding of mosquito habitats, animal sacrificing and preparation practices may have

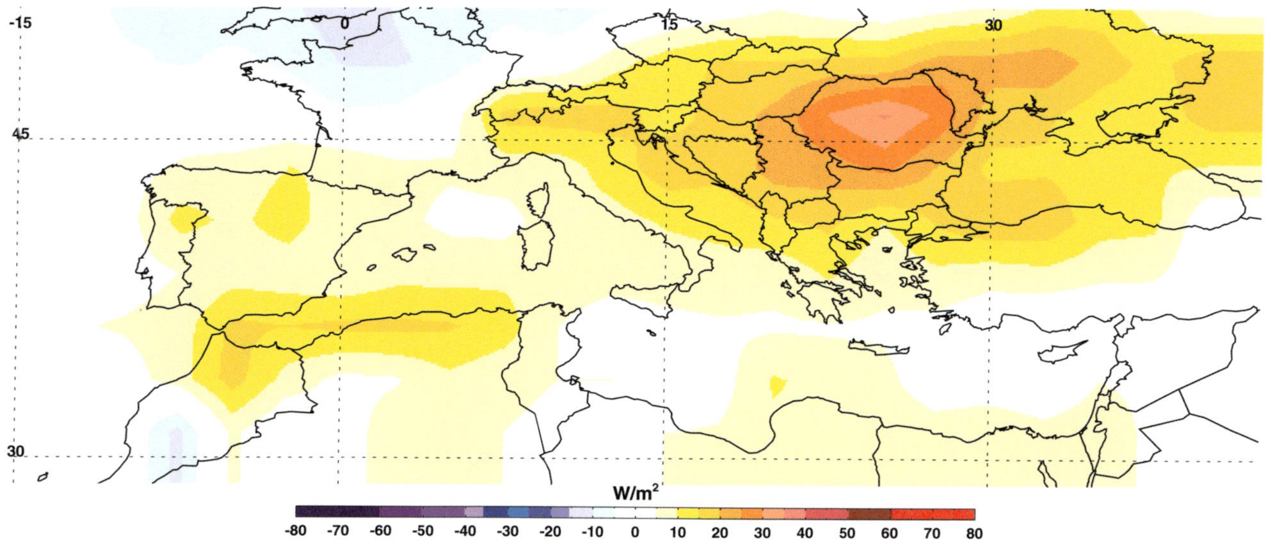

FIGURE 2-8 Outgoing longwave radiation (OLR) anomalies, July 2007, for the Mediterranean region. Positive anomalies (> +10 W/m^2) are indicative of severe drought conditions that persisted during the summer of 2007 across the region. Such severe drought conditions also prevailed during the Chikungunya outbreak in coastal East Africa and the Indian Ocean island during the 2004-2005 period.

TABLE 2-2 Factors in Emergence and Spread of Rift Valley Fever and Chikungunya Fever

	Rift Valley Fever	Chikungunya Fever
Climatic factors	• Flooding	• Warm, dry conditions
Biological factors	• Broad host and vector species range • Abundant livestock hosts	• Genetic adaptation to *Aedes albopictus* • Association of *Aedes albopictus* and *Aedes aegypti* to humans
Physical environment factors	• Dambos, other ground pools	• Vector breeding sites
Social, political, and economic factors	• Livestock trade • Herder and religious practices	• Travel • Delayed notification and control • Previous introduction of *Aedes albopictus* to Indian Ocean islands and Italy by trade

SOURCE: Adapted from Chretien and Linthicum (2007), IOM (2003), and Peters and Linthicum (1994).

contributed to the RVF epidemic in East Africa in 2006. In coastal Kenya in 2004, the availability of vector breeding sites (i.e., unprotected domestic water stores) appears to have facilitated the emergence of chikungunya fever. In developing early warning systems for outbreaks linked to extreme weather, consideration of the nonclimatic facilitating factors may enable more precise identification of populations at risk, with better targeting of risk communication.

The RVF and chikungunya fever outbreaks also suggest the need for infectious disease early warning systems to integrate with other natural disaster prediction and response programs. In both of these epidemics, climatic conditions facilitating disease emergence and transmission had other public health effects as well. Flooding in the Horn of Africa during late 2006-early 2007 affected more than 1 million people (WHO, 2007b), destroying homes, livestock, and crops; displacing families; causing hygiene breakdown and water-borne disease epidemics; and obstructing delivery of aid (Save the Children, 2007). Drought in Kenya during 2004 contributed to massive crop failure and food shortages. Coastal areas (where the chikungunya fever epidemics occurred) were particularly affected, since rainfall was well below normal during 2003 and the areas lacked community-based mechanisms for emergency intervention because they had not recently experienced severe drought (UN, 2004).

There are few operational early warning systems for climate-linked epidemics (WHO, 2004). But there is potential for developing such systems—WHO

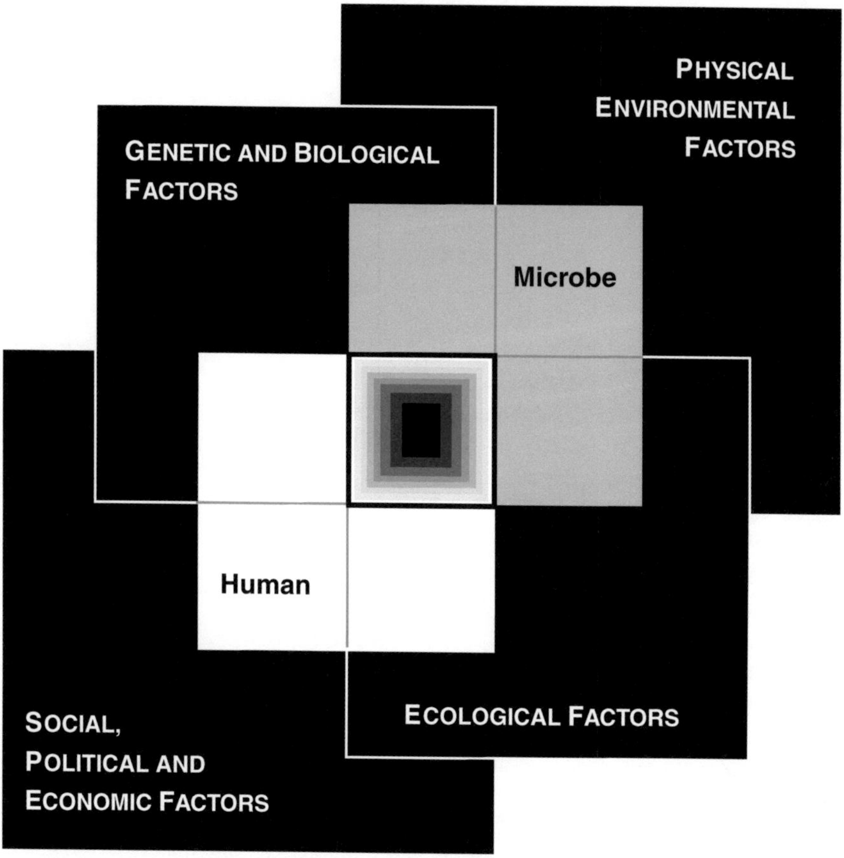

FIGURE 2-9 The Convergence Model.
SOURCE: IOM (2003).

has assessed climate-infectious disease links and recommended development of climate-based predictive models for cholera, malaria, and several other infectious diseases (WHO, 2004); and many countries maintain or are developing early warning systems for natural hazards. Citing the Indian Ocean tsunami of December 26, 2004, as a "wake-up call" about the role that early warning systems could play in reducing the human and physical impacts of natural hazards, United Nations (UN) Secretary General Kofi Annan called for the development of a global early warning system for all natural hazards (UN, 2006). The UN Platform for the Promotion of Early Warning,[7] initiated in 2004, is leading early warning

[7]See http://www.unisdr.org/ppew.

actors toward this goal. Integration of epidemic prediction with such related efforts could speed the development of epidemic prediction systems and facilitate more comprehensive risk communication to communities facing extreme weather events and other natural hazards.

PLAGUE AND CLIMATE

Nils Chr. Stenseth, Dr.philos.[8]
University of Oslo

Plague, caused by the bacterium *Yersinia pestis*, is found on all continents except Antarctica and Australia (Figure 2-10). The plague bacillus causes a rapidly progressing, serious illness that, in its bubonic form, is likely to lead to death by septicemia (40 to 70 percent mortality). Without prompt antibiotic treatment, pneumonic and bubonic plagues are nearly always fatal. For these reasons the plague bacterium *Y. pestis* is considered one of the most pathogenic bacteria for humans (Gage and Kosoy, 2005). Throughout history, it has played a dramatic role, and it continues to be a threat worldwide (Figure 2-10), particularly in Africa (Figure 2-11).

Plague is currently recognized as a reemerging disease increasing in frequency throughout the world (Duplantier et al., 2005; Schrag and Wiener, 1995; Stenseth et al., 2008; WHO, 2003, 2005) as well as being a potential agent of bioterrorism (Inglesby et al., 2000; Koirala, 2006). Throughout its geographic distribution, its main reservoir is composed of a variety of wild (and in some cases commensal) rodents and the bacterium is transmitted between individual hosts primarily by flea vectors (see "The (Full) Plague Eco-Epidemiological System" below). Understanding what determines the dynamics of plague necessitates an understanding of the dynamic rodent-flea-bacterium system in the wild.

The dynamics of the reservoir species are known to be profoundly influenced by climate variation (see Stenseth, 1999; Stenseth et al., 2002, 2006). Here, I summarize our findings from the analysis of long-term data monitoring in Kazakhstan. I both address what might happen should the climate change as expected (IPCC, 2007) and assess whether there has been a climate component underpinning the past plague pandemics.

The Three Big Historical Plague Pandemics

Plague has given rise to at least three major pandemics. The first ("the Justinian plague") spread around the Mediterranean Sea in the sixth century A.D., the second ("the Black Death") started in Europe in the fourteenth century and

[8]Founding Chair of the Centre for Ecological and Evolutionary Synthesis (CEES), Department of Biology, University of Oslo, PO Box 1066 Blindern, N-0316 Oslo, Norway.

FIGURE 2-10 The global distribution of plague. The map shows countries with a known presence of plague in wild reservoir species (black) (after WHO, 2005). For the United States, only the mainland below 50ºN is shown.
SOURCE: Stenseth et al. (2008).

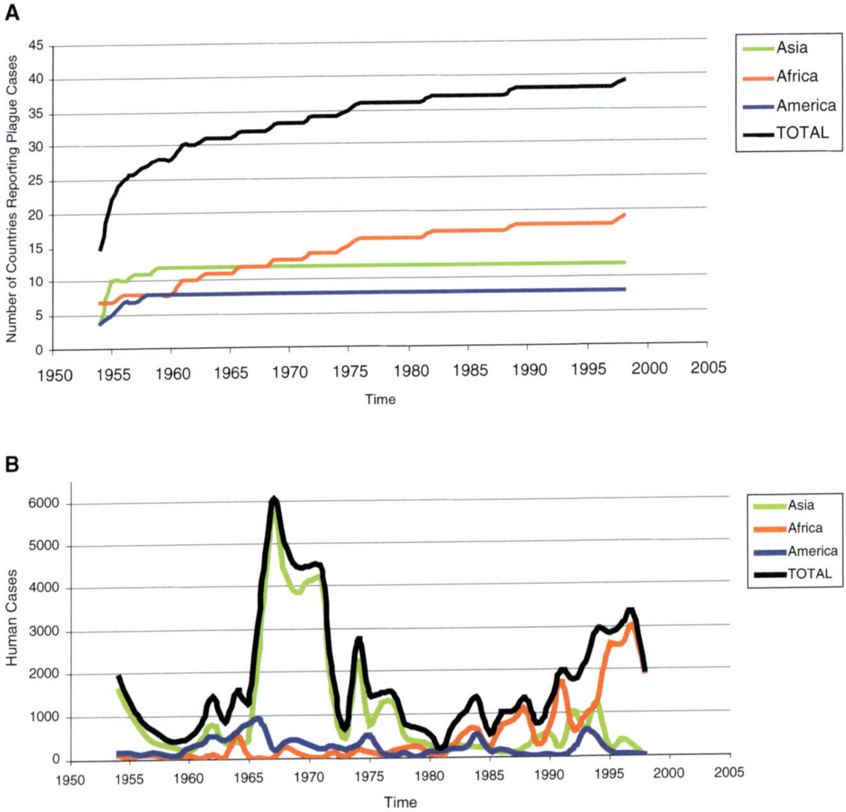

FIGURE 2-11 The global distribution of plague: (A) cumulative number of countries that reported plague to WHO per continent, from 1954-1998; (B) the temporal distribution of plague cases by continent, from 1954-1998, also from WHO. (Panel B is corrected relative to a similar one given in Stenseth et al. (2008): for 1997 and 1998 the numbers have, in dialogue with WHO, been corrected for Madagascar.)
SOURCE: Modified from Stenseth et al. (2008).

recurred intermittently for more than 300 years, and the third started in China during the middle of the nineteenth century and spread throughout the world. Purportedly, each pandemic was caused by a different biovar of *Yersinia pestis*, respectively Antiqua (still found in Africa and central Asia), Medievalis (currently limited to central Asia), and Orientalis (nearly worldwide; Guiyoule et al., 1994; Twigg, 1984; see Figure 2-12).

The Black Death decimated medieval Europe, and as a result, had a major impact on the continent's socioeconomic development, culture, art, religion, and

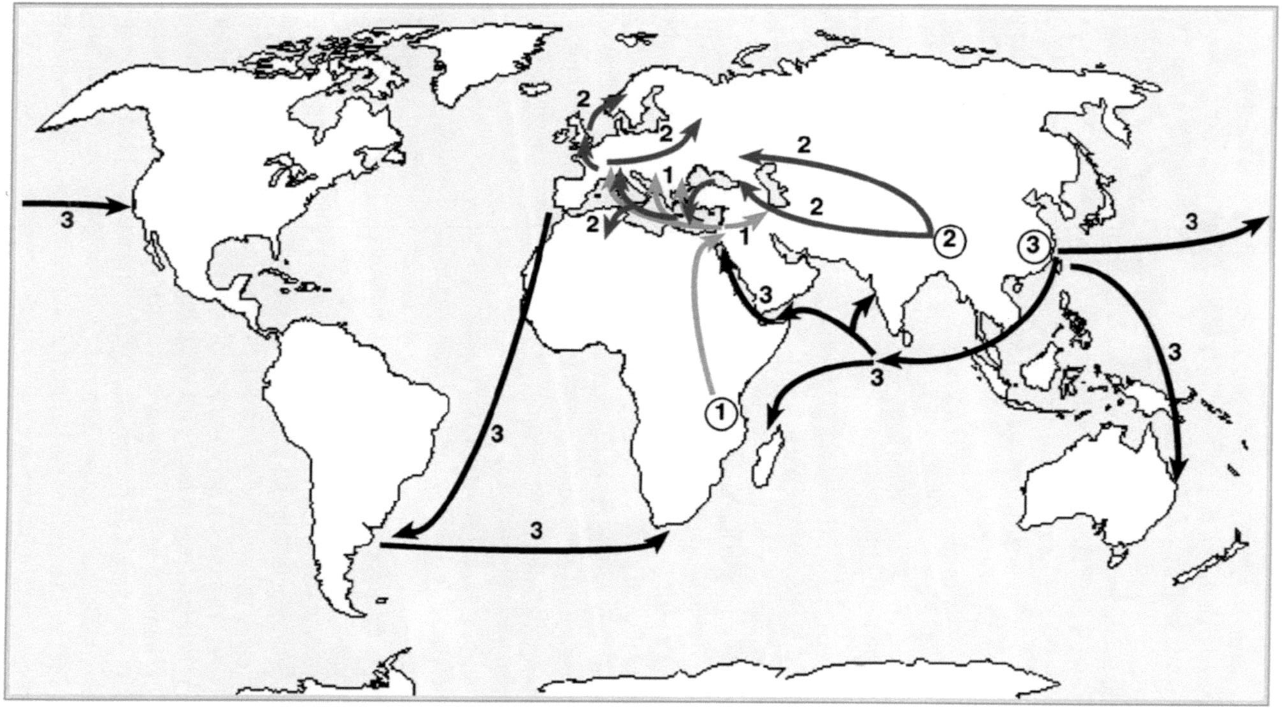

FIGURE 2-12 Routes followed by the three plague pandemic waves (labeled 1, 2, and 3). Circled numbers indicated the regions thought to be the origin of each of the three pandemics: the Justinian plague (541 A.D. to 767 A.D.); the Black Death and subsequent epidemics from 1346 to the early nineteenth century; and the Third Pandemic, in the mid-nineteenth century in the Yunnan region of China, started in 1855.
SOURCE: Achtman et al. (1999).

politics (Twigg, 1984; Ziegler, 1969). Although some have questioned whether the Black Death (as well as the first pandemics) was caused by *Y. pestis* (Cohn, 2002; Scott and Duncan, 2001), it seems settled today (Stenseth et al., 2008). It is generally accepted that the epidemiology of the Black Death plague, as reflected in historical records, does not always match the "classical" rat-flea-human plague cycle, but the reported medical symptoms were very similar during each historical pandemic. It should be appreciated, however, that "classical" plague epidemiology is only one of several possibilities to explain the Black Death and may not even be the most typical of the different plague ecology systems (Drancourt et al., 2006). The discovery of *Y. pestis* genetic material in those who died from the Black Death and are buried in medieval graves (Raoult and Aboudharam, 2000) further supports the view that *Y. pestis* was the causative agent of the Black Death.

The (Full) Plague Eco-Epidemiological System

Soon after Yersin's discovery of the plague bacillus (Yersin, 1894), it became clear that urban outbreaks were linked to transmission among commensal rats and their fleas. In this classic urban plague scenario, infected rats (transported, for example, by ships) arrive in a new city and transmit the infection to local house rats and their fleas, which then serve as sources of human infection. Occasionally, humans develop a pneumonic form of plague, which is then directly transmitted from human to human through respiratory droplets.

The epidemiology of plague, however, is much more complicated than this urban plague scenario suggests, involving several other pathways of transmission. This complicated epidemiology necessitates reconsidering plague ecology within its full ecological web (Figure 2-13).

Maintenance of plague foci depends on a whole suite of rodent hosts and their associated fleas. Under favorable conditions, the plague bacillus might survive in the environment, essentially in rodent burrows (Baltazard et al., 1963). When an infected flea happens to feed on a commensal rodent, the cycle continues in the latter. As commensal rodents die, their fleas are forced to move to alternate hosts (e.g., humans). If humans develop pneumonic plague, the infection may transmit from person to person through exposures to respiratory droplets spread by coughing. Humans may also become infected through handling infected animals (or meat), including rodents, camels, or cats. Cats may also develop pneumonic plague, passing their infection to their owners through coughing. Finally, there is evidence that the human flea, *Pulex irritans*, can be involved in human-to-human transmission (Blanc, 1956; Laudisoit et al., 2007). Mammalian predators, birds of prey, and other birds that use rodent burrows for nesting may move over larger areas than the rodents themselves, spreading the infection over longer distances. Infected commensal rats or humans may also travel over long distances.

Because of its widespread occurrence in wildlife rodent reservoir species one must recognize that plague cannot be eradicated. There is a critical need,

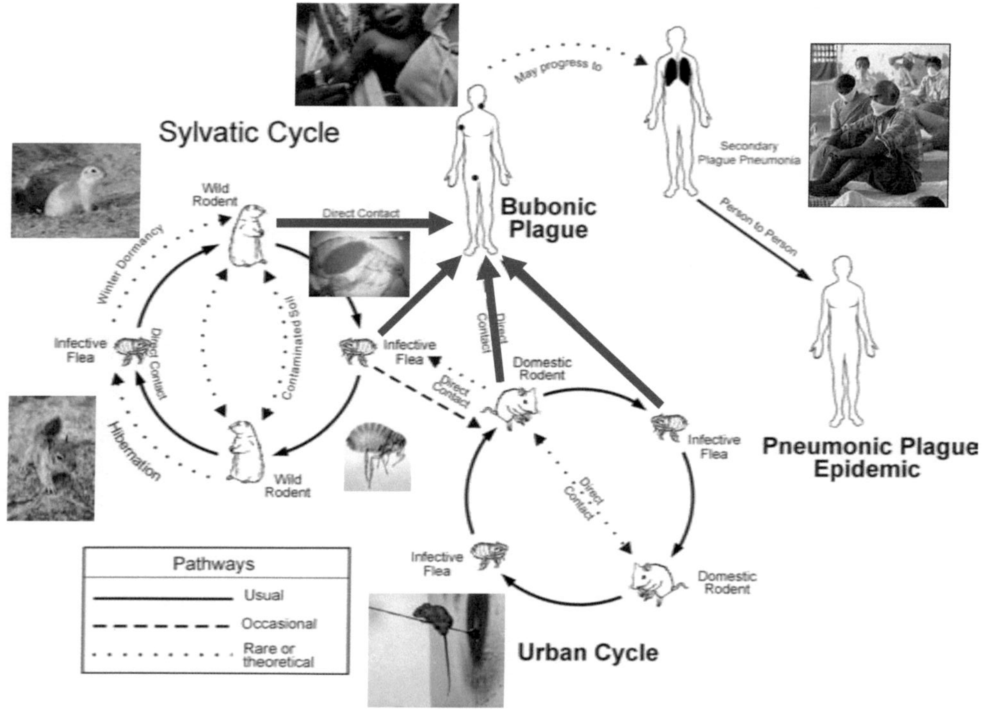

FIGURE 2-13 Possible transmission pathways for the plague bacterium, *Yersinia pestis*. Thick arrows indicate pathways to people.
SOURCE: Adapted from Chamberlain (2004) and printed with permission from Neal R. Chamberlain, Ph.D., A.T. Still University of Health Sciences.

therefore, to understand how human risks are affected by the dynamics of these reservoirs and how people interact with them.

The capacity of the plague bacillus to form permanent foci under highly diverse ecological conditions attests to its extraordinary adaptability. During its emergence in central Asia, *Y. pestis* accumulated copies of insertion sequences rendering its genome highly plastic (Parkhill et al., 2001). The capacity to undergo genomic rearrangements may thus be an efficient means for the plague bacillus to adapt to new ecological niches. *Y. pestis* was recently shown to be able to acquire antibiotic resistance plasmids under natural conditions (Galimand et al., 1997; Guiyoule et al., 2001), probably during its transit in the flea midgut (Hinnebusch et al., 2002). Obviously, the emergence and spread of multidrug-resistant strains of *Y. pestis* would represent a major threat to human health.

Although the number of human cases of plague is relatively low, it would be a mistake to overlook its threat to humanity because of the disease's inherent communicability, rapid spread, rapid clinical course, and high mortality if left untreated. A plague outbreak may also cause widespread panic, as occurred in 1994, when a relatively small outbreak, with 50 deaths, was reported in the city of Surat, India (Mudur, 1995), which led to a nationwide collapse in tourism and trade, with an estimated cost of $600 million U.S. dollars (Fritz et al., 1996).

Studying the Plague Dynamics of Central Asia: The Effect of Climate Variation

Together with colleagues, I have been studying the dynamics of the plague ecological system based on long-term monitoring data from the former Soviet Union (specifically from Kazakhstan), some of which have been published (Davis et al., 2004, 2007; Frigessi et al., 2005; Kausrud et al., 2007; Park et al., 2007; Samia et al., 2007; Stenseth et al., 2006) but much more is to come, including information on human plague cases. Currently, we are expanding our geographic area of interest to include China, India, Madagascar, and the United States.

Our core set of monitoring data comes from southeastern Kazakhstan (74-78°E and 44-47°N; see Figure 2-14). Each spring and autumn, between 1949 and 1995, a proportion of inhabited burrows and site-count observations were done at different locations within the PreBalkhash area (see Figure 2-14; for details, see Stenseth et al., 2006).

For monitoring purposes, the area was divided into 10×10 km^2 sectors. Four sectors constitute a 20×20 km^2 *primary square* (PSQ), and four PSQs constitute a *large square* (LSQ; Figure 2-14). At a given site, the great gerbil population densities were estimated at most twice per year. On approximately 85 percent of these occasions, up to 8,576 gerbils (median = 604) were trapped per LSQ, based on independent plague prevalence data (see Stenseth et al., 2006) and season, and tested for *Y. pestis* infection. The LSQs chosen had the longest regular and continuous time-series data required by our analysis. We also have access to

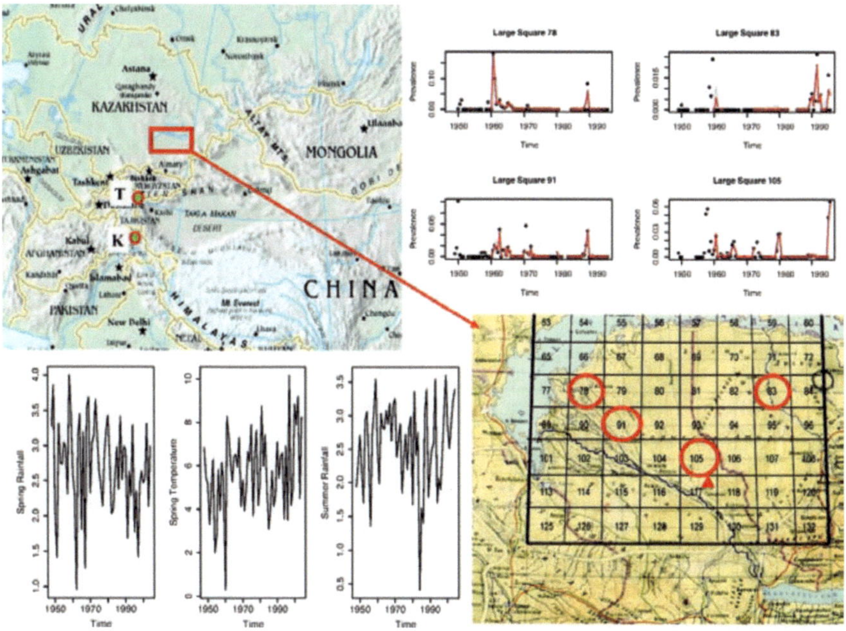

FIGURE 2-14 The field data used in Stenseth et al. (2006) were collected in a natural plague focus in Kazakhstan. The data are plague prevalence in great gerbils, counts of fleas collected from trapped gerbils, and meteorological observations. *Left Upper*: Kazakhstan on a map of Central Asia with the PreBalkhash focus (between 74 and 78°E and 44 and 47°N) marked as a square. The historic climate (tree-ring) measurement sites are circles marked K (Karakorum) and T (Tien Shan). These sites are located approximately 1,000 and 600 km from the research area, respectively. *Lower Right*: The LSQ in the PreBalkhash focus from which we have prevalence. The four LSQs (40 × 40 km) circled in red, namely LSQs 78, 83, 91, and 105, represent key sites where collection of samples for testing the presence of plague was more regular and continuous. The Bakanas meteorological station is located in LSQ 117, marked by a red triangle. *Upper Right*: The time-series plots of the observed prevalence per LSQ. Open and filled circles denote the observed prevalence during the spring and fall, respectively. The time series of the prevalence fitted by using the model defined by the model is shown in red. Using the same model but without any climatic covariates gives the time series shown in gray. Note that owing to the presence of missing values in some covariates (occupancy) and prevalence data, the curves of the fitted values are discontinuous. The fitted values from the model provide a closer fit and reproduce the peaks in prevalence far better than the model without the climatic variables. *Lower Left*: Time-series plots of the climate variables, spring rainfall, spring temperature, and summer rainfall (from left to right).
SOURCE: Stenseth et al. (2006).

plague prevalence data: gerbils caught were tested for plague through isolation of *Y. pestis* from blood, spleen, or liver smears.

Spring climatic variables used were the average monthly temperature during the spring (i.e., March and April) and the log average of the spring rainfall. The fall climatic variable used is the log average of summer rainfall over June, July, and August. Incorporating the climatic effects in the model resulted in fitted values that track the peak occurrences in prevalence more closely than the model without the climatic variables.

Climate variability over the past millennium was estimated by using a large data set of 203 *Juniperus turkestanica* tree-ring width series to reconstruct temperature variations in the Tien Shan Mountains (Kirghizia) (Esper et al., 2003) and a total of 40 stable oxygen isotope ($\delta^{18}O$) series to reconstruct precipitation variations in the Karakorum Mountains (Pakistan) (Treydte et al., 2006). Climatic variations at these sites are found to be correlated with those in the study area.

We also used the NDVI (Hall et al., 2005; Los et al., 2000; see also Pettorelli et al., 2005), which is based on the difference between near-infrared and visible light reflected from the ground, thereby giving an index of light absorbed by chlorophyll on the ground, an index we also extended through proxy data back in time (see Kausrud et al., 2007).

The following discussion summarizes our findings to date. Davis et al. (2004) demonstrated that plague within an area invades, fades out, and reinvades in response to fluctuations in the abundance of its main reservoir host, the great gerbil. Broadly speaking, they found that infection spreads and persists when total abundance is above a single threshold value and fades out when it is below (see Figure 2-15).

Stenseth et al. (2006) reported that a 1°C increase in spring temperatures is predicted to lead to a >50 percent increase in prevalence (see also Samia et al., 2007). Changes in spring temperature were found to be the most important environmental variable determining the prevalence level, leading to the following scenario: Warmer spring conditions result in an elevated vector-host ratio, which leads to a higher prevalence level in the gerbil host population. Moreover, the climatic conditions that support increased prevalence among gerbils, given unchanged gerbil abundance, also favor increased gerbil abundance (see Kausrud et al., 2007), implying that the threshold density (as found by Davis et al., 2004) condition for plague will be reached more often, thereby increasing the frequency with which plague can occur.

Kausrud et al. (2007), focusing on rodent-host dynamics, drew the following five main conclusions from their analyses:

1. Density fluctuations of the great gerbil, the main host, are highly correlated over large areas, suggesting that climate may be a synchronizing agent. This is probably an important factor causing large-scale plague epizootics in the region.

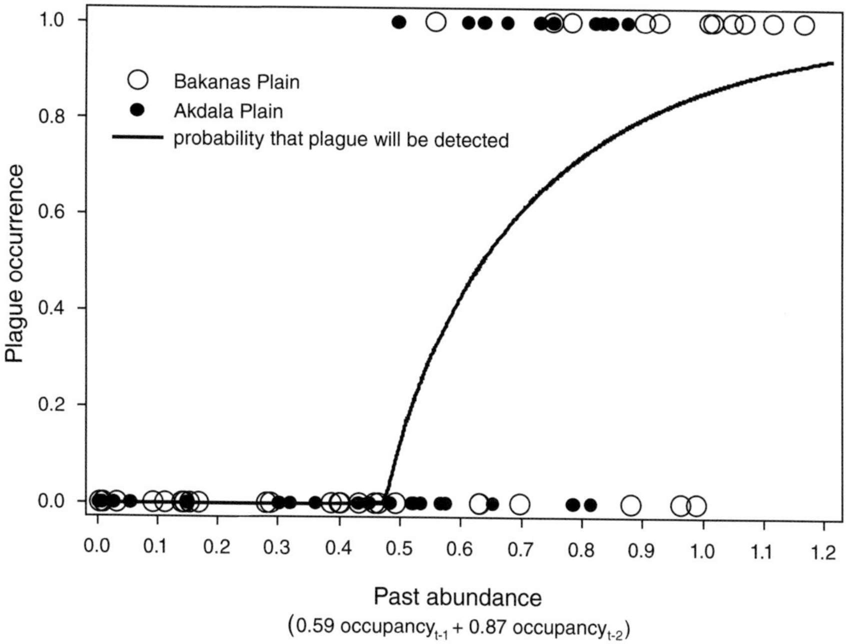

FIGURE 2-15 Relationship between the likelihood of detecting plague (solid line) in gerbils and past burrow occupancy rates together with data on presence or absence of plague at two sites: Bakanas plain (open circles) and Akdala plain (filled circles); see Davis et al. (2004) for details. The likelihood of detecting *Y. pestis* is 0 below a threshold value of 0.476 (95 percent confidence interval: 0.355, 0.572) but rises rapidly once the threshold is attained and continues to increase for even higher values. The seasonal data on abundance and presence of infection are pooled such that presence in a particular year means the disease was detected in either spring or autumn (or both seasons) of that year. Occupancy data represent averages of spring and autumn estimates.
SOURCE: Reprinted from Davis et al. (2004) with permission from AAAS.

2. Great gerbil population densities at large spatial scales can be well predicted 6 to 12 months in advance when combining spatial environmental effects and intrinsic dynamics. This insight is certainly important for predicting plague dynamics.

3. While great gerbil population growth rates exhibit greater variability in areas with low April NDVI index, average population density is not strongly correlated to average vegetation productivity. This suggests that the gerbils will be capable of maintaining population densities where plague can persist over most of their range even if, as predicted, the climate in central Asia becomes increasingly arid.

4. While the presence of plague infection in an area is associated with population decrease over the following months, plague seems unlikely to be the main driving force behind great gerbil density fluctuations.
5. The magnitude of plague epizootics associated with the great gerbil may be expected to increase under predicted effects of ongoing climate change, a confirmation of the conclusion previously drawn by Stenseth et al. (2006).

Altogether, the model reported by Stenseth et al. (2006) suggests that warmer springs (and wetter summers) can trigger a cascading effect on the occurrence and level of plague prevalence, in years with above-threshold great gerbil abundance during the fall 2 calendar-years earlier and in a region that is itself dry and continental (hot summers, cold winters; see IPCC, 2007). Our analyses, moreover, favor the suggestion that enhanced flea survival and reproduction are critical to this effect. Given the multiple routes of plague transmission (flea-borne, direct via several pathways, transfer from other reservoirs), climatic influences on other epidemiological processes cannot be precluded. More generally, it is widely accepted that the distribution and dynamics of vector-borne infections are particularly sensitive to climatic conditions, by virtue of the sensitivity of the (arthropod) vectors themselves to variations in temperature, humidity, and often, quantities of standing water used as breeding sites.

The model reported by Stenseth et al. (2006) may also shed light on the emergence of the Black Death and the plague's Third Pandemic, thought to have spread from an outbreak "core region" in central Asia. Analyses of tree-ring proxy climate data demonstrated that conditions during the period of the Black Death (1280-1350) were both warmer and increasingly wet. The same was true during the origin of the Third Pandemic (1855-1870), when the climate was wetter and underwent an increasingly warm trend. Our analyses are thus in agreement with the hypothesis that the medieval Black Death and the mid-nineteenth-century plague pandemic may have been triggered by favorable climatic conditions in central Asia. Figure 2-16 summarizes the link between climate and the two last plague pandemics.

Such climatic conditions have recently become more common (IPCC, 2007), and whereas regional scenarios suggest a decrease in annual precipitation with increasing variance, mean spring temperatures are predicted to continue increasing (Huntington, 2006). Indeed, during the period from the 1940s, plague prevalence has been high in its host-reservoir in Kazakhstan (see Stenseth et al., 2006). Effective surveillance and control during the Soviet period resulted in few human cases. However, recent changes in the public health systems, coupled with a period of political transition in central Asia and an increased prevalence of plague in its natural reservoir in the region, shadow a future of increased risk of human infections.

In a yet-to-be-published study, Kausrud et al. (2008), using the same surveillance data from 1950 to 1995, together with regional climate indices, have

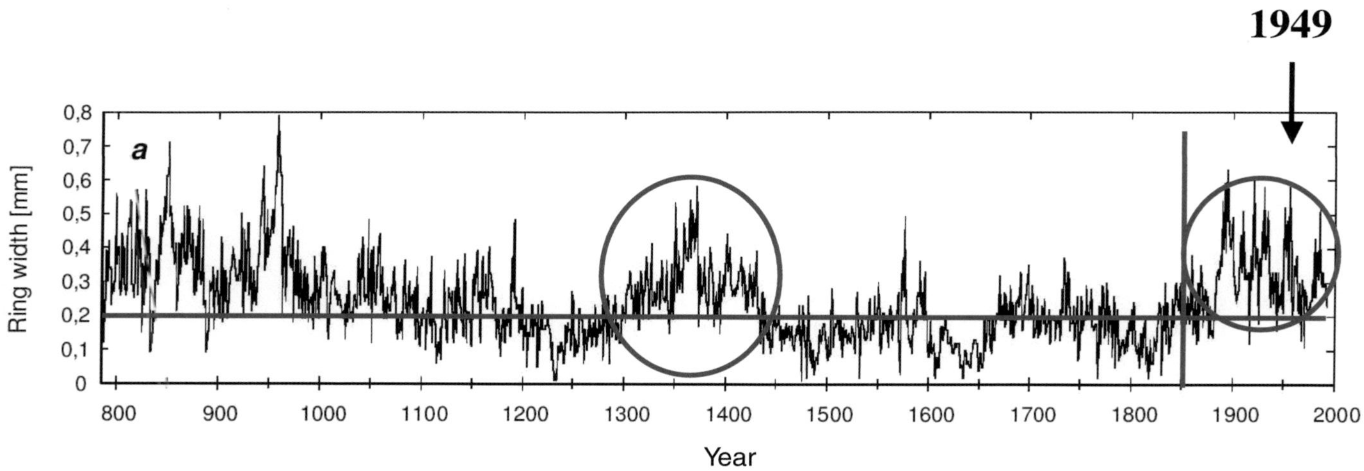

FIGURE 2-16 Tree-ring data suggesting that conditions during the Black Death and the Third Pandemic were similar. The two circles highlight the start of the Black Death and the Third Pandemic; the horizontal line is inserted for the purpose of baseline reference; the vertical gray line indicates the very start of the Third Pandemic (1855); 1949 is the year for which the monitoring and intervention program started in Kazakhstan.

SOURCE: Based on data in Stenseth et al. (2006).

found that climate influences plague dynamics through the rodent-host and flea-vector relationship. Simulating backwards, Kausrud et al. (2008) successfully predicted human plague patterns in Kazakhstan from 1904 to 1950. Using tree-ring data extending back in time to 1000 A.D., this model allows us to compare model predictions with historical plague epidemiology. Analysis suggests an eco-epidemiological basis for considering the Black Death epidemic as having originated in central Asia during climatically favorable conditions (for the plague system). The same model, used for prediction forward, suggests that expected climate change will sustain and possibly increase plague activity in central Asia.

Effect of Climate on Plague Dynamics in Other Regions of the World

Based upon our work on the Kazakh data, we are now extending our interest to other parts of the world. Together with Zhang et al. (2007), I have been involved in some preliminary analysis of data on human plague cases from China. These show a clear effect of large-scale climate influence. Unpublished work that I have done in cooperation with a student of mine (Ben Ari et al., 2008) similarly shows that the number of human plague cases in the western United States is strongly influenced by the Pacific Decadal Oscillation (PDO) and the number of days with above-normal temperatures. In short, a warmer and wetter climate is associated with increased prevalence level of the plague bacterium in the rodent reservoir, which subsequently might lead to an increased number of human cases. These results match up nicely with the previously published cascade model by Parmenter et al. (1999) emphasizing that the climate connection works partly through the rodent-host dynamics and the flea-vector dynamics (see Figure 2-17) in the same region.

Additional Reasons for Being Concerned: Bioterror

As indicated in the introduction, we should not overlook the fact that plague has been weaponized throughout history—from catapulting diseased corpses over city walls, to dropping infected fleas from airplanes, to refined modern aerosol formulations (Inglesby et al., 2000; Koirala, 2006). The weaponization research carried out on plague from the 1930s through the 1990s fueled biological warfare fears that may actually have stimulated research on infectious disease surveillance and response strategies. More recently, however, the fears of small-scale bioterrorism and a desire by government authorities to more fully control all access to plague materials increase the danger of stifling basic research on plague ecology, epidemiology, and pathophysiology that is required to improve its clinical management in endemic areas. Terrorist use of an aerosol released in a confined space could result in significant mortality and widespread panic (Inglesby et al., 2000; Koirala, 2006), and no one would want the knowledge and materials for weaponizing plague to fall into the hands of non-state actors. However, the

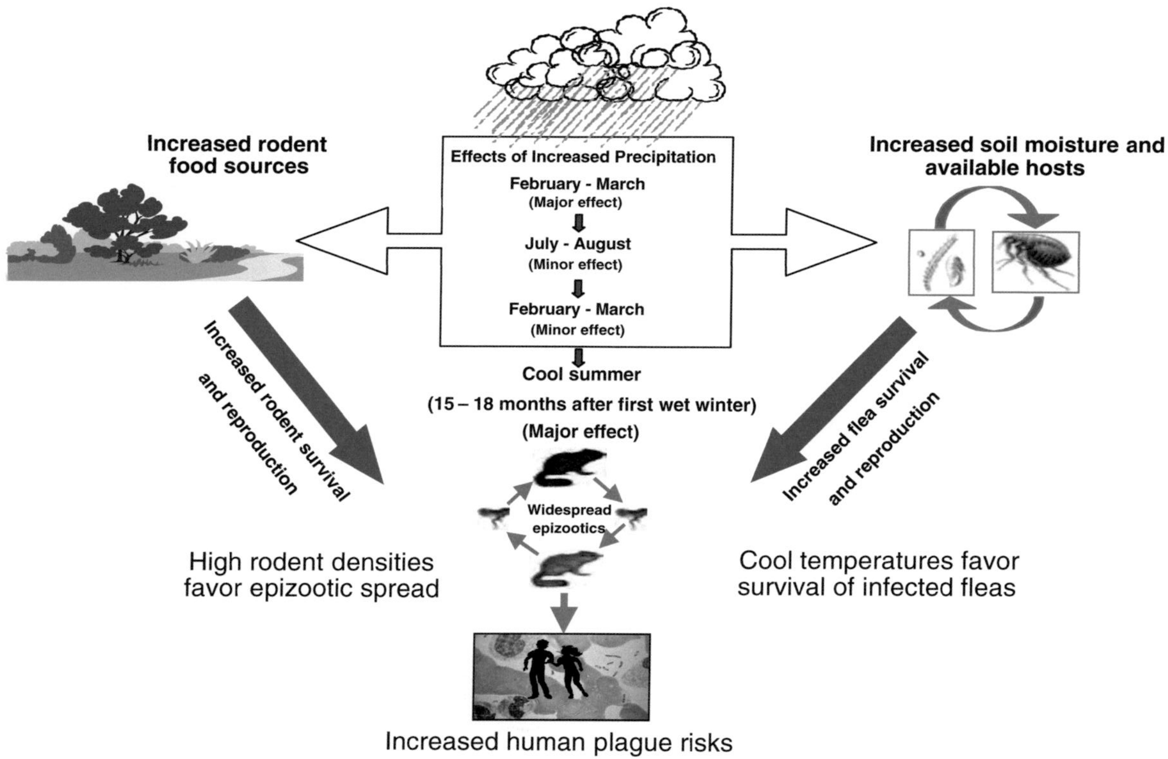

FIGURE 2-17 The modified trophic cascade model of Parmenter et al. (1999).
SOURCE: Adapted from Parmenter et al. (1999) with permission from the *American Journal of Tropical Medicine and Hygiene*.

need for scientifically sound studies of the dynamics of infection, transmission, outbreak management, and improved surveillance and monitoring systems has never been greater.

Conclusion: It Is Unwise to Neglect Plague

In conclusion, it should be noted that although plague may not match the so-called big three diseases (malaria, HIV/AIDS, and tuberculosis; see Hotez et al., 2006) in numbers of human cases, it by far exceeds these diseases in pathogenicity and rapid spread under the right environmental conditions. Plague should be seen not only as a historical curiosity but as a reemerging disease of the twenty-first century. Plague should not continue to be neglected and relegated to the sidelines; it is a disease which should concern us today.

Plague remains a fairly poorly understood threat that we cannot afford to ignore. Nevertheless, much progress has recently been made toward understanding the dynamics of the full plague eco-epidemiological system, and not the least how it responds to climate variation and change. We know that climate does affect the dynamics (and indeed the level) of plague. However, it is difficult at present to say what that effect will be. For example, in central Asia there might be higher levels of plague in the rodent reservoir populations, if current climate prognoses for the region materialize. Also, higher levels in the wildlife reservoir will automatically lead to a greater chance of people being infected by the plague bacillus. In other places of the world there might be lower plague levels in the reservoirs—we simply do not know, but we ought to know if we are to be maximally prepared for what happens should climate change. It is certain, though, that the picture regarding plague might be much more serious than conveyed by Anyamba et al. (2006). Only by knowing more about how the eco-epidemiological plague systems in different parts of the world will respond to given climate scenarios can we take the necessary precautionary measures to reduce the risks associated with human infections. Indeed, knowing how climate is affecting the components of the eco-epidemiological system depicted in Figure 2-13, and subsequently how these climate drivers might change the dynamics of the system, will put us in a greatly improved position for predicting where and under what environmental conditions the risk of human plague infections might increase and where and under what conditions it might decrease (or remain unchanged). Much of the insight derived from studying particular plague systems will be general and applicable to other plague systems—and indeed to other vector-borne infectious disease systems. However, since the involved host and vector species are different from one part of the world to another (indeed, the plague eco-epidemiological system is characterized by a whole suite of rodent host species and their associated fleas, differing from one place to another), studies similar to those that I have summarized for central Asia are greatly needed. Such additional studies may help us to understand which insights derived from the central-Asian studies may or may not be generalized to other places where plague occurs. Such additional studies will further help us develop more region-specific prediction regarding what might happen should climate change in some specific way.

Acknowledgments

I thank Tamara Ben Ari for having read and commented on an earlier version of this paper; furthermore, I thank her and Kyrre Linné Kausrud for allowing me to summarize yet unpublished work. Over the years working on plague dynamics, I have benefited enormously from collaboration with several colleagues, most importantly Herwig Leirs, Hildegunn Viljugrein, Mike Begon, Kung-Sik Chan, Noelle I. Samia, Stephen Davis, Kyrre Linné Kausrud, Tamara Ben Ari, Lise Heier, Elisabeth Carniel, Mark Achtman, Kenneth L. Gage, Vladimir S. Ageyev, Nikolay L. Klassovskiy, and Sergey B. Pole. I have learned a lot from them—any misunderstandings of what they have tried to teach me is due solely to my own shortcomings. On a more administrative side, I would like to thank Dr. M. Pletschette for his stimulating encouragement, which made me start working on plague in the first case. My work on plague has been generously funded over the years through the European Union Projects (ISTC K-159, STEPICA [INCO-COPERNICUS, ICA 2-CT2000-10046], as well as Marie Curie Early Stage Training grant to CEES), the Norwegian Research Council, and my own university and center. Last, but not least, I extend my thanks to the many hundreds of Kazakh plague zoologists who collected so many data over all these years.

CLIMATE CHANGE AND PLANT DISEASE RISK

Karen A. Garrett, Ph.D.[9]
Kansas State University

Plant Disease and Ecosystem Services

One of the most important effects of plant disease is its impact on crop plant productivity. Oerke et al. (1994) estimated that damage by disease and insect pests resulted in a 42 percent loss in the eight most important food and cash crops. Pimentel et al. (2000) estimated that 65 percent of U.S. crop losses, $137 billion, were due to introduced pathogens. The effects of plant disease can also be considered within the broader context of ecosystem services, defined as the benefits provided to humans by ecosystems, including services provided by plants and their pathogens (Daily, 1997). Ecosystem services include the following: (1) provisioning services, such as the more obvious provisioning of food, fiber, fuel, and also the provisioning of genetic resources; (2) supporting services, such as

[9]Associate Professor, Department of Plant Pathology.

soil formation, nutrient cycling, and primary productivity by plants, all of which have great economic value but tend to be appreciated only when there are breakdowns, such as the loss of soil during the U.S. dust bowl; (3) regulating services, such as regulation of climate, disease, and insect herbivory, and water purification; and (4) cultural services, such as opportunities for education, recreation, tourism, and inspiration. The Millennium Ecosystem Assessment[10] provides an example of system evaluation based on ecosystem services. Cheatham et al. (in revision) have synthesized perspectives on plant disease and its management in the context of ecosystem services. In addition to the direct effects of disease on crop production, disease and its management by increased tillage, pesticide use, and other methods may reduce services provided by plants such as soil formation and climate and water regulation. Disease may also remove plants that provide important cultural services in addition to the range of other potential services.

Some examples among the many notorious plant diseases illustrate the issues for disease management and the potential impact when diseases cannot be managed effectively. Chestnut blight has had one of the most definitive effects, essentially removing the once common American chestnut from the landscape of eastern North America (Anagnostakis, 2000). Potato late blight is infamous as the proximate cause of the Irish potato famine and continues as a major constraint to potato production, making the use of pesticides a typical part of potato management in many areas (Hijmans et al., 2000). Karnal bunt of wheat offers an example of a disease that does not cause major yield loss, but has an important economic impact on regions where it is present through limits on trade with Europe and other parts of the world where the pathogen has not been detected (Rush et al., 2005). Sudden oak death has changed the structure of some western U.S. forests and threatens to impact forests throughout a much wider area (Rizzo et al., 2005). Soybean rust is a new pathogen to the United States, with the potential to become established throughout much of the U.S. soybean production areas (Pivonia and Yang, 2004). Wheat stem rust was an important pathogen in the United States in the 1900s, motivating the removal of barberry plants that served as an alternate host and supported sexual reproduction of the pathogen. Disease resistance in U.S. wheat has been effective against this pathogen, but now new pathogen types for which this resistance is not useful have arisen in Africa and are likely to arrive in the United States in the near future (Stokstad, 2007).

The effects of climate on plant disease have been a direct object of study for decades. In contrast to many human diseases, the pathogens causing important plant diseases are often present on and around plants, ready to infect when environmental conditions become conducive. This has motivated the development of plant disease forecasting systems based on climatic variables such as

---

[10]The Millennium Ecosystem Assessment is an evaluation of the effects of ecosystem change on human well-being assembled from the work of more than 1,360 scientists (see http://www.millenniumassessment.org).

temperature and precipitation (De Wolf and Isard, 2007). Although such models are not new, the need to address climate change has placed new demands on these models, the research underpinning them, and policy drawing upon them. This paper begins with a brief introduction to the typical methods for managing disease. The observed and potential effects of climate change on plant disease are then reviewed, with an emphasis on biological thresholds and interactions that may lead to particularly large impacts from climate shifts. The paper concludes with a discussion of research and policy needs for plant disease management in response to climate change.

The Usual Challenges for Managing Plant Disease

Pesticides are a common tool for managing plant disease. For some plant diseases such as potato late blight, crop production without pesticides is currently impractical in many systems. In regions where education about pesticide safety is lacking, some farmers and their families experience chronic pesticide exposure. There are estimated to be between 1 and 5 million cases of pesticide poisoning each year, including many thousands of fatalities (UNEP, 2004). Shifts in pesticide use may thus result in shifts in unmanaged pesticide exposure, so that changes in demand for pesticides due to climate or other factors may have unexpected impacts on human health as well. Other disease management methods may be useful for specific diseases, such as removal of infected plant materials, introduction of biocontrol agents, management for disease-suppressive soils, or use of certified seed to avoid introduction of pathogens.

Deployment of disease resistance genes is often the most attractive option for disease management in agricultural systems. For some diseases, resistance offers completely effective management, whereas for others, effective resistance is not known although partial resistance may still be a useful management component. There is little cost from use of resistance genes to growers or consumers, except that in some cases it may be challenging for plant breeders to combine desired resistance genes with other desirable plant characteristics. Breeding crops for disease resistance also offers challenges in terms of identifying resistance that is durable. The deployment of resistance genes is much more efficient if the genes are useful against pathogen populations for long periods of time even if exposed to large pathogen populations under disease-conducive environmental conditions. Pathogen adaptation to overcome disease resistance is an ongoing problem for the management of many diseases (McDonald and Linde, 2002).

The use of cultivar mixtures is one method of resistance gene deployment that may increase the useful life of resistance genes in some cases. The management of rice blast in China offers a particularly dramatic example of the utility of mixtures for disease management, applied to over a million hectares. Higher-value susceptible rice varieties were grown in strips mixed with strips of lower-value resistant varieties. Both resistant and susceptible varieties experienced a decrease in disease pressure compared to test plots where single varieties were grown for comparison (Zhu et al., 2000). In this case, it seems that microclimate was an important factor, such that the taller susceptible varieties experience relatively drier conditions when surrounded by the shorter resistant varieties (Zhu et al., 2005).

The fact that agriculturalists have the ability to manipulate crop plant genetics makes plant disease management in agriculture much easier, in some respects, than human disease management. Problems can also arise from this ability, however, as particularly successful crop varieties become widespread. Thus, a common challenge for plant disease management is the general homogeneity of cropping systems in the United States and trends toward greater crop homogeneity in most regions of the world. This homogeneity makes it easier for plant pathogens adapted to the common crop varieties to spread rapidly throughout crop plant populations. Margosian et al. (in revision) have evaluated the connectivity of the four major crop plants in the United States in terms of availability of the crop host species. The connectivity of a landscape for a particular organism, in this case a plant pathogen, is a measure of the ease with which the organism can move through the landscape. Maize and soybean are strongly connected throughout much of their range. Wheat and cotton production are more fragmented, so that pathogen populations cannot move as readily through all production areas. Conversion to biofuel production has the potential to increase crop homogeneity.

Maps of disease risk based on climate can be generated for diseases with reliable and widely applicable forecasting models. For example, Hijmans et al. (2000) mapped the risk of potato late blight based on climate parameters. Using updated forecasting models for potato late blight risk, Villanueva et al. (in preparation) estimated disease risk in the Altiplano region around Lake Titicaca (Figure 2-18). Such models are available for only very well-studied diseases, but Magarey et al. (2007) have developed a general model of infection risk for application in mapping the risk of new pathogens for which detailed models are not yet available. The combination of maps of current and future climatic conditions with models of pathogen risk can be adapted to evaluate changes in global risk in response to climate change. For example, Bergot et al. (2004) predicted the spread of the host-generalist pathogen *Phytophthora cinnamomi* in Europe.

Implications of Climate Change

Climate change will impact the productivity of agricultural and wildland plant populations through many mechanisms. One method for studying climate change effects on crop productivity is to study the correlation between climate variables and yield to date. Yield is the product of a number of factors, including losses to plant disease; partitioning the effects of these different factors will be necessary to develop a full understanding of the impacts of climate change.

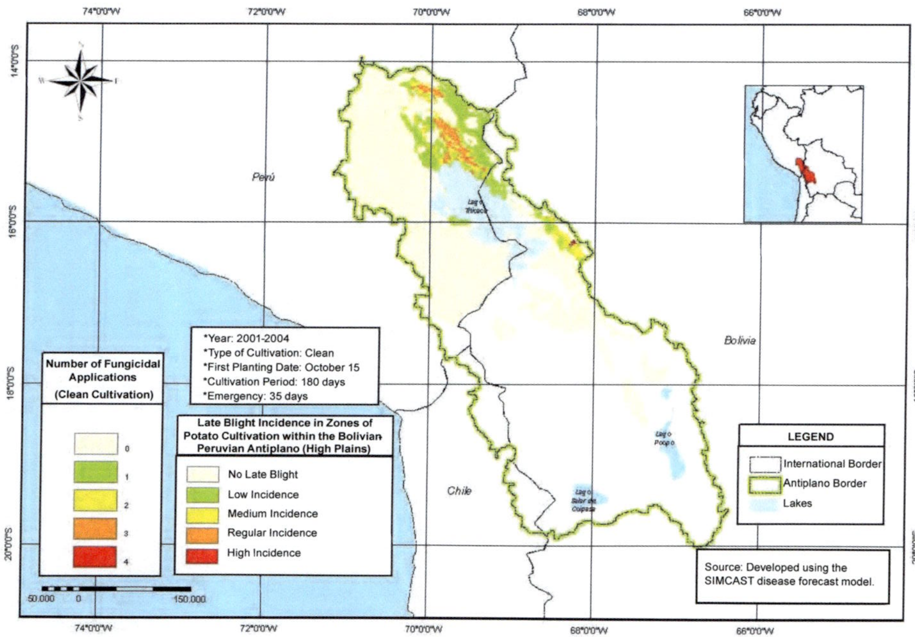

FIGURE 2-18 Estimated potato late blight severity in the Altiplano area of Peru and Bolivia based on weather measures during 2001-2004 used in a late blight forecasting model. The comparable estimates for disease severity in 1995-1998 were for no late blight occurrence in the region. As temperatures increase in the region, the risk of potato late blight may be expected to increase at higher altitudes. Color indicates the level of estimated disease, ranging from green = low to red = high, with a corresponding range in the number of fungicide applications needed for successful potato production. This figure was translated into English from the original (in Spanish) by Mila Gonzalez.
SOURCE: Map courtesy of H. Villanueva, R. Raymundo, H. Juarez, W. Perez, and G. Forbes, International Potato Center.

The general trend toward warmer temperatures in recent years in many regions has facilitated study of the correlation between climate variables and yield. For example, rice productivity in the Philippines has shown a negative correlation with night temperature from 1979 to 2003 (Peng et al., 2004). Of course a challenge for such analyses is to account well for all of the other potential causal factors of the system that may vary along with climatic parameters. Lobell et al. (2008) have analyzed which regions of the world are most likely to be confronted with food security issues resulting from climate change, concluding that south Asia and southern Africa are particularly at risk. In another analysis of global agriculture, Cline (2007) points out that potential benefits to agriculture in some areas in the first decades of global temperature change may give the public a false sense of security and make it more difficult to put policies in place to avoid problems from more extreme changes in later decades.

A first step toward understanding wildland plant responses to climate change and the potential for adaptation to new climatic conditions is to address gene expression and underlying genetic diversity in wild plant populations. Travers et al. (2007) studied the effects of simulated precipitation change on big bluestem, the dominant grass of tallgrass prairie of the U.S. Great Plains. Under the predicted future precipitation patterns with fewer and larger precipitation events leading to longer periods of drought stress, they observed lower expression of a gene associated with the hypersensitive response, a disease resistance reaction. Frank (2007) also studied big bluestem, finding higher infection rates and dampened phytohormonal responses to infection when plants experienced severe drought stress. Studying the diversity of resistance genes in wild plant populations is still challenging because little is known about them and for the moment there are few tools available. Rouse (2007) studied a gene in big bluestem that is related to genes conferring disease resistance in sorghum, finding evidence for historical disease patterns in natural populations that vary in diversity for these genes across a gradient of disease conduciveness.

The effectiveness of disease resistance genes may vary with climatic parameters. For example, Webb et al. (in preparation) found that rice genes conferring resistance to rice blast have different effectiveness depending on temperature. Most resistance genes tested were less effective at higher temperatures, but one of the most effective genes was actually more effective at 35-29°C day-night temperatures than at 29-21°C. These differential responses will influence the selection pressures experienced by pathogen populations as temperatures fluctuate annually and shift over years (Webb et al., in preparation).

Climatic changes and changes in CO_2 concentrations can affect plant physiology, growth, and architecture in several ways that influence plant disease risk. On shorter time scales, stomatal closure in response to drought stress makes it more difficult for some pathogens to enter leaves. If plant canopies close earlier in the season due to changed conditions, the increased humidity in canopy microclimates may favor many pathogens. CO_2 concentrations are expected to impact pathogens directly as well, although a model for this impact is unlikely to be simple. For example, in a study of a set of fungal pathogens, Chakraborty et al. (2000) found that some species reproduced more rapidly under increased CO_2, while other species reproduced more slowly.

In wildland systems, climate change and increased CO_2 concentrations may also have mixed effects. Mitchell et al. (2003) found that the fungal pathogen load in tallgrass prairie increased overall in response to higher ambient CO_2. In montaine prairie, Roy et al. (2004) found mixed effects of simulated temperature increases, with some pathogens increasing in abundance and others decreasing. Desprez-Loustau et al. (2007) predicted that the effect of climate change on a set of forest pathogens in Europe will be to increase favorability for the majority of pathogens. In general, rising temperatures may favor soil fungi that cause damping-off in seedlings, sometimes with high rates of mortality, a trend unlikely to be observed in the short term unless studies are designed specifically to look for such effects.

Range shifts in pathogens are frequently observed. As others have discussed at this workshop, such range shifts can be difficult to interpret. For example, needle blight is moving northward in North America as temperature and precipitation patterns shift (Woods et al., 2005). It is reasonable to think that such range shifts may be driven by changing climatic conditions, but the correlative nature of the data makes it impossible to determine this conclusively. Ultimately these relationships will have to be addressed in projects that combine the full range of factors in field studies as well as more limited and controlled experiments that allow clear conclusions about the effects of factors to partition effects.

The potential importance of extreme weather events is illustrated by the introduction of soybean rust to the United States. It is likely that spores of soybean rust entered the United States via Hurricane Ivan (Isard et al., 2005).[11] If such extreme weather events become more common, global movement of pathogens will be accelerated. Soybean rust also offers an interesting example of the potential interactions between two invasive species. The widely introduced and problematic kudzu vine is another host of this pathogen and has the potential to play an important role as a pathogen reservoir during seasons when soybeans are not available for infection. Until now, however, movement of soybean rust has been slower than expected based on some predictions, probably due to environmental conditions that have not been conducive to disease. If the public becomes too complacent about the slower-than-predicted progress of soybean rust across the United States, this may result in more substantial problems if there is not support for needed research and if soybean growers do not prepare adequately.

The ultimate impact of changes in plant disease pressure, in either agricultural systems or wildland systems, will be determined in part by what plant

[11] Of course, spores of this pathogen may well have entered the United States previously but been unsuccessful in establishing infection. Entry of large numbers of spores may be necessary for an invasive pathogen to "beat the odds."

genotypes or species replace those that have experienced more damage by disease. Eviner and Likens (2008) summarize factors important for predicting the effects of disease on ecosystems, where one of the most important factors may be the functional similarities of infected host individuals versus the species that replace them. Through a broad ecosystem science lens, plant species may be generalized as composing "a single giant photosynthesizing leaf." From this standpoint, damage to one or a few plant species may not be important if other species can play the same role. In eastern U.S. forests, while other tree species increased in abundance to photosynthesize in the place of American chestnuts, they did not provide other important ecosystem services such as production of chestnuts as food for humans and wildlife. Likewise, most agricultural systems are not diverse enough to readily accommodate removal of an important species such as soybeans, if soybean production were to become uneconomical due to a new disease such as soybean rust.

Potential Interactions, Thresholds, and Positive Feedback Loops

If a small change in average temperature or precipitation patterns results in a small change in plant disease risk, this may be relatively easy to accommodate in agricultural disease management and may have little impact on wildland systems. Climate change is a greater concern when interactions serve to amplify the effects on biological systems or when systems are currently near thresholds such that small changes in abiotic drivers may push them beyond the threshold and thus have important effects. Effects may also be exacerbated if positive feedback loops are in place so that increased disease pressure further increases disease risk.

Abiotic environmental conditions are understood to be critically important in plant disease epidemiology, as commonly represented in the "plant disease triangle" (Figure 2-19). The three components of this triangle are a susceptible host, a virulent pathogen (and effective vector, as needed), and a conducive abiotic environment. For example, many fungal and oomycete pathogens benefit from higher levels of humidity. Surprising new disease problems may occur if the susceptible host and virulent pathogen have been present all along and the environment shifts to become more conducive. For example, potato late blight became an extreme problem for Irish food security during the potato famine when wetter years supported rapid disease development. The further interaction between high losses to disease and widespread reliance on potatoes as a primary food led to a disastrous situation.

Allee effects represent one type of threshold. An Allee effect occurs when a species experiences greater limitations on per capita reproduction for small population sizes. Quorum sensing provides an interesting potential mechanism for this type of phenomenon, where bacterial populations may become pathogenic only when intraspecific signaling indicates that a sufficiently large population is present for infection. Smaller population sizes may also make it less likely that

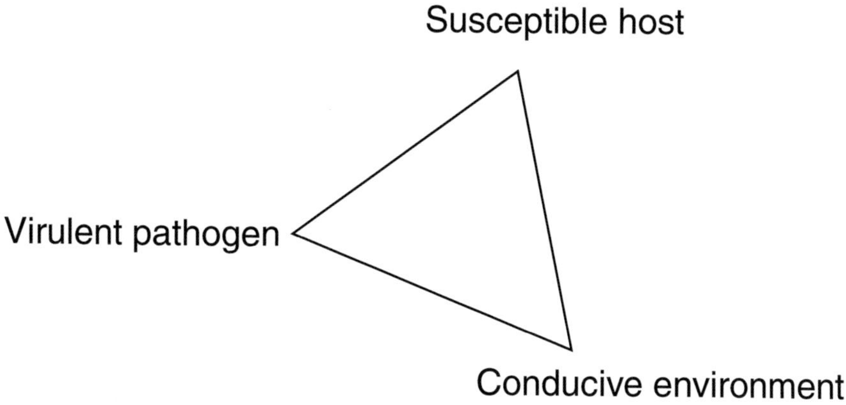

FIGURE 2-19 The plant disease triangle, illustrating the relationship between host, pathogen, and environment necessary for disease to occur.

individuals encounter mates. For example, the Karnal bunt pathogen requires encounters between two mating types for reproduction to occur, yet its propagules are wind dispersed, making encounters between individuals of different mating types unlikely when populations are small. The resulting Allee effect may help to explain why this species has not been more invasive, since encounters between mating types will be even less likely when dispersed by wind over larger areas (Garrett and Bowden, 2002). For species that experience them, such Allee effects interact with disease nonconducive environmental conditions to reduce the chance of infection still further. As a result, if climatic conditions become more conducive to disease so that pathogens are released from the constraint of the Allee effect, pathogen populations may increase much more rapidly than anticipated.

The typical "compound interest" development of plant disease epidemics for pathogens with multiple generations per season can also result in important threshold structures. Infection levels can often increase by orders of magnitude toward the end of the season. If the length of the growing season increases, regional production of particular crop species may expand over time, with the longer season length allowing for huge increases in pathogen populations toward the end of the growing season. These populations may reduce yields during that season and also serve as large sources of inoculum for upcoming cropping seasons. Such higher regional inoculum loads may produce positive feedback loops, rendering local application of some management techniques less useful. For example, local application of techniques such as sanitation (removal of infected plant materials), use of cultivar mixtures, and use of disease resistance based on lower inoculum production all rely, at least to some extent, on an ability to control local inoculum loads. If regional inoculum loads are too high, the contributions of these methods will be diminished (Garrett et al., in revision). Likewise, in wildland systems, plant diversity probably provides baseline regulation of plant disease that is unappreciated but may be diminished if systems become saturated with inoculum. Conversely, if seasons become shorter or if climatic conditions during significant parts of growing seasons become less disease conducive, greater benefits may be obtained from some management techniques.

Pathogen range shifts may occur as climatic conditions change to allow infection and overwintering or oversummering in new areas. The effects of climatic shifts may interact with other phenomena, such as the introduction of new pathogen species or pathogen genotypes. For example, overwintering of the potato late blight pathogen is facilitated by the presence of different mating types, which allows sexual production of a much hardier oospore and the potential for adaptation through production of new genotypes (Widmark et al., 2007). The combination of milder winters and introduction of new mating types may greatly increase problems with such pathogens. Range shifts and pathogen introductions will also result in new encounters between pathogen species, with the potential for hybridization to produce new pathogens (Brasier, 2001). Likewise, the introduction or range shift of new vector species may make diseases much more important, such as in the case of the movement of the glassy-winged sharpshooter and resulting increased risk of Pierce's disease of grapevines (Redak et al., 2004).

Phenological shifts and range shifts in response to climate change may not follow the same patterns for plant hosts and pathogens. Some pathogens can only infect particular plant growth stages or organs, such as flowers. For example, the *Fusarium* head blight pathogen infects wheat anthers or other floral organs (Bai and Shaner, 2004). Shifts in flowering time phenology in response to climate change may not match shifts in pathogen phenology such that infection rates may unexpectedly rise or fall. Different patterns of geographic range shifts may result in new pathogen-host combinations (Parker and Gilbert, 2004). The genetic potential for adaptability of pathogen populations will be important in determining whether any resulting reductions in infection will be short term or lasting. In general, the timeline of pathogen adaptation is likely to be much shorter than the timeline for plant adaptation. This will be especially true for long-lived plant species in wildlands, but also for annual crop species even with the full attention of agricultural scientists.

Policy may also interact in important ways with abiotic conditions. Along with the Irish potato famine, another dramatic example is the dustbowl in the central United States. Policies that supported extensive plowing of lands in this area coincided with climatic conditions favoring wind erosion of soils. Either factor alone might have caused problems, but the combination of the two led to conditions devastating to the region. The interaction of biological and sociological factors may also result in amplified effects of climate change. For example, if plant pathogens are intentionally introduced (Fletcher and Stack, 2007; Madden and Wheelis, 2003), bioterrorists using them might seek out the most environmentally conducive conditions for their establishment in vulnerable cropping systems. Regions where local food security is closely tied to local food production will be particularly vulnerable to changes in crop disease pressure. Yet societies in these regions may also tend to rely on crop species that are less well supported by research and development. These "orphan crops," such as millet, quinoa, cassava, and teff, need more research support to buffer the vulnerability of societies to which they are important (Nelson et al., 2004).

Responding to Climate Change

The good news for formulation of strategies for plant disease management under changing climate conditions is that much of what needs to be done is the same with or without climate change. Even if there were no long-term trend in climatic parameters, climatic variation from season to season, year to year, and region to region requires knowledge and tools for adapting to the different scenarios. However, the potential for new combinations of climatic variables, along with the potential for interactions and for more rapid variation in conditions, reinforces the need for research and policy responses to plant disease risk (Coakley et al., 1999; Garrett et al., 2006). Research directed explicitly toward understanding the complexity of system responses to climate change is needed.

A mechanistic understanding of plant and pathogen responses to climate change will be based on characterizing current populations and their potential for adaptation. New genomic tools make it possible to characterize gene expression and genotypic diversity much more readily in both wildland and agricultural plant communities. These tools can be applied in concert with other "-omics" approaches to link responses in gene expression (transcriptomics), lipidomics, and metabolomics for a fuller mechanistic understanding of adaptive potential. These approaches will have to be applied in multifactor studies of climate change effects, so that the interactions between the effects of changes in temperature, precipitation, CO_2, and other environmental factors can be understood, along with the potential for adaptation.

Tools for the study of pathogen population and community structure, gene expression, and other responses are evolving rapidly. Advances in sequencing technologies make the routine characterization of microbial communities feasible (Riesenfeld et al., 2004; Roesch et al., 2007) and will eventually make it inexpensive. Microarrays, such as the GeoChip (He et al., 2007), are being designed to study microbial gene function in soils. New microarrays are needed to study the presence and expression of microbial genes related to plant disease. It will be important to collect baseline information about microbial community structure and function soon, so that changes in microbial communities under new climatic conditions can be studied. Experiments to compare responses of microbial communities to new environments will also be important (e.g., Waldrop and Firestone, 2006). Undoubtedly there are many forms of disease suppressiveness provided by microbial communities in soils that offer benefits to agriculture and regulate disease in natural systems and are currently underappreciated.

Research to clarify the effects of host landscape structures will help to improve strategies and will be necessary for studying changes at regional, continental, and global scales. Current regional analyses of climatic effects on disease risk tend to be calculated for disease risk in individual "pixels," important for developing a first-approximation estimate of risk. The next stage for such models will be to incorporate risk neighborhoods to improve estimates, where the risk for any given location will increase with proximity to higher-risk areas. Finally, regional and global models will need to incorporate pathogen evolution. Formulating and parameterizing these models will require advances in epidemiological theory and experimentation. For example, better data and models related to pathogen and host dispersal, current levels of intraspecific diversity, and the strength of selection under different climate change scenarios are needed.

Long-term geographically representative records of disease occurrence and the distribution of pathogens and hosts are rare, despite their importance for understanding epidemiology and trends in epidemics (Jeger and Pautasso, 2008). Global networks supporting the analysis of epidemics are needed. Progress toward this goal is in place; for example, the United States has developed a National Plant Diagnostic Network to facilitate data collection and analysis (Stack and Fletcher, 2007). To be most effective, this network ultimately needs to be linked with comparable national networks in other countries. It is to the advantage of the United States to assist other countries in setting up such networks for gathering and analyzing data, so that we can all benefit from more complete information. The use of model predictions for modifying agricultural management has proven useful in many parts of the world, including applications by resource-poor farmers based on climate predictions in Zimbabwe (Patt et al., 2005).

One of the most important investments we can make is in conservation, characterization, and the development of strategies for optimal use of plant genetic resources. In wildland systems, conservation is necessary to increase the chances that plant populations are large enough to include individual genotypes adapted to new climate scenarios. In agricultural systems, conservation of diversity in crop species and their wild relatives is necessary to increase the chances that genes needed for resistance and tolerance to new biotic and abiotic stresses are maintained (Johnson, 2008). *In situ* conservation allows natural selection to continue acting on these species. *Ex situ* conservation is a useful backup strategy and simplifies some analyses of accessions. International networks for conservation of crop genetic diversity, such as the institutions in the Consultative Group for International Agricultural Research (CGIAR), are critical for ensuring conservation and analysis of accessions. The funding currently available for such programs is very low compared to the importance of their mission. While investments such as the Svalbard Global Seed Vault provide a last resort, active investigation of plant resources is needed.

Ultimately our best response to the challenge of climate change in agriculture will be to develop diverse, flexible, and resilient agricultural systems that can adapt more readily to new climatic conditions. These systems will have to include well-prepared and well-funded agricultural scientists working globally to develop new strategies. In wildland systems, replacing plant species or genotypes at risk is a less attractive option. Since invasive pathogens can have the most important effects and have the potential to exacerbate the effects of climate change, policies to better reduce the spread of exotic pathogens will be important (Anderson et al., 2004; Burdon et al., 2006).

Acknowledgments

I appreciate valuable comments from members and staff of the Forum on Microbial Threats, R. Bowden, D. Rotenberg, and P. Garfinkel. It is also a pleasure to acknowledge support by the U.S. National Science Foundation (NSF) through Grant DEB-0516046 and NSF Grant EF-0525712 as part of the joint NSF-National Institutes of Health (NIH) Ecology of Infectious Disease program; by the U.S. Agency for International Development (USAID) to the Office of International Research, Education, and Development (OIRED) at Virginia Tech for the Sustainable Agriculture and Natural Resource Management (SANREM) Collaborative Research Support Program (CRSP) under Award No. EPP-A-00-04-00013-00 and for the Integrated Pest Management (IPM) CRSP under Award No. EPP-A-00-04-00016-00; and by the Kansas State Experiment Station (Contribution No. 08-308-B).

CLIMATE CHANGE AND INFECTIOUS DISEASE: IMPACT ON HUMAN POPULATIONS IN THE ARCTIC[12]

Alan J. Parkinson, Ph.D.[13]
Centers for Disease Control and Prevention

Introduction: The Arctic Environment

The circumpolar region is defined as the region that extends above 60°N latitude, borders the Arctic Ocean, and includes all of or the northern parts of eight nations: the United States (Alaska), Canada, Greenland, Iceland, Norway,

[12]The findings and conclusions in this report are those of the author and do not necessarily represent the official position of the Centers for Disease Control and Prevention.

[13]Arctic Investigations Program, Division of Emerging Infections and Surveillance Services, National Center for Preparedness Detection and Control of Infectious Disease, Anchorage, AK.

Finland, Sweden, and the Russian Federation (see Figure 2-20). The climate in the Arctic varies geographically from severe cold in arid uninhabited regions to temperate forests bordering coastal agrarian regions. Approximately 4 million people live in the Arctic and almost half reside in northern regions of the Russian Federation. Peoples of the Arctic and sub-Arctic regions live in social and physical environments that differ substantially from those of their more southern dwelling counterparts. These populations are comprised of varying proportions of indigenous and nonindigenous peoples (Stephansson Arctic Institute, 2004; see Figure 2-21).

The indigenous populations of northern Canada (Northwest Territories, Yukon, Nunavut, northern Quebec, and Labrador), Alaska, and Greenland generally reside in small communities in remote regions. They have little economic infrastructure and depend on subsistence hunting, fishing, and gathering of food for a significant proportion of their diet. In these remote areas, access to public health and acute care systems is often marginal and poorly supported. Life expectancy of the indigenous peoples of Alaska, northern Canada, and Greenland is lower than that of the general populations of the United States, Canada, and Nordic countries (Young, 2008). Similarly the infant morality rate for the indigenous segments of these populations is higher than that of the comparable national populations. Mortality rates for heart disease and cancer, once much lower among the indigenous populations of the United States, Canada, and northern European countries, are now similar to their respective national rates. The indigenous populations of Alaska, Canada, and Greenland have higher mortality rates for unintentional injury and suicide. Other health concerns of the indigenous peoples of the Arctic include the high prevalence of certain infectious diseases, such as hepatitis B, *Helicobacter pylori*, respiratory syncytial virus (RSV) infections in infants, and sexually transmitted diseases, as well as heath impacts associated with exposures to environmental pollutants, rapid economic change and modernization, and climate change (Bjerregaard et al., 2004).

Climate Change and the Arctic Environment

The Arctic, like most other parts of the world, warmed substantially over the twentieth century, principally in recent decades. Arctic climate models project continued warming with a 3-5°C mean increase by 2100. The winters will warm more than summers, the mean annual precipitation is projected to increase, and continued melting of land and sea ice is expected to increase river discharge and contribute to rising sea levels. These changes will be accompanied by greater overall climate variability and an increase in extreme weather events (Arctic Council, 2005).

The rapid warming in the Arctic is already bringing about substantial ecological and socioeconomic impacts, many of which result from the thawing of permafrost, flooding, and shoreline erosion resulting from storm surges and

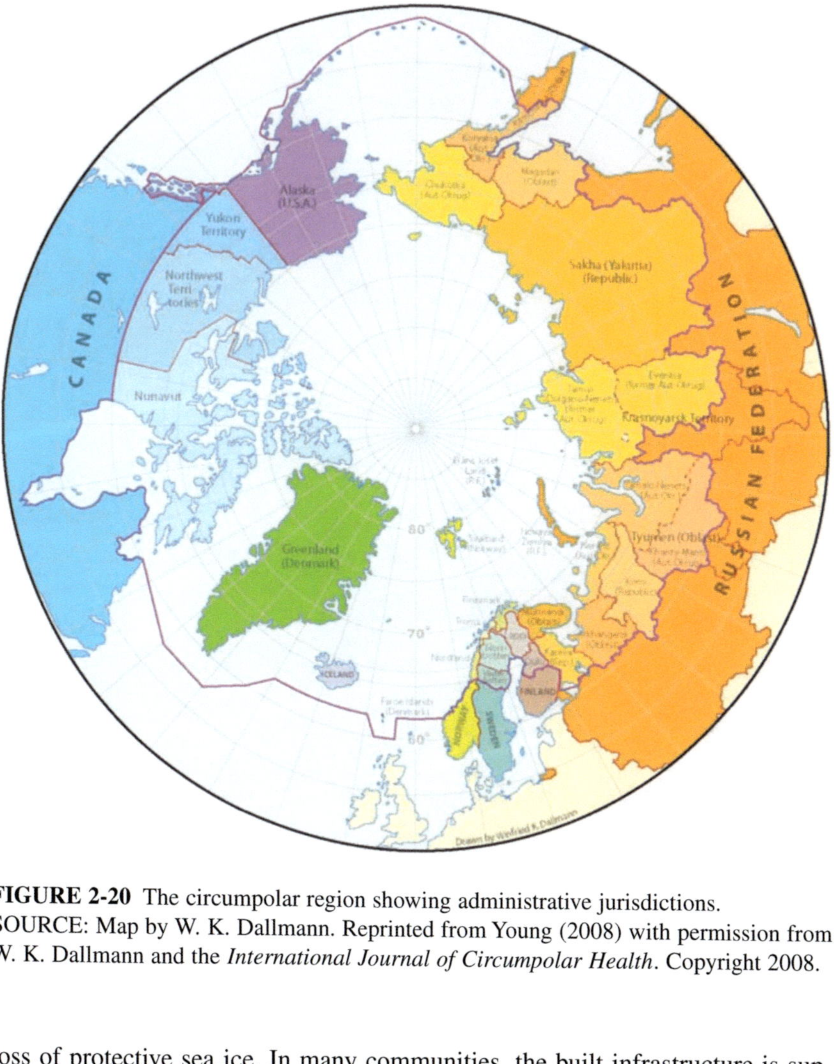

FIGURE 2-20 The circumpolar region showing administrative jurisdictions.
SOURCE: Map by W. K. Dallmann. Reprinted from Young (2008) with permission from W. K. Dallmann and the *International Journal of Circumpolar Health*. Copyright 2008.

loss of protective sea ice. In many communities, the built infrastructure is supported by permafrost. Loss of this permafrost foundation will result in damage to water intake systems and pipes, and may result in contamination of the community water supply. In addition, loss of foundation support for access roads, boardwalks, water storage tanks, and wastewater treatment facilities will render water distribution and wastewater treatment systems inoperable. Several villages already face relocation because village housing, water system, and infrastructure are being undermined (Warren et al., 2005).

Rapid warming has resulted in the loss of annual Arctic sea ice. On September 11, 2007, the Arctic sea ice cover reached the lowest extent recorded since

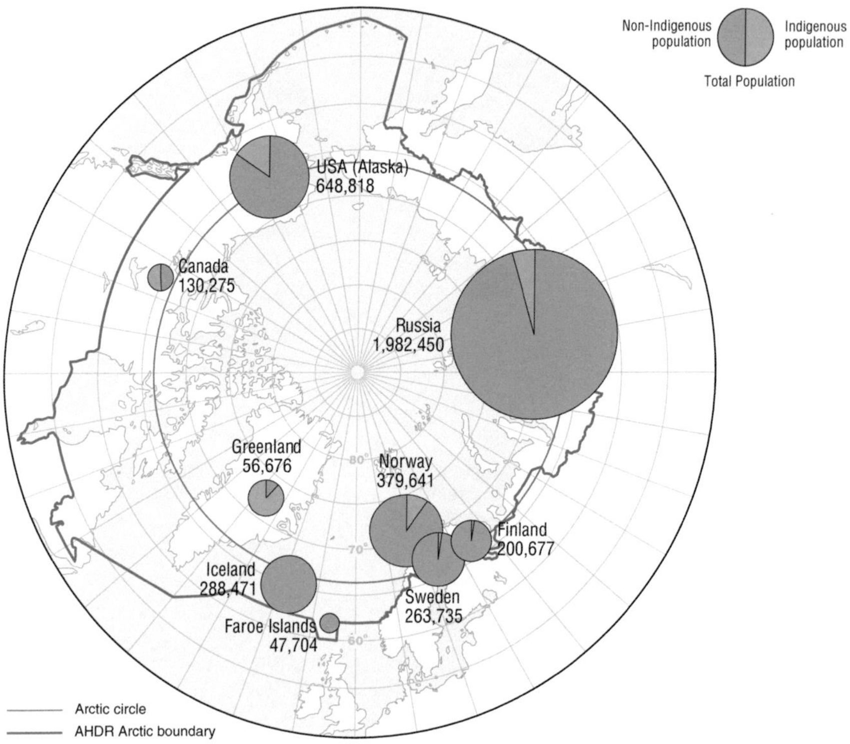

FIGURE 2-21 The circumpolar region showing indigenous and nonindigenous population distributions.
SOURCE: Reprinted from Stefansson Arctic Institute (2004) with permission from W. K. Dallmann, Norwegian Polar Institute and the Stefansson Arctic Institute. Copyright 2004.

observations began in the 1970s, exceeding the most pessimistic model predictions of an ice-free Arctic by 2050 (Richter-Menge et al., 2008; Figure 2-22). This dramatic reduction in sea ice will have widespread effects on marine ecosystems, coastal climate, human settlements, and subsistence activities. For the first time the reduction in annual sea ice has created ice-free shipping lanes to the northwest, from northern Labrador through the Arctic archipelago in northern Canada, to the Bering Strait, and has almost completely cleared a passage to the northeast, from the Bering Strait along the northern coast of the Russian Federation to Norway (see Figure 2-23). Both routes represent time- and fuel-saving shortcuts between the Pacific and Atlantic Oceans and will bring an increase in marine transport and access to vast oil, gas, and mineral reserves once inaccessible to exploration and exploitation.

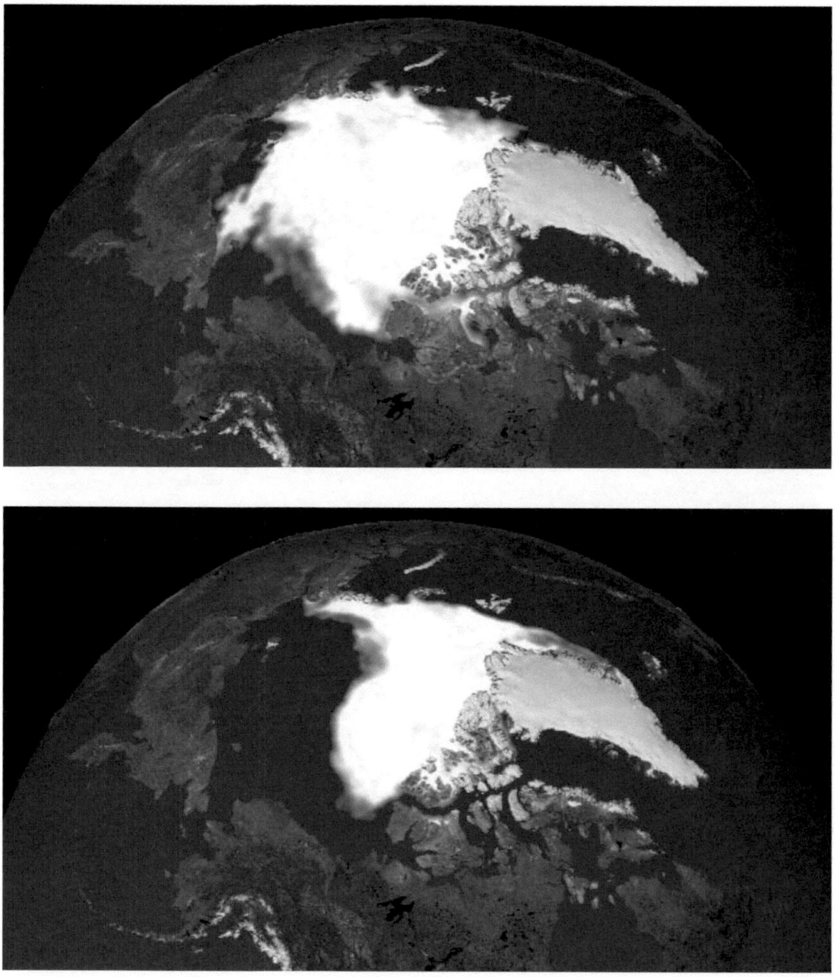

FIGURE 2-22 The Arctic ice cap, September 2001 (Top) and September 2007 (Bottom).
SOURCE: NASA, as printed in Borgerson (2008).

Such access will bring many benefits as well as risks to once isolated Arctic communities. Construction of new coast guard or military bases and other industrial ventures will bring employment opportunities to local populations, but will also affect population distribution, dynamics, culture, and local environments. Tourism will most likely increase. Public sector and government services will then increase to support the new emerging economies. These events will greatly

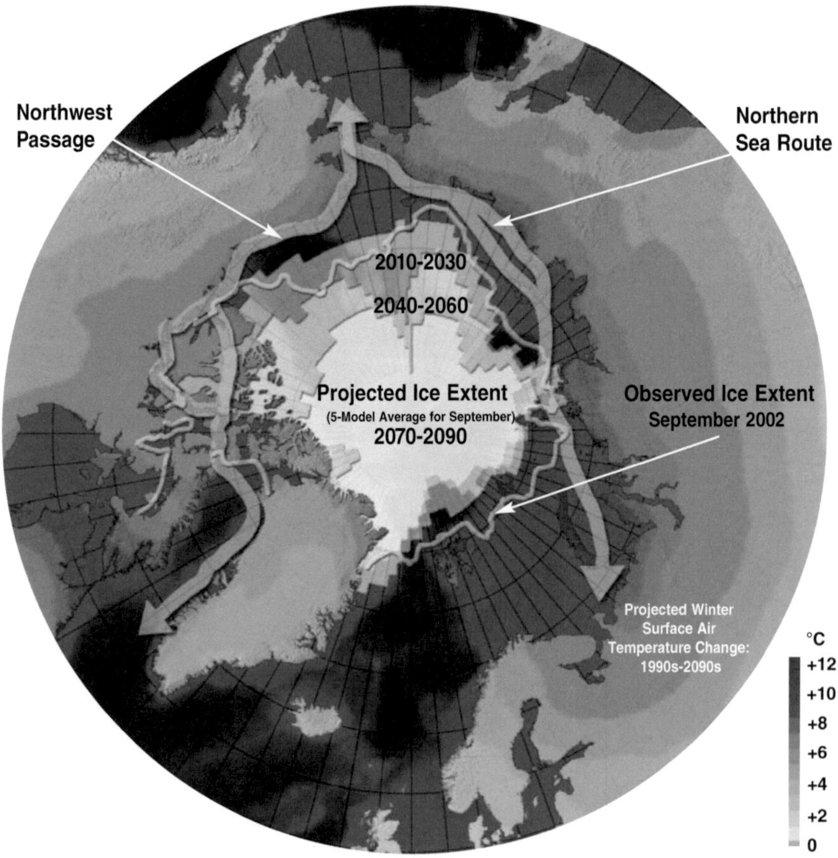

FIGURE 2-23 Proposed northwest and northeast shipping lanes through the Arctic Ocean joining the Atlantic and Pacific Oceans.
SOURCE: Map by C. Grabhorn Reprinted from ACIA (2004) with permission from Cambridge University Press and C. Grabhorn.

challenge the traditional subsistence way of life for many communities and lead to rapid and long-term cultural change, which will create additional stress on an already vulnerable population (Curtis et al., 2005).

Climate Change and Human Health

The direct health effects of climate change will result from changes in ambient temperature, altered patterns of risk from outdoor activities, and changes in the incidence of infectious diseases. As ambient temperature increases, the incidence of hypothermia and associated morbidity and mortality may decrease. Conversely hyperthermia may increase, particularly among the very young and the elderly (Nayha, 2005). However, because of the low mean temperature in many Arctic regions, the likelihood of such events having large impacts on public health for the general population is low. More significantly, unintentional injury, mostly related to subsistence hunting and fishing—already a significant cause of mortality among Arctic residents—may increase (Arctic Council, 2005). The reduction in river and sea ice thickness, curtailed ice season, reduced snow cover, and permafrost thawing will make hunting and gathering more difficult, dangerous, and less successful, thereby increasing the risk of injuries and death by drowning.

Permafrost thawing erosion or flooding can force relocation. Communities and families undergoing relocation will have to adapt to new ways of living, may face unemployment, and will have to integrate and create new social bonds. Relocation may also lead to rapid and long-term cultural change and loss of traditional culture, which will increase individual and community stress, leading to mental and behavioral health challenges (Hess et al., in press).

Climate change already poses a serious threat to the food security of many Arctic communities because of their reliance on traditional subsistence hunting and fishing for survival. Populations of marine and land mammals, fish, and waterfowl may be reduced or displaced by changing habitats and migration patterns, further reducing the traditional food supply. Release of environmental contaminants from the atmosphere and melting glaciers and sea ice may increase the levels of these pollutants entering the food chain, making traditional foods less desirable (AMAP, 2003). Reduction in traditional food supply will force indigenous communities to depend increasingly on nontraditional and often less healthy Western foods. This will most likely result in increasing rates of modern diseases associated with processed foods, such as obesity, diabetes, cardiovascular diseases, and outbreaks of food-borne infectious diseases associated with imported fresh and processed foods (Bjerregaard et al., 2004; Orr et al., 1994).

Many host-parasite systems are particularly sensitive to climate change. Specific stages of the life cycles of many helminths may be greatly affected by temperature. For example, small increases in temperature can substantially increase the transmission of lung worms and muscle worms pathogenic to wildlife that are important as a food source for many northern communities (Hoberg et al., 2008).

Climate Change and Infectious Diseases in the Arctic

It is well known that climate and weather affect the distribution and risk of many vector-borne diseases, such as malaria, RVF, plague, and dengue fever in tropical regions of the globe. Weather also affects the distribution of food- and water-borne diseases and emerging infectious diseases, such as West Nile virus, hantavirus, and Ebola hemorrhagic fever (Haines et al., 2006). Less is known about the impact of climate change and the risk and distribution of infectious diseases in Arctic regions. It is known that Arctic populations have a long history of both endemic and epidemic infectious diseases (Parkinson et al., 2008). However, with the introduction of antimicrobial drugs, vaccines, and public health systems, morbidity and mortality due to infectious diseases have been greatly reduced. Despite these advances, high rates of invasive diseases caused by *Streptococcus pneumoniae, Haemophilus influenzae,* and *Mycobacterium tuberculosis* persist (Bruce et al., 2008a,b; Christensen et al., 2004; Dawar et al., 2002; Degani et al., 2008; Gessner et al., 1998; Meyer et al., 2008; Netesov and Conrad, 2001; Nguyen et al., 2003; Singleton et al., 2006; Søborg et al., 2001). Sharp seasonal epidemics of viral respiratory infections also commonly occur (Bulkow et al., 2002; Karron et al., 1999; Van Caeseele et al., 2001). The overuse of antimicrobial drugs in some regions has led to the emergence of multidrug-resistant *S. pneumoniae*, *Helicobacter pylori*, and methicillin-resistant *Staphylococcus aureus* (Baggett et al., 2003, 2004; McMahon et al., 2007; Rudolph et al., 1999, 2000).

The impact of climate on the incidence of these existing infectious disease challenges is unknown. In many Arctic regions, however, inadequate housing and sanitation are already important determinants of infectious disease transmission. The cold northern climate keeps people indoors amplifying the effects of household crowding, smoking, and inadequate ventilation. Crowded living conditions increase person-to-person spread of infectious diseases and favor the transmission of respiratory and gastrointestinal diseases and skin infections. Many homes in communities across the Arctic lack basic sanitation services (e.g., flush toilet, shower or bath, kitchen sink). Providing these services is difficult in remote villages where small isolated populations live in a harsh cold climate. A recent study in western Alaska demonstrated two to four times higher hospitalization rates among children less than 3 years of age for pneumonia, influenza, and childhood RSV infections in villages where the majority of homes had no in-house piped water, compared with villages where the majority of homes had in-house piped water service. Likewise, outpatient *Staphylococcus aureus* infections and hospitalization for skin infections among persons of all ages were higher in villages with no in-house piped water service compared to villages without water service (Hennessy et al., 2008). Damage to the sanitation infrastructure by melting permafrost or flooding may therefore result in increased rates of hospitalization among children for respiratory infections, as well as an increased rate of skin infections and diarrheal diseases caused by bacterial, viral, and parasitic pathogens.

Some infectious diseases are unique to the Arctic and lifestyles of the indigenous populations and may increase in a warming Arctic. For example, many Arctic residents depend on subsistence hunting, fishing, and gathering for food, and on a predictable climate for food storage. Food storage methods often include above ground air-drying of fish and meat at ambient temperature, below ground cold storage on or near the permafrost, and fermentation. Changes in climate may prevent the drying of fish or meat, resulting in spoilage. Similarly, loss of the permafrost may result in spoilage of food stored below ground. Outbreaks of food-borne botulism occur sporadically in communities in the United States, Canadian Arctic, and Greenland and are caused by ingestion of improperly prepared fermented traditional foods (CDC, 2001; Proulx et al., 1997; Sobel et al., 2004; Sørensen et al., 1993; Wainwright et al., 1988). Because germination of *Clostridium botulinum* spores and toxin production will occur at temperatures greater than 4°C, it is possible that warmer ambient temperatures associated with climate change may result in an increased rate of food-borne botulism in these regions. Preliminary studies have shown that fermentation of aged seal meat challenged with *C. botulinum* at temperatures above 4°C results in toxin production (Leclair et al., 2004).

Outbreaks of gastroenteritis caused by *Vibrio parahaemolyticus* have been related to the consumption of raw or inadequately cooked shellfish collected from seawater at temperatures of higher than 15°C. Prior to 2004, the most northerly outbreak occurred in northern British Columbia in 1997. However, in July 2004, an outbreak of gastroenteritis caused by *V. parahaemolyticus* was documented among cruise ship passengers consuming raw oysters while visiting an oyster farm in Prince William Sound, Alaska (McLaughlin et al., 2005). The outbreak investigation documented an increase of 0.21°C per year in the July-August water temperature since 1997, and reported that 2004 was the first year that the oyster farm water temperature exceeded 15°C in July. This event provides direct evidence of an association between rising seawater temperature and the onset of illness.

Warmer temperatures may allow infected host animal species to survive winters in larger numbers, increase in population, and expand their range of habitation, thus increasing the opportunity to pass infections to humans. For example, milder weather and less snow cover may have contributed to a large outbreak of Puumala virus infection in northern Sweden in 2007. Puumala virus is endemic in bank voles, and in humans causes hemorrhagic fever with renal syndrome (Pettersson et al., 2008). Similar outbreaks have been noted in the Russian Federation (Revich, 2008). The climate-related northern expansion of the boreal forest in Alaska and northern Canada has favored the steady northward advance of the beaver, extending the range of *Giardia lamblia*, a parasitic infection of beaver that can infect other mammals, including humans who use untreated surface water (Arctic Council, 2005). Similarly, warmer temperatures in the Arctic and sub-Arctic regions could support the expansion of the geographical range and populations of foxes and voles, common carriers of *Echinococcus multilocularis,* the cause of alveolar echinococcus in humans (Holts et al., 2005). The prevalence of alveolar echinococcus has risen in Switzerland as fox populations have increased in size and expanded their geographic ranges into urban areas (Schweiger et al., 2007). Alveolar echinococcus was common in two regions of northwestern Alaska prior to 1997. Disease in humans was associated with contact with dogs; however, improvements in housing and dog lot management have largely eliminated dog-to-human transmission in Alaska. This may not be the case, however, in other parts of the Arctic where human infections with *Echinococcus granulosis*, and *E. multilocularis* are still reported, particularly in association with communities dependent on reindeer herding and dog use (Castrodale et al., 2002; Rausch, 2003).

Climate change may also influence the density and distribution of animal hosts and mosquito vectors, which could result in an increase in human illness or a shift in the geographical range of disease caused by these agents. The impact of these changes on human disease incidence has not been fully evaluated, but there is clearly potential for climate change to shift the geographical distribution of certain vector-borne and other zoonotic diseases. For example, West Nile virus entered the United States in 1999 and in subsequent years infected human, horse, mosquito, and bird populations across the United States and as far north as northern Manitoba, Canada (Parkinson and Butler, 2005). In the Russian Federation infected birds and humans have been detected as far north as the region of Novosibirsk (Revich, 2008). Although there is, at present, insufficient information about the relationship between climate and the spread of West Nile virus, a number of factors may contribute to its further northward migration. Milder winters could favor winter survival of infected *Culex spp.* mosquitoes, the predominant vector of West Nile virus, which since the 1970s have migrated as far north as Prince Albert, Saskatchewan in Canada. Longer, hotter summers increase the transmission season leading to higher numbers of infected mosquitoes and greater opportunities for human exposure. Climate change may alter the disease ecology and migration patterns of other reservoirs such as birds. These factors may affect disease incidence and result in expansion of the range of other arthropod vector-borne diseases.

A number of mosquito-borne viruses that cause illness in humans circulate in the U.S. Arctic and northern regions of the Russian Federation (Walters et al., 1999). Jamestown Canyon and Snowshoe Hare viruses are considered emerging threats to the public health in the United States, Canada, and the Russian Federation, causing flu-like symptoms and central nervous system diseases, such as aseptic meningitis and encephalitis (Walters et al., 1999). Sindbis virus also circulates in northern Europe. The virus is carried northward and amplified by migratory birds. In the late summer, ornithophilic mosquitoes pass the virus onto humans causing epidemics of Pogosta disease in northern Finland, an illness characterized by a rash and arthritis (Kurkela et al., 2008). In Sweden, the incidence of tick-borne encephalitis (TBE) has substantially increased since the mid-1980s (Lindgren and Gustafson, 2001). This increase corresponds to a trend of milder winters and an earlier onset of spring, resulting in an increase in the tick population (*Ixodes ricinus*) that carries the virus responsible for TBE and other potential pathogens (Skarphédinsson et al., 2005). Similarly in northeastern Canada, climate change is projected to result in a northward shift in the range of *Ixodes scapularis,* a tick that carries *Borrelia burgdorferi,* the etiologic agent of Lyme disease. The current northern limit of *Ix. scapularis* is southern Ontario including the shoreline of Lake Erie and southern coast of Nova Scotia. Some temperature-based models show the potential for a northward expansion of *Ix. scapularis* above 60°N latitude and into the Northwest Territories by 2080 (Ogden et al., 2005). However, it should be noted that tick distribution is influenced by additional factors such as habitat suitability and dispersal patterns which can affect the accuracy of these predictions. Whether or not disease in humans is a result of these climate change-induced alterations in vector range depends on many other factors, such as land-use practices, human behavior (suburban development in wooded areas, outdoor recreational activities, use of insect repellents, etc.), human population density, and adequacy of the public health infrastructure.

Response to Climate Change in the Arctic

In 1992, the IOM published a report titled *Emerging Infections: Microbial Threats to Health in the United States*. This report uncovered major challenges for public health in the medical community primarily related to detecting and managing infectious disease outbreaks and monitoring the prevalence of endemic infectious diseases. It stimulated a national movement to reinvigorate the U.S. public health system to address the HIV/AIDS epidemic, the emergence of new diseases, the resurgence of old diseases, and the persistent evolution of antimicrobial resistance. In a subsequent report, the IOM provided an assessment of the capacity of the public health system to respond to emerging threats and made recommendations for addressing infectious disease threats to human health (IOM, 2003).

Because climate change is expected to exacerbate many of the factors contributing to infectious disease emergence and reemergence, the recommendations of the 2003 IOM report can be applied to the prevention and control of emerging infectious disease threats resulting from climate change. A framework for public health response to climate change in the United States has recently been proposed (Frumkin et al., 2008; Hess et al., in press). The framework emphasizes the need to capitalize on and enhance existing essential public health services and to improve coordination efforts between government agencies (federal, state, and local), academia, the private sector, and nongovernmental organizations.

Applying this framework to Arctic regions requires enhancing the public health capacity to monitor diseases with potentially large public health impacts, including respiratory diseases in children, skin infections, and diarrheal diseases, particularly in communities with failing sanitation systems. Monitoring certain vector-borne diseases, such as West Nile virus, Lyme disease, and TBE, should be priorities in areas at the margins of focal regions known to support both animal and insect vectors, and where climate change may promote the geographic expansion of vectors. Because Arctic populations are relatively small and widely dispersed over a large area, region-specific detection of significant trends in emerging climate-related infectious diseases may be delayed. This difficulty may be overcome by linking regional monitoring systems together for the purposes of sharing standardized information on climate-sensitive infectious diseases of mutual concern. Efforts should be made to harmonize notifiable disease registries, laboratory methods, and clinical surveillance definitions across administrative jurisdictions to allow comparable disease reporting and analysis. An example of such a network is the International Circumpolar Surveillance system for emerging infectious diseases. This network links hospital and public health laboratories together for the purposes of monitoring invasive bacterial diseases and tuberculosis in Arctic populations (Parkinson et al., 2008).

Public health capacity should be enhanced to respond to infectious disease food-borne outbreaks (e.g., botulism, gastroenteritis caused by *Giardia lamblia* or *Vibro parahaemolyticus*). Public health research is needed to determine the baseline prevalence of potential climate-sensitive infectious diseases (e.g., West Nile virus, *Borrelia burgdorferi*, *Brucella spp.*, *Echinococcus spp.*, *Toxoplasma spp.*) in both human and animal hosts in regions where emergence may be expected. Such studies can be used to accumulate additional evidence of the effect of climate change or weather on infectious disease emergence, to guide early detection and public health intervention strategies, and to provide science-based support for public health actions on climate change. The circumpolar coordination of research efforts will be important not only to harmonize research protocols, laboratory methods, data collection instruments, and data analysis, but also to maximize the impact of scarce resources and to minimize the impact of research on affected communities. Coordination can be facilitated through existing international cooperatives, such as the Arctic Council,[14] the International Union for Circumpolar Health,[15] and the newly formed International Network of Circumpolar Health Researchers.[16]

The challenge in the Arctic, however, will be to ensure sufficient public health capacity to allow the detection of disease outbreaks and monitor infectious disease trends most likely to be influenced by climate. The remoteness of many communities from clinical or public health facilities, and the harsh weather conditions of Arctic regions, often preclude appropriate specimen and epidemiologic data collection during an outbreak investigation, research, or ongoing surveillance activities. Staffing shortages are frequent in many in local clinics and regional hospitals that are already overwhelmed by routine and urgent care priorities, leaving little capacity for existing staff to assist public health personnel in outbreak investigations, research, or maintenance of routine surveillance activities.

[14]See http://www.arctic-council.org.

[15]See http://www.iuch.org.

[16]See http://www.inchr.org.

Additional resources and training may be needed to ensure adequate staffing at these facilities, to address existing gaps between regional clinics and hospitals and public health departments, and to ensure a sufficiently trained staff to address the emerging public health impacts posed by climate change.

A key aspect of the public health response to climate change in Arctic regions will be the formation of community-based partnerships with tribal governments to identify potential threats to the community and develop strategies to address those threats. Communities at greatest risk should be targeted for education, outreach, and assessment of existing or potential health risks, vulnerabilities, and engagement in the design of community-based monitoring and the formulation of intervention strategies. The identification, selection, and monitoring of basic indicators for climate change and community health will be important for any response to climate change at the community level (Furgal, 2005). The selection of site- or village-specific indicators should be guided by local concerns and may include activities such as the surveillance of a key wildlife or insect species in a region where climate changes may contribute to the emergence of new zoonotic diseases or the measurement of weather (i.e., precipitation and temperature), water quality (i.e., turbidity, pathogens), and gastrointestinal illness (i.e., clinic visits) in a community. Linking communities across regions and internationally should facilitate the sharing of standard protocols, data collection instruments, and data for analysis. These linkages will be important for the detection of trends over larger geographic regions, should enhance a community's ability to detect changes that impact health, and will allow the development of strategies to minimize the negative health impacts of climate change on Arctic residents in the future.

Conclusion

Resident indigenous populations of the Arctic are uniquely vulnerable to climate change because of their close relationship with, and dependence on, the land, sea, and natural resources for their cultural, social, economic, and physical well-being. The increasing mean ambient temperature may lead to an increase in food-borne diseases, such as botulism and gastrointestinal illnesses. An increase in mean temperature may also influence the incidence of zoonotic and arboviral infectious diseases by changing the population density and range of animal hosts and insect vectors. The public health response to these emerging microbial threats should include enhancing the public health capacity to monitor climate-sensitive infectious diseases with potentially large public health impacts; the prompt investigation of infectious disease outbreaks that may be related to climate change; and research on the relationship between climate and infectious disease emergence to guide early detection and public health interventions. The development of community-based monitoring networks with links to regional and national public health agencies as well as circumpolar health organizations will facilitate method standardization, data-sharing, and the detection of infectious disease trends over a larger geographic area. This capacity is essential for the development of strategies to minimize the negative effects of climate change on the health of Arctic residents in the future.

REFERENCES

Overview References

Borgerson, S. G. 2008. Arctic meltdown: the economic and security implications of global warming. *Foreign Affairs* 87(2):63-77.

Chretien, J. P., A. Anyamba, S. A. Bedno, R. F. Breiman, R. Sang, K. Sergon, A. M. Powers, C. O. Onyango, J. Small, C. J. Tucker, and K. J. Linthicum. 2007. Drought-associated chikungunya emergence along coastal East Africa. *American Journal of Tropical Medicine and Hygiene* 76(3):405-407.

Fritz, C. L., D. T. Dennis, M. A. Tipple, G. L. Campbell, C. R. McCance, and D. J. Gubler. 1996. Surveillance for pneumonic plague in the United States during an international emergency: a model for control of imported emerging diseases. *Emerging Infectious Diseases* 2(1):30-36.

IPCC (Intergovernmental Panel on Climate Change). 2007. *Climate change 2007: the physical science basis*. Contribution of Working Group I to the fourth assessment report of the IPCC. Cambridge, UK: Cambridge University Press.

Linthicum, K. J., A. Anyamba, C. J. Tucker, P. W. Kelley, M. F. Myers, and C. J. Peters. 1999. Climate and satellite indicators to forecast Rift Valley fever epidemics in Kenya. *Science* 285(5426):397-400.

Stenseth, N. C., N. I. Samia, H. Viljugrein, K. L. Kausrud, M. Begon, S. Davis, H. Leirs, V. M. Dubyanskiy, J. Esper, V. S. Ageyev, N. L. Klassovskiy, S. B. Pole, and C. Kung-Sik. 2006. Plague dynamics are driven by climate variation. *Proceedings of the National Academy of Sciences* 103(35):13110-13115.

Colwell References

Colwell, R. R. 1996. Global climate and infectious disease: the cholera paradigm. *Science* 274(5295):2025-2031.

Colwell, R. R., and A. Huq. 1994. Vibrios in the environment: viable but nonculturable *Vibrio cholerae*. In: Vibrio cholerae *and cholera: molecular to global perspectives*, edited by I. K. Wachsmuth, O. Olsvik, and P. A. Blake. Washington, DC: American Society for Microbiology. Pp. 117-133.

Gil, A. I., V. R. Louis, I. N. Rivera, E. Lipp, A. Huq, C. F. Lanata, D. N. Taylor, E. Russek-Cohen, N. Choopun, R. B. Sack, R. R. Colwell. 2004. Occurrence and distribution of *Vibrio cholerae* in the coastal environment of Peru. *Environmental Microbiology* 6(7):699-706.

Rawlings, T., G. M. Ruiz, and R. R. Colwell. 2007. Association of *Vibrio cholerae* O1 El Tor and O139 Bengal with the copepods *Acartia tonsa* and *Eurytemora affinis*. *Applied Environmental Microbiology* 73(24):7926-7933.

WHO (World Health Organization). 2005. *Weekly epidemiological record* 80(31):261-268, http://www.who.int/wer/2005/wer8031.pdf (accessed May 1, 2008).

Chretien et al. References

Anyamba, A., K. J. Linthicum, R. Mahoney, C. J. Tucker, and P. W. Kelley. 2002. Mapping potential risk of Rift Valley fever outbreaks in African savannas using vegetation index time series data. *Photogrammetric Engineering and Remote Sensing* 68(2):137-145.

Anyamba, A., J. P. Chretien, J. Small, C. J. Tucker, and K. J. Linthicum. 2006. Developing global climate anomalies suggest potential disease risks for 2006-2007. *International Journal of Health Geographics* 5:60.

Bedno, S. A., C. O. Onyango, C. Njugana, R. Sang, S. Gaydos, K. Sergon, and R. F. Breiman. 2006. Outbreak of chikungunya in Lamu, Kenya, 2004. Paper presented at the International Conference on Emerging Infectious Diseases, Atlanta, GA.

CDC (Centers for Disease Control and Prevention). 1998. Rift Valley fever—East Africa, 1997-1998. *Morbidity and Mortality Weekly Report* 47(13):261-264.

Chretien, J, P., and K. J. Linthicum. 2007. Chikungunya in Europe—what's next? *Lancet* 370(9602):1805-1806.

Chretien, J. P., A. Anyamba, S. A. Bedno, R. F. Breiman, R. Sang, K. Sergon, A. M. Powers, C. O. Onyango, J. Small, C. J. Tucker, and K. J. Linthicum. 2007. Drought-associated chikungunya emergence along coastal East Africa. *American Journal of Tropical Medicine and Hygiene* 76(3):405-407.

FAO (Food and Agriculture Organization). 2006. Possible RVF activity in the Horn of Africa. *EMPRES Watch.*

IOM (Institute of Medicine). 2003. *Microbial threats to health: emergence, detection, and response.* Washington, DC: The National Academies Press.

IPCC (Intergovernmental Panel on Climate Change). 2007a. *Climate change 2007: the physical science basis.* Contribution of Working Group I to the fourth assessment report of the IPCC. Cambridge, UK: Cambridge University Press. Chapter 3.

———. 2007b. *Climate change 2007: the physical science basis.* Contribution of Working Group I to the fourth assessment report of the IPCC. Cambridge, UK: Cambridge University Press. Chapter 10.

Kovats, R. S., M. J. Bouma, S. Hajat, E. Worrall, and A. Haines. 2003. El Niño and health. *Lancet* 362(9394):1481-1489.

Linthicum, K. J., F. G. Davies, C. L. Bailey, and A. Kairo. 1984. Mosquito species encountered in a flooded grassland dambo in Kenya. *Mosquito News* 44:228-232.

Linthicum, K. J., A. Anyamba, C. J. Tucker, P. W. Kelley, M. F. Myers, and C. J. Peters. 1999. Climate and satellite indicators to forecast Rift Valley fever epidemics in Kenya. *Science* 285(5426):397-400.

Mavalankar, D., P. Shastri, and P. Raman. 2007. Chikungunya epidemic in India: a major public-health disaster. *Lancet Infectious Disease* 7(5):306-307.

Peters, C. J., and K. J. Linthicum. 1994. Rift Valley fever. In *Handbook of zoonoses*, Second edition, edited by G. B. Beran. Boca Raton, FL: CRC Press, Inc.

Rezza, G., L. Nicoletti, R. Angelini, R. Romi, A. C. Finarelli, M. Panning, P. Cordioli, C. Fortuna, S. Boros, F. Magurano, G. Silvi, P. Angelini, M. Dottori, M. G. Ciufolini, G. C. Majori, and A. Cassone. 2007. Infection with chikungunya virus in Italy: an outbreak in a temperate region. *Lancet* 370(9602):1840-1846.

Save the Children Alliance. 2007 (January 11). *Horn of Africa emergency statement*, http://www.savethechildren.net/alliance/media/newsdesk/2007-01-01.html (accessed March 4, 2008).

Sergon, K., C. Njuguna, R. Kalani, V. Ofula, C. Onyango, L. S. Konongoi, S. Bedno, H. Burke, A. M. Dumilla, J. Konde, M. K. Njenga, R. Sang, and R. F. Breiman. 2008. Seroprevalence of chikungunya virus (CHIKV) infection on Lamu Island, Kenya, October 2004. *American Journal of Tropical Medicine and Hygiene* 78(2):333-337.

Tsetsarkin, K. A., D. L. Vanlandingham, C. E. McGee, and S. Higgs. 2007. A single mutation in Chikungunya virus affects vector specificity and epidemic potential. *PLoS Pathogens* 3(12):e201.
UN (United Nations). 2004. *Kenya flash appeal*, http://www.un.org/depts/ocha/cap/kenya.html (accessed March 4, 2008).
———. 2006. *Global survey of early warning systems: an assessment of capacities, gaps, and opportunities towards building a comprehensive global early warning system for all natural hazards*, http://www.unisdr.org/ppew/info-resources/ewc3/Global-Survey-of-Early-Warning-Systems.pdf (accessed March 4, 2008).
Watts, D. M., D. S. Burke, B. A. Harrison, R. E. Whitmire, and A. Nisalak. 1987. Effect of temperature on the vector efficiency of *Aedes aegypti* for dengue 2 virus. *American Journal of Tropical Medicine and Hygiene* 36(1):143-152.
WHO (World Health Organization). 2004. *Using climate to predict infectious disease outbreaks: a review*, http://www.who.int/globalchange/publications/oeh0401/en/ (accessed March 4, 2008).
———. 2006. *Chikungunya and dengue in the south west Indian Ocean*, http://www.who.int/csr/don/2006_03_17/en/ (accessed March 4, 2008).
———. 2007a. Outbreaks of Rift Valley fever in Kenya, Somalia and United Republic of Tanzania, December 2006-April 2007. *Weekly Epidemiological Record* 82(20):169-178.
———. 2007b. *Health action in crises. Highlights No 140—8 to 14 January 2007*, http://www.who.int/hac/donorinfo/highlights/highlights_140_08_14jan2007.pdf (accessed March 4, 2008).

Stenseth References

Achtman, M., K. Zurth, G. Morelli, G. Torrea, A. Guiyoule, and E. Carniel. 1999. *Yersinia pestis*, the cause of plague, is a recently emerged clone of *Yersinia pseudotuberculosis*. *Proceedings of the National Academy of Sciences* 96(24):14043-14048.
Anyamba, A., J. P. Chretien, J. Small, C. J. Tucker, and K. J. Linthicum. 2006. Developing global climate anomalies suggest potential disease risks for 2006-2007. *International Journal of Health Geographics* 5:60.
Baltazard, M., Y. Karimi, M. Eftekhari, M. Chamsa, and H. H. Mollaret. 1963. La conservation interépizootique de la peste en foyer invétéré hypothèses de travail. *Bulletin de la Société de Pathologie Exotique* 56:1230-1241.
Ben Ari, T., A. Gershunov, K. L. Gage, T. Snäll, P. Ettestad, K. L. Kausrud, and N. C. Stenseth. 2008. Human plague in U.S.: the importance of regional and local climate. *Biology Letters* (in review).
Blanc, G. 1956. Une opinion non conformiste sur le mode de transmission de la peste. *Revue d'Hygiéne et de Médecine Sociale* 4(6):532-562.
Chamberlain, N. 2004. *Plague*, http://www.kcom.edu/faculty/chamberlain/Website/lectures/lecture/plague.htm (accessed July 1, 2008).
Cohn, S. K., Jr. 2002. *The Black Death transformed: disease and culture in early Renaissance Europe*. London, UK: Edward Arnold Publishers.
Davis, S., M. Begon, L. De Bruyn, V. S. Ageyev, N. L. Klassovskiy, S. B. Pole, H. Viljugrein, N. C. Stenseth, and H. Leirs. 2004. Predictive thresholds for plague in Kazakhstan. *Science* 304(5671):736-738.
Davis, S., H. Leirs, H. Viljugrein, N. C. Stenseth, L. De Bruyn, N. Klassovskiy, V. Ageyev, and M. Begon. 2007. Empirical assessment of a threshold model for sylvatic plague. *Journal of the Royal Society Interface* 4(15):649-657.
Drancourt, M., L. Houhamdi, and D. Raoult. 2006. *Yersinia pestis* as a telluric, human ectoparasite-borne organism. *Lancet Infectious Diseases* 6(4):234-241.

Duplantier, J. M., J. B. Duchemin, S. Chanteau, and E. Carniel. 2005. From the recent lessons of the Malagasy foci towards a global understanding of the factors involved in plague reemergence. *Veterinary Research* 36(3):437-453.

Esper, J., S. G. Shiyatov, V. S. Mazepa, R. J. S. Wilson, D. A. Graybill, and G. Funkhouser. 2003. Temperature-sensitive Tien Shan tree ring chronologies show multi-centennial growth trends. *Climate Dynamics* 21(7/8):8p.

Frigessi, A., M. Holden, C. Marshall, H. Viljugrein, N. C. Stenseth, L. Holden, V. Ageyev, and N. L. Klassovskiy. 2005. Bayesian population dynamics of interacting species: great gerbils and fleas in Kazakhstan. *Biometrics* 61(1):230-238.

Fritz, C. L., D. T. Dennis, M. A. Tipple, G. L. Campbell, C. R. McCance, and D. J. Gubler. 1996. Surveillance for pneumonic plague in the United States during an international emergency: a model for control of imported emerging diseases. *Emerging Infectious Diseases* 2(1):30-36.

Gage, K. L., and M. Y. Kosoy. 2005. Natural history of plague: perspectives from more than a century of research. *Annual Review of Entomology* 50(1):505-528.

Galimand, M., A. Guiyoule, G. Gerbaud, B. Rasoamanana, S. Chanteau, E. Carniel, and P. Courvalin. 1997. Multidrug resistance in *Yersinia pestis* mediated by a transferable plasmid. *New England Journal of Medicine* 337(10):677-680.

Guiyoule, A., F. Grimont, I. Iteman, P. A. Grimont, M. Lefevre, and E. Carniel. 1994. Plague pandemics investigated by ribotyping of *Yersinia pestis* strains. *Journal of Clinical Microbiology* 32(3):634-641.

Guiyoule, A., G. Gerbaud, C. Buchrieser, M. Galimand, L. Rahalison, S. Chanteau, P. Courvalin, and E. Carniel. 2001. Transferable plasmid-mediated resistance to streptomycin in a clinical isolate of *Yersinia pestis*. *Emerging Infectious Diseases* 7(1):43-48.

Hall, F. G., G. Collatz, S. Los, E. Brown de Colstoun, and D. Landis, eds. 2005. *ISLSCP Initiative II*. DVD/CD-ROM.

Hinnebusch, B. J., M.-L. Rosso, T. G. Schwan, and E. Carniel. 2002. High-frequency conjugative transfer of antibiotic resistance genes to *Yersinia pestis* in the flea midgut. *Molecular Microbiology* 46(2):349-354.

Hotez, P. J., D. H. Molyneux, A. Fenwick, E. Ottesen, S. Ehrlich Sachs, and J. D. Sachs. 2006. Incorporating a rapid-impact package for neglected tropical diseases with programs for HIV/AIDS, tuberculosis, and malaria. *PLoS Medicine* 3(5):e102.

Huntington, T. G. 2006. Evidence for intensification of the global water cycle: review and synthesis. *Journal of Hydrology* 319(1-4):83-95.

Inglesby, T. V., D. T. Dennis, D. A. Henderson, J. G. Barlett, M. S. Ascher, E. Eitzen, A. D. Fine, A. M. Friedlander, J. Hauer, J. F. Koerner, M. Layton, J. McDade, M. T. Osterholm, T. O'Toole, G. Parker, T. M. Perl, P. K. Russell, M. Schoch-Spana, and K. Tonat. 2000. Plague as a biological weapon. *Journal of the American Medical Association* 283(17):2281-2290.

IPCC (Intergovernmental Panel on Climate Change). 2007. *Climate change 2007: impacts, adaptation, and vulnerability*. Contribution of Working Group II to the fourth assessment report of the Intergovernmental Panel on Climate Change. Cambridge, UK: Cambridge University Press. Chapter 8.

Kausrud, K., H. Viljugrein, A. Frigessi, M. Begon, S. Davis, H. Leirs, V. Dubyanskiy, and N. C. Stenseth. 2007. Climatically driven synchrony of gerbil populations allows large-scale plague outbreaks. *Proceedings: Biological Sciences* 274(1621):1963-1969.

Kausrud, K. L., H. Viljugrein, A. Frigessi, M. Begon, S. Davis, H. Leirs, T. Ben Ari, and N. C. Stenseth. 2008. The epidemiological history of plague in Central Asia: a paleoclimatic modelling study *Proceedings of the National Academy of Sciences* (in review).

Koirala, J. 2006. Plague: disease, management, and recognition of act of terrorism. *Infectious Disease Clinics of North America* 20(2): viii, 273-287.

Laudisoit, A., H. Leirs, R. H. Makundi, S. Van Dongen, S. Davis, S. Neerinckx, J. Deckers, and R. Libois. 2007. Plague and the human flea, Tanzania. *Emerging Infectious Diseases* 13(5):687-693.

Los, S., G. Collatz, P. Sellers, C. Malmström, N. Pollack, R. Defries, L. Bounoua, M. Parris, C. Tucker, and D. Dazlich. 2000. A global 9-year biophysical land surface data set from NOAA AVHRR data. *Journal of Hydrometeorology* 1:183-199.

Mudur, G. 1995. India's pneumonic plague outbreak continues to baffle. *British Medical Journal* 311(7007):706.

Park, S., K. S. Chan, H. Viljugrein, L. Nekrassova, B. Suleimenov, V. S. Ageyev, N. L. Klassovskiy, S. B. Pole, and N. C. Stenseth. 2007. Statistical analysis of the dynamics of antibody loss to a disease-causing agent: plague in natural populations of great gerbils as an example. *Journal of the Royal Society Interface* 4(12):57-64.

Parkhill, J., B. W. Wren, N. R. Thomson, R. W. Titball, M. T. G. Holden, M. B. Prentice, M. Sebaihia, K. D. James, C. Churcher, K. L. Mungall, S. Baker, D. Basham, S. D. Bentley, K. Brooks, A. M. Cerdeno-Tarraga, T. Chillingworth, A. Cronin, R. M. Davies, and P. Davis. 2001. Genome sequence of *Yersinia pestis*, the causative agent of plague. *Nature* 413(6855):523-527.

Parmenter, R. R., E. P. Yadav, C. A. Parmenter, P. Ettestad, and K. L. Gage. 1999. Incidence of plague associated with increased winter-spring precipitation in New Mexico. *American Journal of Tropical Medicine and Hygiene* 61(5):814-821.

Pettorelli, N., J. O. Vik, A. Mysterud, J.-M. Gaillard, C. J. Tucker, and N. C. Stenseth. 2005. Using the satellite-derived NDVI to assess ecological responses to environmental change. *Trends in Ecology and Evolution* 20(9):503-510.

Raoult, D., and G. Aboudharam. 2000. Molecular identification by "suicide PCR" of *Yersinia pestis* as the agent of medieval Black Death. *Proceedings of the National Academy of Sciences* 97(23):12800-12803.

Samia, N. I., K.-S. Chan, and N. C. Stenseth. 2007. A generalized threshold mixed model for analyzing nonnormal nonlinear time series, with application to plague in Kazakhstan. *Biometrika* 94(1):101-118.

Schrag, S. J., and P. Wiener. 1995. Emerging infectious diseases: what are the relative roles of ecology and evolution? *Trends in Ecology and Evolution* 10(8):319-324.

Scott, S., and C. J. Duncan. 2001. *Biology of plagues: evidence from historical populations.* Cambridge, UK: Cambridge University Press.

Stenseth, N. C. 1999. Population cycles in voles and lemmings: density dependence and phase dependence in a stochastic world. *Oikos* 87(3):427-460.

Stenseth, N. C., A. Mysterud, G. Ottersen, J. W. Hurrell, C. Kung-Sik, and M. Lima. 2002. Ecological effects of climate fluctuations. *Science* 297(5585):1292-1296.

Stenseth, N. C., N. I. Samia, H. Viljugrein, K. L. Kausrud, M. Begon, S. Davis, H. Leirs, V. M. Dubyanskiy, J. Esper, V. S. Ageyev, N. L. Klassovskiy, S. B. Pole, and C. Kung-Sik. 2006. Plague dynamics are driven by climate variation. *Proceedings of the National Academy of Sciences* 103(35):13110-13115.

Stenseth, N. C., B. B. Atshabar, M. Begon, S. R. Belmain, E. Bertherat, E. Carniel, K. L. Gage, H. Leirs, and L. Rahalison. 2008. Plague: past, present, and future. *PLoS Medicine* 5(1):e3.

Treydte, K. S., G. H. Schleser, G. Helle, D. C. Frank, M. Winiger, G. H. Haug, and J. Esper. 2006. The twentieth century was the wettest period in northern Pakistan over the past millennium. *Nature* 440(7088):1179-1182.

Twigg, G. 1984. *The Black Death: a biological reappraisal.* London, UK: Batsform Academic and Educational.

WHO (World Health Organization). 2003. Plague, Algeria. *Weekly Epidemiological Record* 78(29):253.

———. 2005. Plague. *Weekly Epidemiological Record* 80(15):138-140.

Yersin, A. 1894. La peste bubonique à Hong-Kong. *Annales de l'Institut Pasteur* 8:662-667.

Zhang, Z., Z. Li, Y. Tao, M. Chen, X. Wen, L. Xu, H. Tian, and N. C. Stenseth. 2007. Relationship between increase rate of human plague in China and global climate index as revealed by cross-spectral analysis and cross-wavelet analysis. *Integrative Zoology* 2(3):144-153.

Ziegler, P. 1969. *The Black Death*. Wolfeboro Falls, NH: Alan Sutton Publishing Inc.

Garrett References

Anagnostakis, S. L. 2000. Revitalization of the majestic chestnut: chestnut blight disease. *APSnet*, http://www.apsnet.org/online/feature/chestnut/ (accessed March 28, 2008).

Anderson, P. K., A. A. Cunningham, N. G. Patel, F. J. Morales, P. R. Epstein, and P. Daszak. 2004. Emerging infectious diseases of plants: pathogen pollution, climate change and agrotechnology drivers. *Trends in Ecology and Evolution* 19(10):535-544.

Bai, G., and G. Shaner. 2004. Management and resistance in wheat and barley to *Fusarium* head blight. *Annual Review of Phytopathology* 42:135-161.

Bergot, M., E. Cloppet, V. Pérarnaud, M. Déqué, B. Marçais, and M.-L. Desprez-Loustau. 2004. Simulations of potential range expansion of oak disease caused by *Phytophthora cinnamomi* under climate change. *Global Change Biology* 10:1539-1552.

Brasier, C. M. 2001. Rapid evolution of introduced plant pathogens via interspecific hybridization. *BioScience* 51(2):123-133.

Burdon, J. J., P. H. Thrall, and L. Ericson. 2006. The current and future dynamics of disease in plant communities. *Annual Review of Phytopathology* 44:19-39.

Chakraborty, S., A. V. Tiedemann, and P. S. Teng. 2000. Climate change: potential impact on plant diseases. *Environmental Pollution* 108(3):317-326.

Cheatham, M. R., M. N. Rouse, P. D. Esker, S. Ignacio, W. Pradel, R. Raymundo, A. H. Sparks, G. A. Forbes, T. R. Gordon, and K. A. Garrett. Beyond yield: plant disease in the context of ecosystem services. *Phytopathology* (in revision).

Cline, W. R. 2007. *Global warming and agriculture: impact estimates by country*. Washington, DC: Center for Global Development and Peterson Institute for International Economics.

Coakley, S. M., H. Scherm, and S. Chakraborty. 1999. Climate change and plant disease management. *Annual Review of Phytopathology* 37:399-426.

Daily, G. C., ed. 1997. *Nature's services: societal dependence on natural ecosystems*. Washington, DC: Island Press.

De Wolf, E. D., and S. A. Isard. 2007. Disease cycle approach to plant disease prediction. *Annual Review of Phytopathology* 45:203-220.

Desprez-Loustau, M.-L., C. Robin, G. Reynaud, M. Déqué, V. Badeau, D. Piou, C. Husson, and B. Marçais. 2007. Simulating the effects of a climate-change scenario on the geographical range and activity of forest-pathogenic fungi. *Canadian Journal of Plant Pathology* 29:101-120.

Eviner, V. T., and G. E. Likens. 2008. Effects of pathogens on terrestrial ecosystem function. In *Infectious disease ecology: effects of ecosystems on disease and of disease on ecosystems*, edited by R. Ostfeld, F. Keesing, and V. Eviner. Princeton, NJ: Princeton University Press. Pp. 260-283.

Fletcher, J., and J. P. Stack. 2007. Agricultural biosecurity: threats and impacts for plant pathogens. In *Global infectious disease surveillance and detection*. Washington, DC: The National Academies Press. Pp. 86-94.

Frank, E. E. 2007. Rust and drought effects on gene expression and phytohormone concentration in big bluestem. M.S. Thesis, Kansas State University, Manhattan, Kansas.

Garrett, K. A., and R. L. Bowden. 2002. An Allee effect reduces the invasive potential of *Tilletia indica*. *Phytopathology* 92:1152-1159.

Garrett, K. A., S. P. Dendy, E. E. Frank, M. N. Rouse, and S. E. Travers. 2006. Climate change effects on plant disease: genomes to ecosystems. *Annual Review of Phytopathology* 44:489-509.

Garrett, K. A., L. N. Zúñiga, E. Roncal, G. A. Forbes, C. C. Mundt, Z. Su, and R. J. Nelson. Intraspecific functional diversity in hosts and its effect on disease risk across a climatic gradient. *Ecological Applications* (in revision).

He, Z. L., T. J. Gentry, C. W. Schadt, L. Y. Wu, J. Liebich, S. C. Chong, Z. J. Huang, W. M. Wu, B. H. Gu, P. Jardine, C. Criddle, and J. Zhou. 2007. GeoChip: a comprehensive microarray for investigating biogeochemical, ecological and environmental processes. *ISME Journal* 1:67-77.

Hijmans, R. J., G. A. Forbes, and T. S. Walker. 2000. Estimating the global severity of potato late blight with GIS-linked disease forecast models. *Plant Pathology* 49(6):697-705.

Isard, S. A., S. H. Gage, P. Comtois, and J. M. Russo. 2005. Principles of the atmospheric pathway for invasive species applied to soybean rust. *BioScience* 55(10):851-861.

Jeger, M. J., and M. Pautasso. 2008. Plant disease and global change—the importance of long-term data sets. *New Phytologist* 177(1):8-11.

Johnson, R. C. 2008. Gene banks pay big dividends to agriculture, the environment, and human welfare. *PLoS Biology* 6(6):e148.

Lobell, D. B., M. B. Burke, C. Tebaldi, M. D. Mastrandrea, W. P. Falcon, and R. L. Naylor. 2008. Prioritizing climate change adaptation needs for food security in 2030. *Science* 319(5863):607-610.

Madden, L., and M. Wheelis. 2003. The threat of plant pathogens as weapons against U.S. crops. *Annual Review of Phytopathology* 41:155-176.

Magarey, R. D., G. A. Fowler, D. M. Borchert, T. B. Sutton, and M. Colunga-Garcia. 2007. NAPPFAST: an Internet system for the weather-based mapping of plant pathogens. *Plant Disease* 91(4):336-345.

Margosian, M. L., K. A. Garrett, J. M. S. Hutchinson, and K. A. With. Connectivity of the American agricultural landscape: assessing the national risk of crop pest and disease spread. *BioScience* (in revision).

McDonald, B. A., and C. Linde. 2002. Pathogen population genetics, evolutionary potential, and durable resistance. *Annual Review of Phytopathology* 40:349-379.

Mitchell, C. E., P. B. Reich, D. Tilman, and J. V. Groth. 2003. Effects of elevated CO_2, nitrogen deposition, and decreased species diversity on foliar fungal plant disease. *Global Change Biology* 9(3):438-451.

Nelson, R. J., R. L. Naylor, and M. M. Jahn. 2004. The role of genomics research in improvement of "orphan" crops. *Crop Science* 44:1901-1904.

Oerke, E. C., H. W. Dehne, F. Schönbeck, and A. Weber. 1994. *Crop production and crop protection*. Amsterdam, The Netherlands: Elsevier Science, B.V.

Parker, I. M., and G. S. Gilbert. 2004. The evolutionary ecology of novel plant-pathogen interactions. *Annual Review of Ecology, Evolution and Systematics* 35:675-700.

Patt, A., P. Suarez, and C. Gwata. 2005. Effects of seasonal climate forecasts and participatory workshops among subsistence farmers in Zimbabwe. *Proceedings of the National Academy of Sciences* 102(35):12623-12628.

Peng, S., J. Huang. J. E. Sheehy, R. C. Laza, R. M. Visperas, X. H. Zhong, G. S. Centeno, G. S. Khush, and K. G. Cassman. 2004. Rice yields decline with higher night temperature from global warming. *Proceedings of the National Academy of Sciences* 101(27):9971-9975.

Pimentel, D., L. Lach, R. Zuniga, and D. Morrison. 2000. Environmental and economic costs of nonindigenous species in the United States. *BioScience* 50(1):53-65.

Pivonia, S., and X. B. Yang. 2004. Assessment of the potential year-round establishment of soybean rust throughout the world. *Plant Disease* 88(5):523-529.

Redak, R. A., A. H. Purcell, J. R. S. Lopes, M. J. Blua, R. F. Mizell, and P. C. Andersen. 2004. The biology of xylem fluid-feeding insect vectors of *Xylella fastidiosa* and their relation to disease epidemiology. *Annual Review of Entomology* 49:243-270.

Riesenfeld, C. S., P. D. Schloss, and J. Handelsman. 2004. Metagenomics: genomic analysis of microbial communities. *Annual Review of Genetics* 38:525-552.

Rizzo, D. M., M. Garbelotto, and E. A. Hansen. 2005. *Phytophthora ramorum*: integrative research and management of an emerging pathogen in California and Oregon forests. *Annual Review of Phytopathology* 43:309-335.

Roesch, L. F., R. R. Fulthorpe, A. Riva, G. Casella, A. K. M. Hadwin, A. D. Kent, S. H. Daroub, F. A. O. Camargo, W. G. Farmerie, and E. W. Triplett. 2007. Pyrosequencing enumerates and contrasts soil microbial diversity. *ISME Journal* 1:283-290.

Rouse, M. N. 2007. Diversity of a disease resistance gene homolog in *Andropogon gerardii* (Poaceae) is correlated with precipitation. M.S. Thesis, Kansas State University, Manhattan, Kansas.

Roy, B. A., S. Gusewell, and J. Harte. 2004. Response of plant pathogens and herbivores to a warming experiment. *Ecology* 85:2570-2571.

Rush, C. M., R. Riemenschneider, J. M. Stein, T. Boratynski, R. L. Bowden, and M. H. Royer. 2005. Status of karnal bunt of wheat in the United States 1996-2004. *Plant Disease* 89(3):212-223.

Stack, J. P., and J. Fletcher. 2007. Plant biosecurity infrastructure for disease surveillance and diagnostics. In *Global infectious disease surveillance and detection*. Washington, DC: The National Academies Press. Pp. 95-102.

Stokstad, E. 2007. Plant pathology: deadly wheat fungus threatens world's bread baskets. *Science* 315(5820):1786-1787.

Travers, S. E., M. D. Smith, J. Bai, S. H. Hulbert, J. E. Leach, P. S. Schnable, A. K. Knapp, G. A. Milliken, P. A. Fay, A. Saleh, and K. A. Garrett. 2007. Ecological genomics: making the leap from model systems in the lab to native populations in the field. *Frontiers in Ecology and the Environment* 5:19-24.

UNEP (United Nations Environment Programme). 2004. *Childhood pesticide poisoning: information for advocacy and action*. Châtelaine, Switzerland: United Nations Environment Programme.

Villanueva, H., R. Raymundo, H. Juarez, W. Perez, and G. Forbes. In preparation. The article and journal titles were not available at the time of publication.

Waldrop, M. P., and M. K. Firestone. 2006. Response of microbial community composition and function to soil climate change. *Microbial Ecology* 52:716-724.

Webb, K. M., J. Bai, I. Oña, K. A. Garrett, T. W. Mew, C. M. Vera Cruz, and J. E. Leach. In preparation. The article and journal titles were not available at the time of publication.

Widmark, A.-K., B. Andersson, A. Cassel-Lundhagen, M. Sandström, and J. E. Yuen. 2007. *Phytophthora infestans* in a single field in southwest Sweden early in spring: symptoms, spatial distribution and genotypic variation. *Plant Pathology* 56:573-579.

Woods, A., K. D. Coates, and A. Hamann. 2005. Is an unprecedented *Dothistroma* needle blight epidemic related to climate change? *BioScience* 55(9):761-769.

Zhu, Y., H. Chen, J. Fan, Y. Wang, Y. Li, J. Chen, J. Fan, S. Yang, L. Hu, H. Leung, T. W. Mew, P. S. Teng, Z. Wang, and C. C. Mundt. 2000. Genetic diversity and disease control in rice. *Nature* 406(6797):718-722.

Zhu, Y., H. Fang, Y. Wang, J. X. Fan, S. Yang, T. W. Mew, and C. Mundt. 2005. Panicle blast and canopy moisture in rice cultivar mixtures. *Phytopathology* 95(4):433-438.

Parkinson References

AMAP (Arctic Monitoring and Assessment Programme). 2003. *AMAP Assessment 2002: human health in the Arctic*. Oslo, Norway: Arctic Monitoring and Assessment Program.

Arctic Council. 2005. *Arctic climate impact assessment scientific report*. New York: Cambridge University Press. Pp. 863-960.

Baggett, H. C., T. W. Hennessy, R. Leman, C. Hamlin, D. Bruden, and A. Reasonover. 2003. An outbreak of community-onset methicillin resistant *Staphylococcus aureus* skin infections in southwestern Alaska. *Infection Control and Hospital Epidemiology* 24(6):397-402.

Baggett, H. C., T. W. Hennessy, K. Rudolph, D. Bruden, A. Reasonover, and A. J. Parkinson. 2004. Community-onset methicillin-resistant *Staphylococcus aureus*, associated with antibiotic use and cytotoxin Panton-Valentine leukocidin during a furnunculosis outbreak in rural Alaska. *Journal of Infectious Diseases* 189(9):1565-1573.

Bjerregaard, P., K. T. Young, E. Dewailly, and S. O. E. Ebbesson. 2004. Indigenous health in the Arctic: an overview of the circumpolar Inuit population. *Scandinavian Journal of Public Health* 32(5):390-395.

Borgerson, S. G. 2008. Arctic meltdown: the economic and security implications of global warming. *Foreign Affairs* 87(2):63-77.

Bruce, M. G., S. L. Deeks, T. Zulz, D. Bruden, C. Navarro, M. Lovegren, L. Jette, K. Kristinsson, G. Sigmundsdottir, K. Brinklov Jensen, O. Lovoll, J. P. Nuorti, E. Herva, A. Nystedt, A. Sjostedt, A. Koch, T. W. Hennessy, and A. J. Parkinson. 2008a. International Circumpolar Surveillance for invasive pneumococcal disease, 1999-2005. *Emerging Infectious Diseases* 14(1):25-33.

Bruce, M. G., S. L. Deeks, T. Zulz, C. Navarro, C. Palacios, C. Case, C. Hemsley, T. W. Hennessy, A. Corriveau, B. Larke, I. Sobel, M. Lovegren, C. DeByle, R. Tsang, and A. J. Parkinson. 2008b. Epidemiology of *Haemophilus influenzae* serotype A, North American Arctic 2000-2005. *Emerging Infectious Diseases* 14(1):48-55.

Bulkow, L. R., R. J. Singleton, R. A. Karron, L. H. Harrisson, and Alaska RSV Study Group. 2002. Risk factors for severe respiratory syncytial virus infection among Alaska Native children. *Pediatrics* 109(2):210-216.

Castrodale, L. J., M. Beller, J. F. Wilson, P. M. Schantz, D. P. McManus, L. H. Zhang, F. G. Fallico, and F. D. Sacco. 2002. Two atypical cases of cystic echinococcosis (*Echinococcus granulosis*) in Alaska 1999. *American Journal of Tropical Medicine and Hygiene* 66(3):325-327.

CDC (Centers for Disease Control and Prevention). 2001. Botulism outbreak associated with eating fermented food—Alaska. *Morbidity and Mortality Weekly Report* 50(32):680-682.

Christensen, J., P. Poulsen, and K. Ladefoged. 2004. Invasive pneumococcal disease in Greenland. *Scandinavian Journal of Infectious Diseases* 36(5):325-329.

Curtis, T., S. Kvernmo, and P. Bjerregaad. 2005. Changing living conditions, life style and health. *International Journal of Circumpolar Health* 64(5):442-450.

Dawar, M., L. Moody, J. D. Martin, C. Fung, J. Isaac-Renton, and D. M. Patrick. 2002. Two outbreaks of botulism associated with fermented salmon roe—British Columbia, August 2001. *Canadian Communicable Disease Reports* 28(6):45-49.

Degani, N., C. Navarro, S. Deeks, and M. Lovegren. 2008. Invasive bacterial diseases in northern Canada. *Emerging Infectious Diseases* 14(1):34-40.

Frumkin, H., J. Hess, G. Luber, J. Maliay, and M. McGeehin. 2008. Climate change: the public health response. *American Journal of Public Health* 98(3):435-445.

Furgal, C. 2005. Monitoring as a community response for climate change and health. *International Journal of Circumpolar Health* 64(5):440-441.

Gessner, B. D., N. S. Weiss, and C. M. Nolan. 1998. Risk factors for pediatric tuberculosis infection and disease after household exposure to adult index cases in Alaska. *Jornal de Pediatria* 132(3):509-513.

Haines, A., R. S. Kovars, D. Campbell-Lendrun, and C. Corvalan. 2006. Climate change and human health: impacts, vulnerability, and mitigation. *Lancet* 360(9528):2101-2109.

Hennessy, T., T. Ritter, R. C. Holman, D. L. Bruden, K. L. Yorita, L. Bulkow, J. E. Cheek, R. J. Singleton, and J. Smith. 2008. Relationship between in-home water service and the risk of respiratory tract, skin, and gastrointestinal tract infections among Alaska Natives. *American Journal of Public Health* 98(5):1-8.

Hess, J., J. Malilay, and A. J. Parkinson. In press. Climate change: the importance of place and places of special risk. *American Journal of Preventive Medicine*.

Hoberg, E. P., L. Polley, E. J. Jenins, S. J. Kutz, A. M. Vetch, and B. T. Elkin. 2008. Integrated approaches and empiric models for investigation of parasitic diseases in northern wildlife. *Emerging Infectious Diseases* 14(1):10-17.

Holts, D. W., C. Hanns, T. O. O'Hara, J. Burek, and R. Franz. 2005. New distribution records of *Echinococcus multilocularis* in the brown lemming from Barrow, Alaska. *Journal of Wildlife Diseases* 41(1):257-259.

IOM (Institute of Medicine). 1992. *Emerging infections: microbial threats to health in the United States*. Washington, DC: National Academy Press.

———. 2003. *Microbial threats to health: emergence, detection, and response*. Washington, DC: The National Academies Press.

Karron, R. A., R. J. Singleton, L. Bulkow, A. J. Parkinson, D. Kruse, I. DeSmet, C. Indorf, K. M. Petersen, D. Leombruno, D. Hurlburt, M. Santosham, and L. H. Harrison. 1999. Severe respiratory syncytial virus disease in Alaska Native children. *Journal of Infectious Diseases* 180(1):41-49.

Kurkela, S., O. Rätti, E. Huhtamo, N. Y. Uzcátegui, P. J. Nuorti, J. Laakkonen, T. Manni, P. Helle, A. Vaheri, and O. Vapalahti. 2008. Sindbis virus infection in resident birds, migratory birds, and humans, Finland. *Emerging Infectious Diseases* 14(1):41-47.

Leclair, D., J. W. Austin, J. Faber, B. Cadieux, and B. Blanchfield. 2004. Toxicity of aged seal meat challenged with *Clostridium botulinum* type E. Federal Food Safety and Nutrition Research Meeting, Ottawa, Ontario, October 4-5.

Lindgren, E., and R. Gustafson. 2001. Tick-borne encephalitis in Sweden and climate change. *Lancet* 358(9275):16-18.

McLaughlin, J. B., A. Depoala, C. A. Bopp, K. A. Martinek, N. Napolilli, C. Allison, S. Murry, E. C. Thompson, M. M. Bird, and T. P. Middaugh. 2005. Emergence of *Vibrio parahaemolyticus* gastroenteritis associated with consumption of Alaskan oysters and its global implications. *New England Journal of Medicine* 353(14):1463-1470.

McMahon, B. J., M. G. Bruce, T. W. Hennessy, D. L. Bruden, F. Sacco, H. Peters, D. A. Hurlburt, J. M. Morris, A. L. Reasonover, G. Dailde, D. E. Berg, and A. J. Parkinson. 2007. Reinfection after successful eradication of *Helicobacter pylori*—a 2 year prospective study in Alaska Natives. *Alimentary Pharmacology and Therapeutics* 23(8):1215-1223.

Meyer, A., K. Ladefoged, P. Poulsen, and A. Kock. 2008. Population-based survey of invasive bacterial diseases, Greenland, 1995-2004. *Emerging Infectious Diseases* 14(1):76-79.

Nayha, S. 2005. Environmental temperature and mortality. *International Journal of Circumpolar Health* 64(5):451-458.

Netesov, S. V., and L. J. Conrad. 2001. Emerging infectious diseases in Russia 1990-1999. *Emerging Infectious Diseases* 7(1):1-5.

Nguyen, D., J. F. Proulx, J. Westley, L. Thibert, S. Dery, and M. A. Behr. 2003. Tuberculosis in the Inuit community of Quebec, Canada. *American Journal of Respiratory and Critical Care Medicine* 168(11):1353-1357.

Ogden, N. H., A. Maarouf, I. K. Barker, M. Bigras-Poulin, L. R. Lindsay, M. G. Morshed, C. J. O'Callaghan, F. Ramay, D. Waltner-Twews, and D. F. Charron. 2005. Climate change and the potential for range expansion of the Lyme disease vector *Ixodes scapularis* in Canada. *International Journal of Parasitology* 36(1):63-70.

Orr, P., B. Lorencz, R. Brown, R. Kielly, B. Tan, D. Holton, H. Clugstone, L. Lugtig, C. Pim, S. McDonald, G. Hammond, M. Moffatt, J. Spika, D. Manuel, W. Winther, D. Milley, H. Lior, and N. Sinuff. 1994. An outbreak of diarrhea due to verotoxin-producing *Esherichia coli* in the Canadian Northwest Territories. *Scandinavian Journal of Infectious Diseases* 26(6):675-684.

Parkinson, A. J., and J. C. Butler. 2005. Potential impact of climate change on infectious disease emergence in the Arctic. *International Journal of Circumpolar Health* 64(5):478-486.

Parkinson, A. J., M. Bruce, and T. Zultz. 2008. International circumpolar surveillance, and arctic network for surveillance of infectious diseases. *Emerging Infectious Diseases* 14(1):18-24.

Pettersson, L., J. Boman, P. Juto, M. Evander, and C. Ahlm. 2008. Outbreak of Puumala virus infection, Sweden. *Emerging Infectious Diseases* 14(5):808-810.

Proulx, J. F., V. Milor-Roy, and J. Austin. 1997. Four outbreaks of botulism in Ungava Bay Nunavik, Quebec. *Canadian Communicable Disease Report* 23(4):30-32.

Rausch, R. 2003. Cystic echinococcosis in the Arctic and sub-Arctic. *Parasitology* 127(suppl): S73-S85.

Revich, B. A. 2008. Climate change alters human health in Russia. *Studies on Russian Economic Development* 19(3):311-317.

Richter-Menge, J., S. Nghiem, D. Perovich, and I. Rigor. 2008. Sea ice cover. In *Arctic report card, 2007*, www.arctic.noaa.gov//reportcard/seaice.html (accessed April 4, 2008).

Rudolph, K. M., M. J. Crain, A. J. Parkinson, and M. C. Roberts. 1999. Characterization of a multidrug-resistant clone of invasive *Streptococcus pneumoniae* serotype 6B in Alaska by using pulsed-field gel electrophoresis and PspA typing. *Journal of Infectious Diseases* 180(5):1577-1583.

Rudolph, K. M., A. J. Parkinson, A. L. Reasonover, L. R. Bulkow, D. J. Parks, and J. C. Butler. 2000. Serotype distribution and antimicrobial resistance patterns of invasive isolates of *Streptococcus pneumoniae*: Alaska 1991-1998. *Journal of Infectious Diseases* 182(2):490-496.

Schweiger, A., R. W. Ammann, D. Candinas, P. A. Clavien, J. Eckert, B. Gottstein, N. Halkic, B. Muellhaupt, J. Reichen, P. E. Tarr, P. R. Torgerson, and P. Deplazes. 2007. Human alveolar echinococcus after fox population increase Switzerland. *Emerging Infectious Diseases* 13(6):878-882.

Singleton, R., L. Hammitt, T. Hennessy, L. Bulkow, D. DeByle, A. Parkinson, T. E. Cottle, H. Peters, and J. C. Butler. 2006. The Alaska *Haemophilus influenzae* type b experience: lessons in controlling a vaccine-preventable disease. *Pediatrics* 118(2):421-429.

Skarphédinsson, S., M. Jensen, and K. Kristiansen. 2005. Survey of tickborne infections in Denmark. *Emerging Infectious Diseases* 11(7):1055-1061.

Sobel, J., N. Tucker, A. Sulka, J. McMaughlin, and S. Maslanka. 2004. Foodborne botulism in the United States, 1990-2000. *Emerging Infectious Diseases* 10(9):1606-1611.

Søborg, C., B. Søborg, S. Pouelsen, G. Pallisgaard, S. Thybo, and J. Bauer. 2001. Doubling of tuberculosis incidence in Greenland over an 8 year period (1990-1997). *International Journal of Tuberculosis and Lung Disease* 5(3):257-265.

Sørensen, H. C., K. Albøge, and J. C. Misfeldt. 1993. Botulism in Ammassalik. *Ugeskrift for Laeger* 115(2):108-109.

Stefansson Arctic Institute. 2004. *Arctic human development report*. Akureyri, Iceland: Stefansson Arctic Institute.

Van Caeseele, P., A. Macaulay, P. Orr, F. Aoki, and B. Martin. 2001. Rapid pharmacotherapeutic intervention for an influenza A outbreak in the Canadian Arctic: lessons from Sanikiluaq experience. *International Journal of Circumpolar Health* 60(4):640-648.

Wainwright, R. B., W. L. Heyward, J. P. Middaugh, C. L. Hatheway, A. P. Harpster, and T. R. Bender. 1988. Foodborne botulism in Alaska 1947-1985: epidemiology and clinical findings. *Journal of Infectious Diseases* 157(6):1158-1162.

Walters, L. L., S. J. Tyrrell, and R. E. Shope. 1999. Seroepidemiology of California and Bunyamwera (Bunyaviridae) serogroup virus infections in native populations of Alaska. *American Journal of Tropical Medicine and Hygiene* 60(5):806-821.

Warren, J. A., J. E. Berner, and T. Curtis. 2005. Climate change and human health: infrastructure impacts to small remote communities in the North. *International Journal of Circumpolar Health* 64(5):487-497.

Young, T. K. 2008. Circumpolar health indicators: sources, data, and maps. *Circumpolar Health Supplements* 3:55-78.

3

Historical, Scientific, and Technological Approaches to Studying the Climate-Disease Connection

OVERVIEW

A variety of methods are employed to identify, measure, evaluate, and predict the direct and indirect effects of climate change on infectious diseases. As illustrated in the contributions to this chapter, these include the following:

- Analyses of historical records to discern long-term or ancient patterns of climate and infectious disease
- Monitoring programs that track disease in wild animals, which are especially sensitive environmental sentinels
- Comparisons of environmental measurements obtained from satellite imagery with epidemiological data
- Climate-driven predictive models of infectious disease transmission

Each of these approaches contributes to an expanding interdisciplinary effort to understand the influence of climate change and extreme weather events on infectious disease distribution and transmission dynamics.

Historical analysis provides a perspective on climate and infectious disease far more sweeping than can be obtained from scientific monitoring, as demonstrated in this chapter's first paper, which chronicles the association between drought and epidemic disease and its influence on Mexican civilizations over the past millennium. Searching the historical record of the Valley of Mexico for evidence of famines and epidemics, speaker Rodolfo Acuña-Soto of the Universidad Nacional Autónoma de México, and coauthors identified several drought-associated epidemics of hemorrhagic fevers that had swept the region, causing massive mortality. Among these, the authors describe four especially destructive epidemics that appear to have killed between 20 and 90 percent of the entire population, leading to societal collapse: the epidemics of 1003-1011, 1545-1548, 1576-1578, and 1736-1737. The authors also compare circumstances in contemporary Mexico with those associated with apparent past episodes of infectious disease emergence, when increasing human connectivity (roads then, globalization today), and the emergence of new pathogens transmitted by aerosols (smallpox and measles in the past, severe acute respiratory syndrome [SARS] and influenza today), proved to be a very dangerous combination.

Emerging infectious diseases of wildlife arise when the delicate balance of host, pathogen, and environment is disturbed. Therefore, these events represent a critical target for infectious disease monitoring efforts of all sorts, including those that seek to track the influence of climate change, according to speaker William Karesh of the Wildlife Conservation Society. In the chapter's second paper, he and coauthors provide several examples of studies that illustrate the direct and indirect influences of climate on infectious diseases of wildlife. They make the case that such interactions can serve as the basis for monitoring the ecological effects of climate change on emergent diseases that threaten not only wildlife, but also livestock and humans, because wild animals often serve as reservoirs for microbes that may cause pathogenic diseases in humans; these microbes are not necessarily pathogenic in their animal hosts. Moreover, the authors note, wild animals offer a number of advantages for disease monitoring programs: their comparatively short generation times reflect environmental changes more quickly than do humans; the great variety of wild species offers an equally wide range of life histories for the observation of disease dynamics; and they provide sensitive sentinels for changes in the environments to which they are specifically adapted.

As discussed by Chretien and coauthors in Chapter 2 and as first described in Linthicum et al. (1999), efforts to predict risk for Rift Valley fever (RVF) demonstrated that trends in environmental variables detected from satellite imagery can be compared with epidemiological data to reveal relationships between climate and infectious disease transmission and geographic distribution. In his workshop presentation, speaker Compton Tucker of the National Aeronautics and Space Administration (NASA)—who coauthored both of the previously mentioned papers—described how remote sensing data are collected and analyzed, and presented two additional examples of the use of this tool in examining links between climate and infectious disease.

The first involved a search for significant environmental factors common to sporadic outbreaks of Ebola hemorrhagic fever (EHF). Ebola virus also affects nonhuman primates, which have been implicated as the source of several—but not all—human outbreaks through contact with the meat of infected apes (Pinzon et al., 2004). Tucker and colleagues chose to investigate the possibility that Ebola outbreaks occur independently of human cases, in nonhuman primates, and to identify environmental factors that precipitate these outbreaks, which can then spread to humans. Analyzing satellite data—the monthly Normalized Difference Vegetation Index (NDVI), a proxy for precipitation—that had been collected continuously in tropical Africa since 1981, Tucker and coworkers found that the majority of documented EHF outbreaks (in humans) occurred toward the end of rainy seasons, when a sharply drier period was followed by a sudden return to very wet conditions. They hypothesize that these apparent "trigger events" enhance viral transmission from reservoir species—which remain unknown[1]—to nonhuman primates and humans. Today, satellite data from eastern equatorial Africa are screened routinely for the Ebola-triggering weather pattern, Tucker said. The results guide targeted testing for the virus in local primates, which may provide an early warning of future human outbreaks.

Tucker also described the use of satellite imagery to investigate an unusual outbreak of RVF in southwestern Arabia, the first ever recorded there. Records of a satellite-derived index of photosynthetic capacity, the NDVI, showed that significant precipitation had fallen in the region prior to the outbreak and that the RVF outbreak was thus due to natural causes. The origin of the outbreak has since been attributed to infected cattle that were imported into southwestern Arabia from the Horn of Africa (Tucker et al., in press).

As these examples and others in Chapter 2 illustrate, relatively simple correlations between remotely-sensed measurements of climatic variables (e.g., precipitation; sea surface temperature, height) and disease incidence have proven to be useful indicators of risk for a variety of infectious diseases. However, as speaker William Reisen noted, such correlations are not universally applicable and may have to be interpreted in light of other important environmental influences on infectious disease transmission (see also Summary and Assessment section "Predictive Models"). In the chapter's final paper, Reisen and Christopher Barker (both at the University of California, Davis) describe the design, implementation, and limitations of climate-driven predictive models of mosquito-borne encephalitis transmission used by the State of California to estimate disease risk and inform public health interventions.

Under the auspices of this disease surveillance and control program, mosquito abundance and encephalitis virus activity have been actively monitored for more than 50 years throughout many of California's diverse biomes and across wide gradients of latitude (north-south) and elevation (east-west). Early in the

[1]In a personal communication on June 20, 2008, Dr. William Karesh stated that LeRoy et al. (2005) is probably the best work done to demonstrate (1) the presence of Ebola Zaire strain viral particles or viral fragments in 3 species of fruit bats, and (2) serological evidence of immune response to filovirus in those same species of bats. Earlier work by Swanepoel et al. showed viral shedding with no pathology for up to 28 days after fruit bats were experimentally infected with Ebola Zaire in a laboratory setting. To date, there is nothing published on live Ebola virus being isolated from naturally occurring, free-ranging bats. That, in addition to showing that those bats can serve to infect other animals, would help determine if one or more species serve as an effective reservoir.

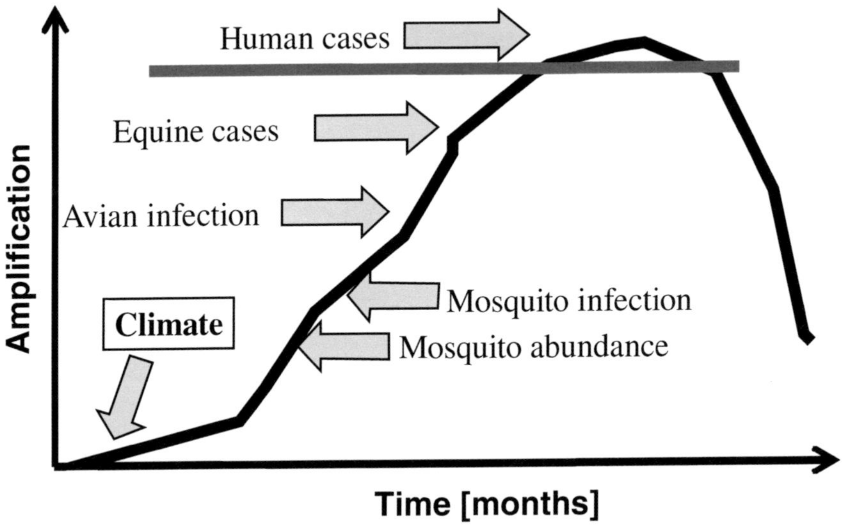

FIGURE 3-1 Sequence of surveillance data collected during seasonal virus amplification.
SOURCE: Reisen and Barker (2008).

season, before insect and wildlife testing become feasible, climate-based forecasts inform disease control measures (Figure 3-1). Surveillance activities begin in the spring, with the goal of arresting viral amplification and avoiding the need for adult mosquito control. In the case of West Nile virus (WNV), early-season temperature measurements are paramount, because the effects of precipitation on viral transmission have been found to vary among regions (Reisen et al., in press) and vector species (e.g., urban *Culex pipiens* mosquitoes do well under hot, dry conditions, whereas rural *Culex tarsalis* do well under wet conditions in many areas).

Although these early-season predictions enable response activities (such as equine vaccination, larval mosquito control, and public education) that can reduce the public health consequences of mosquito-borne disease, Reisen and Barker note that the rationale for applying insecticides in advance of an epidemic is not always understood by the public. They also warn that while WNV "provided a wake-up call for public health agencies and clearly delineated the inability of current control programs to contain an invading, mosquito-borne, zoonosis," waning of the epidemic has led to a loss of funding for WNV research and surveillance and, more importantly, for more general surveillance and detection programs capable of spotting "the next invading pathogen."

DROUGHT, EPIDEMIC DISEASE, AND MASSIVE POPULATION LOSS: 1,000 YEARS OF RECORD IN MEXICO

Rodolfo Acuña-Soto, M.D., M.Sc., D.Sc.[2]
Universidad Nacional Autónoma de México

David W. Stahle, Ph.D.[3]
University of Arkansas

Matthew D. Therrell, Ph.D.[4]
Southern Illinois University

José Villanueva Diaz, Ph.D.[5]
Centro Nacional de Investigación Disciplinaria

Introduction

The Valley of Mexico with its benign climate, rich soil, and once abundant water has been a preferred population center for centuries. Today, with 20 million inhabitants, Mexico City's metropolitan area is one of the largest human conglomerates in history (Yu-ping and Heligman, 1994). While this has been the result of constant growth for the last 85 years, history has not always been this benign. Over the past 1,000 years the Valley of Mexico went through three periods of catastrophic population losses (Clavijero, 1945; Cook and Simpson, 1948; Hugh, 1993).

Founded only 675 years ago, Mexico City is located in the same region where the once magnificent cities Teotihuacán and Tula collapsed 1,255 and 1,000 years ago, respectively. Similar catastrophic events occurred during the sixteenth century, when the Valley of Mexico, as well as the whole country, lost 80 to 90 percent of its inhabitants due to highly lethal epidemics. During the seventeenth to twentieth centuries, the population again went through several calamitous periods of high mortality, droughts, famines, and epidemics (Gerhard, 1986; León, 1982; Ocaranza, 1933; Therrell et al., 2004; Yu-ping and Heligman, 1994).

[2]Corresponding author. Departamento de Microbiología y Parasitología, Facultad de Medicina, Universidad Nacional Autónoma de México, Edificio "A" de Investigación, 2o. Piso, Facultad de Medicina, Ciudad Universitaria, C.P. 04510, México, D.F. Phone (55) 56 23 23 81; Fax: (55) 56 23 23 82; E-mail: yvonne@ibt.unam.mx

[3]Tree-Ring Laboratory, Department of Geosciences, Ozark Hall 113, Fayetteville, AR 72701.

[4]Geograpy and Environmental Resources, Carbondale, Illinois.

[5]Relación Agua-Suelo-Planta-Atmósfera CENID-RASPA, Km. 6.5 Margen Derecha Canal de Sacramento, C. P. 35140, Gómez Palacio, Durango, México.

In spite of the importance of this topic, the formal study of famines and epidemics in Mexico has been primarily descriptive and remains largely incomplete. The aims of this work are to present a chronology of famines and epidemics and to review some of the major events of massive population loss in the Valley of Mexico over the last 1,000 years.

For this study, previously published chronologies of epidemics and famines in Mexico were reviewed. This was complemented with an exhaustive multiyear review of epidemiological, environmental, and demographic information available in archives and libraries in Mexico and the United States. The search included chronicles, old medical books, diaries, newspapers, and official documentation. Quantitative data from censuses and burial records were also obtained. In addition, events indicative of social distress, such as special religious acts, urgent government measures, or local officials asking for help, were recorded. For all documents, priority was given to descriptions written by eye witnesses. Drought was considered as such when firsthand witnesses indicated the absence or drastic reduction of rain, normally associated with crop failure. Information was complemented with data from a previous publication (Therrell et al., 2004). Only materials related to the Valley of Mexico were taken into consideration for this study.[6]

The Chronology of Famines and Epidemics

Over the last 1,500 years, a total of 119 major epidemics and 38 famines were identified (see Table 3-1). Drought was the main cause of 28 (73 percent) famines. The epidemics of smallpox of 1520-1521 and 1538-1539 induced famine by generalized social disruption. Other historic famines were caused by the particularly disastrous combination of summer frost followed by drought. Such was the case of the legendary famine of 1542-1545, when early frost killed all the corn plants in 1542 and was followed by prolonged drought during 1543-1544 when no rain was registered for 20 months. With no new harvest, reserves ran out, creating a very stressful situation that paved the way to a major famine (Therrell et al., 2004). This series of events recurred in 1784-1786, the infamous "year of the hunger" (Cook and Sherburne, 1985). After the Conquest in 1520, famines were recorded with decreasing frequency in the following centuries: 10 in the sixteenth century, 8 in the seventeenth century, and 5 in the eighteenth century; no major famines were recorded during the nineteenth or twentieth centuries.

For the 1,500-year period, a total of 119 epidemics were identified (see Table 3-2). Of these, viral diseases caused 55 (46.2 percent) and bacteria were

---

[6]For representative documents, see Acuña, 1982; Anonymous, 1978, 1980a,b, 1981, 1995, 1998; Carmago, 1999; Chimalpahin, 1998, 2001; Cook and Sherburne, 1985; de Alva Ixtlixóchitl, 1975; de Cárdenas, 1945; de Grijalva, 1924; de Lorenzana, 1992; de Obregón, 1988; de Ortega Montañez, 1955; de Torquemada, 1969; Farfan, 1592; Keber, 1995; Lopez de Hinojoso, 1578, 1595; Mendieta, 1997; Sahagún, 1997; Somolinos, 1956.

TABLE 3-1 Famines in the Valley of Mexico

	Years	Causes	Remarks
1	1003-1011	Drought, war	Famine, epidemic, high mortality
2	1029-1035	Drought	Famine, high mortality
3	1253	Drought	Famine, epidemic
4	1332-1335	Drought	Famine
5	1382-1385	Floods	Famine
6	1403	Locust	Famine
7	1430	Drought	Famine
8	1434	Drought	Famine
9	1446-1456	Locust, floods, frost, snow, drought	Famine, probably associated with epidemic of hemorrhagic fever, high mortality
10	1464	Heat waves, strong winds	Famine
11	1488	Locust	Famine
12	1492	Floods	Famine
13	1498-1499	Floods	Famine
14	1503-1505	Drought	Famine, high mortality
15	1514	Drought	Famine
16	1520-1521	War, first smallpox epidemic	Famine
17	1538-1539	Second smallpox epidemic	Famine
18	1541-1548	Drought, frost, strong winds, crop disease.	Famine, epidemic of hemorrhagic fevers, high mortality
19	1550-1555	Drought, mumps epidemic	Famine
20	1558-1559	Frost, locust	Famine, epidemic of hemorrhagic fevers
21	1562-1564	Drought, epidemics	Famine, epidemics of measles, smallpox, and hemorrhagic fevers; high mortality
22	1571-1573	Drought, epidemic	Famine, epidemic of unknown origin
23	1576-1579	Drought	Famine, epidemic of hemorrhagic fever, high mortality
24	1584-1588	Drought	Famine, epidemic of hemorrhagic fever, high mortality
25	1594	Drought, frost, heat waves	Famine
26	1610-1613	Drought, frost, strong winds	Famine
27	1615-1517	Drought, frost	Famine, epidemic of measles
28	1621-1623	Drought	Famine
29	1629	Floods	Famine
30	1634-1635	Drought	Famine, epidemic of whooping cough
31	1639-1642	Drought	Famine, epidemics of measles, whooping cough, and hemorrhagic fever
32	1658-1663	Drought	Famine, epidemic of measles
33	1691-1697	Drought, frost, floods, speculation	Famine, riots, epidemics of measles and other unknown diseases

continued

TABLE 3-1 Continued

	Years	Causes	Remarks
34	1713-1714	Drought, frost	Famine, epidemic of unknown disease, high mortality
35	1718	Drought	Famine
36	1736-1739	Drought	Famine, epidemic of hemorrhagic fever, high mortality
37	1746-1749	Drought	Famine, epidemics of smallpox and typhus
38	1784-1789	Drought	Famine, epidemic of hemorrhagic fever, high mortality

SOURCE: Adapted from Chimalpahin (2001); Cooper (1965); de Alva Ixtlixóchitl (1975); Flores and Malvido (1985); Garcia et al. (2003); Gerhard (1986); Gibson (1964); Therrell et al. (2004).

TABLE 3-2 Major Epidemics in the Valley of Mexico

	Diseases (n = 15)	Number	Duration (years)	Number of Epidemics That Caused High Mortality[a]
1	Hemorrhagic fevers: cocoliztli, matlazahuatl	24	1.78 ± .62	4
2	Smallpox	20	1.22 ± .42	5
3	Typhus	18	1.25 ± 1.94	3
4	Unknown	19	1.30 ± 1.41	3
5	Measles	13	1.21 ± .42	3
6	Influenza	7	1.0 ± 0	1
7	Typhoid fever	4	1.25 ± .50	0
8	Poliomyelitis	3	1.4 ± .54	0
9	Cholera	3	3.0 ± 2	2
10	Whooping cough	2	1.0 ± 0	0
11	Scarlet fever	2	1.5 ± .70	0
12	Mumps	1	1.0 ± 0	1
13	Chickenpox	1	1.0 ± 0	1
14	Meningitis	1	1.0 ± 0	0
15	Croup	1	1.0 ± 0	0

[a] >1% of the total population.
SOURCE: Adapted from Chimalpahin (1998); Cooper (1965); de Torquemada (1969); Flores (1888); Flores and Malvido (1985); Garcia et al. (2003).

involved in 31 (26.05 percent) of the cases. For the remaining 33 (27.73 percent) epidemics, including 24 described as hemorrhagic fevers, the cause remains to be identified. All 13 known diseases that led to epidemics were caused by exclusive human pathogens with known human-to-human transmission (smallpox, measles, typhus, etc.). Aerosols were the most common mechanism for person-to-person contagion, accounting for 48 (40 percent) of all the epidemics. This was followed by insect vectors, with 18 (8.4 percent) epidemics, and water-borne diseases accounting for 10 (8.4 percent). For the remaining 43 epidemics the mechanism of transmission remains unknown. As in many parts of the world, five diseases were emerging infections of their respective times. All of them were imported and became permanently established in the country (AIDS, chickenpox, measles, mumps, and smallpox). For reasons that remain unclear, cholera disappeared after each introduction. Influenza behaved with the same periodic outbreaks as it does in the rest of the world, and hemorrhagic fevers reemerged locally from a distant past. The four most destructive epidemics (see Table 3-3), with mortality rates ranging from 20 to 90 percent of the entire population, were associated with a sequence of climate extremes, with drought in the years preceding the epidemic followed by wetness during the year of the epidemic (Acuña-Soto et al., 2000, 2002).

Drought and the Collapse of Teotihuacán

The city of Teotihuacán, located about 40 km north of Mexico City, was one of the largest and most sophisticated human conglomerates of the preindustrial world. With a complex urban design, the city was the cultural, religious, and military center of a vast area in Mesoamerica. Following its splendor between the years 300 A.D. and 600 A.D, the city went into decline between 650 A.D. and 750 A.D. (Millon, 1970). Undeniably severe and sustained drought occurred in the eighth and ninth centuries in North America (Acuña-Soto et al., 2005).

TABLE 3-3 Deadliest Epidemics in Central Mexico

	Diseases	Year	Mortality of the Total Population (%)
1	Unknown	1003-1011	90 (?)
2	Hemorrhagic fevers (cocoliztli)	1545-1548	70-80
3	Hemorrhagic fevers (cocoliztli)	1576-1578	50
4	Hemorrhagic fevers (matlazahuatl)	1736-1737	~20-30
5	Smallpox	1520	~15
6	Hemorrhagic fevers (mysterious fevers)	1813-1815	~10
7	Influenza	1918	~5
8	Cholera	1833	~5
9	Smallpox-typhus	1763-1764	~5
10	Smallpox	1779	~4
11	Smallpox	1797-1798	~4
12	Cholera	1850-1854	4
13	Typhus	1915	1
14	Smallpox	1840	1

SOURCE: Adapted from Chimalpahin (1998); Cooper (1965); de Torquemada (1969); Flores (1888); Flores and Malvido (1985); Garcia et al. (2003).

Based on the similarities of the climate (drought) and demographic (large population loss) events of the sixteenth century in the same area, it has been proposed recently that drought-associated epidemics of hemorrhagic fever may have contributed to the massive population loss during the collapse of Teotihuacán (Acuña-Soto et al., 2005). The specific co-occurrence of drought and abandonment of Teotihuacán has not been proved, but it is an attractive hypothesis given the well-documented occurrence of megadrought in adjacent areas (Hodell et al., 1995; Metcalfe and Hales, 1994).

The Fall of Tula

As a result of the meticulous labor of certain educated Indian nobility and Spanish friars during the sixteenth century, pre-Hispanic Mexico's written history—lost as a consequence of the Spanish order to set fire to the library of the city of Texcoco—was partially recovered. Indian authors such as Fernando Alvarado Tezozómoc (Tezozómoc, 1975), Domingo Chimalpahin (Chimalpahin, 1998, 2001), and Fernando de Alva Ixtlixóchitl (de Alva Ixtlixóchitl, 1975) interviewed many elders and studied some of the texts that survived the fire but have since disappeared. Europeans, such as Friar Bernardino de Sahagún (Sahagún, 1997), worked with informants. Fernando de Alva Ixtlixóchitl related the fall of Tula (de Alva Ixtlixóchitl, 1975). He described a series of climatic disasters that plagued the Toltecs for 20 years before a huge epidemic that resulted in a dramatic population loss. Beginning in the year 984 A.D., heavy rains "that destroyed most buildings and lasted for 100 days" were followed by a year of intense heat that "dried all plants and trees." The next year came with frost "that took all the land without leaving anything." The year after, heavy rains came again "with great hailstorms and lightning, so abundant that all the surviving trees were destroyed." This period was followed by an intermission of 12 years of normal weather, but 4 years before the epidemic, a plague of worms infested the grain.

Fernando de Alva Ixtlixóchitl (1975) describes what appears to have been an epidemic in the year 1003, in the style of pre-Hispanic legends:

> In the year 1003, when in the first days, a little boy that was white, blond and beautiful, that had to have been a demon, was on a hill. They took him to the City to show him to the king. When the king saw him, he demanded that they bring him back from where they had taken him, because it did not seem to be a good sign. And then the little demon boy's head began to rot, and many people died from the horrible smell. The Toltecs decided to kill him when one of them was able to reach him, because every one who arrived near the boy died soon after. With this horrible smell, disease spread all over the land and out of the 1,000 Toltecs, 900 died. . . . From this time forward, there was a law that on its fifth birthday, any blond creature would be sacrificed, and this lasted up until the arrival of the Spanish.

The collapse of Tula ended with an extremely violent war, as evidenced by archeological data. During the sixteenth century, several Indian historians wrote about the history of the fall of Tula. Other authors, using independent sources of information, narrate the same events (Chimalpahin, 1998, 2001; Tezozómoc, 1975). Yet the climatic history of the collapse of the Tula Empire is waiting for high-resolution proxy evidence of rainfall.

Drought-Associated Epidemics of Hemorrhagic Fevers of the Sixteenth Century

The post-Conquest collapse of the Mexican population occurred predominantly during the sixteenth century megadrought (Acuña-Soto et al., 2000, 2002, 2004; Stahle et al., 2000). According to all witnesses (Farfan, 1592; Lopez de Hinojoso, 1578, 1595; Somolinos, 1956), the events that caused the highest mortality were a series of epidemics of hemorrhagic fevers referred to as cocoliztli (Nahuatl word for lethal "pestilence") that probably began in 1536 (see Table 3-4). The epidemics of 1545 and 1576-1580 were particularly lethal. Together, they were responsible for approximately 12 million to 15 million deaths in the highlands of Mexico. During the epidemics, a large proportion of the population was incapacitated. Some witnesses described whole families dying of starvation rather than disease, even when not severely ill. Cocoliztli epidemics evolved as an expanding wave originating in central Mexico that radiated outward over the highlands of central Mexico and caused severe social and economic disintegration.

The cause of cocoliztli remains elusive. A brief consensus description, based on contemporaneous attending physicians, is the following: The disease had a very short course, started abruptly with high fever, vertigo, severe headache, insatiable thirst, red eyes, and weak pulse. Shortly after, patients became intensely jaundiced, demented, and restless. Then hard and painful nodules appeared behind one or both ears, sometimes so large that they occupied the entire neck and half of the face. This process was accompanied by intense chest and abdominal pain, as well as dysentery. Toward the end, blood flowed from the ears, anus, vagina, mouth, and nose. The disease was almost inevitably fatal for the native population. The Spaniards were minimally affected, and when they occasionally acquired the disease, it had a benign course (Farfan, 1592; Lopez de Hinojoso, 1578, 1595; Somolinos, 1956).

Drought was particularly important for the epidemics of hemorrhagic fevers in Mexico. Using tree-ring reconstructions of rainfall over central Mexico, cocoliztli epidemics were found to occur in years of abundant rain embedded in the midst of the sixteenth-century megadrought (Acuña-Soto et al., 2002, 2004; Stahle et al., 2000). The 1736 and 1813 epidemics were also highly lethal and were associated with a similar sequence of drought and dryness. The hemorrhagic fevers mysteriously disappeared after 1815 (Acuña-Soto et al., 2000; Cooper, 1965). At

TABLE 3-4 Epidemics of Hemorrhagic Fevers in the Valley of Mexico

	Years	Drought	Mortality (%)	Other
1	1545-1548	Yes	70-80%	Famine
2	1559	Yes	Unknown	Famine
3	1563-1564	Yes	Unknown	Measles, smallpox, famine
4	1566	Yes	Unknown	
5	1570	Yes	Unknown	
6	1576-1578	Yes	50%	Famine
7	1587-1588	Yes	Unknown	Famine
8	1602-1603	Yes	Unknown	
9	1606-1607	Yes	Unknown	
10	1613-1614	Yes	Unknown	Famine
11	1616	Yes	Unknown	Famine and "other" diseases
12	1629-1631	Yes	High	Famine
13	1633	Yes	Unknown	
14	1634	Yes	Unknown	Whooping cough, famine
15	1641-1642	Yes	Unknown	Whooping cough, famine
16	1651	Yes	Unknown	
17	1731	Yes	Unknown	
18	1736-1737	Yes	20-30%	Famine
19	1742	Yes	Unknown	
20	1761-1762	Yes	25,000 in Mexico City	
21	1768-1769	Yes	High infant mortality	Measles
22	1772-1773	Yes	Unknown	Measles and whooping cough
23	1784-1786	Yes	300,000 in the country	Typhus, famine
24	1813-1814	Yes	10%, 25,000 in Mexico City	

SOURCE: Adapted from Chimalpahin (1998); Cooper (1965); de Torquemada (1969); Flores (1888); Flores and Malvido (1985); Garcia et al. (2003).

this time, the specific factors that relate hemorrhagic fevers with drought, as well as the cause of cocoliztli, remain unknown.

Communication Networks Promote Epidemics

Data presented here indicate that the frequency of famines decreased and eventually disappeared in the centuries following the Conquest. Obviously, droughts, the main cause of famines, did not become less severe or less frequent after the Conquest. A possible explanation is that the effects of droughts were reduced by the system of roads developed by the Spanish. In pre-Hispanic times, the native population was fragmented over a vast country; ethnic groups lived in relative isolation, surrounded by enemies and fighting for space and resources. The network of roads built under Spanish rule unified the territory for commercial (collection of silver) and military (control of Indian revolts) purposes. As the road system expanded, the traffic of food, silver, and people was set in motion. Development of the colonial road network had an unintended impact on the introduction and movement of infectious diseases.

Cocoliztli, smallpox, and measles are the diseases that contributed the most to the population collapse of the sixteenth century (Acuña-Soto et al., 2000, 2004; Flores, 1888; Flores and Malvido, 1985; Marr and Kiracofe, 2000). Since the cause of cocoliztli remains unknown, this discussion is centered on the impact of improving communication networks on smallpox and measles, the then-emerging infectious diseases in Mexico. Both diseases are transmitted exclusively by human pathogens and require between 500,000 and 600,000 individuals to remain indefinitely in circulation. Because of the lack of immunity, the first epidemic events of either disease were invariably catastrophic for the native population. After the first wave of infection, the viruses behaved differently depending on the size of the population and the proportion of immune individuals. In large populations, both viruses probably circulated at low frequency during the interepidemic periods. Waiting until a new generation of unprotected children reached a critical number, they then reappeared as epidemics that preferentially affected children. In small and isolated human groups, the situation was different. After a fast outbreak, the infection disappeared quickly because very soon, nonimmune individuals were unavailable; the diseases returned only when they were reintroduced years or decades later. The fragmentation and isolation of Indian groups offered protection from smallpox and measles only until their settlements were connected by the advancing network of roads and missions. Once these settlements were linked by a system of roads, exposure to the repeated introductions of smallpox and measles had devastating effects and correlates with the order in which local Indian populations collapsed. The population loss in Mexico due to smallpox and measles epidemics was not uniform in space and time. In the Valley of Mexico, it occurred during the sixteenth century, with high mortality registered in the north until the second half of the seventeenth century. In the Baja California peninsula, it occurred until the eighteenth century, shortly after the first mission was founded in 1697—178 years after the arrival of the Spaniards in Mexico (Gerhard, 1996).

The Past and the Present Compared

Droughts have been a central factor for at least three of the major population collapses that occurred in central Mexico during the last 1,000 years. Droughts have been associated with both famines and epidemics of hemorrhagic fevers that caused the death of millions of people. The emergence of new pathogens and the increased connectivity of native populations also appear to explain much of the temporal and spatial patterns of famine and epidemic disease in colonial Mexico. To some degree, the globalization of the modern world resembles the colonization of Mexico in the sixteenth to eighteenth centuries. The increased connectivity of populations and the emergence of aerosol-borne pathogens have proven to be a dangerous combination. New roads and trade leveraged the impact of measles and smallpox in colonial Mexico, as globalization has leveraged SARS and influenza in recent times.

Acknowledgments

The authors wish to thank Yvonne Rosenstein for her helpful comments on the manuscript.

WILDLIFE HEALTH AS AN INDICATOR OF CLIMATE CHANGE

Pablo M. Beldomenico, M.V., M.P.V.M., Ph.D.[7]
Wildlife Conservation Society

Damien O. Joly, Ph.D.[7]
Wildlife Conservation Society

Marcela M. Uhart, M.V.[7]
Wildlife Conservation Society

William B. Karesh, D.V.M.[7]
Wildlife Conservation Society

Introduction

The changes in climate we are experiencing as global warming and disturbance in precipitation regimes (IPCC, 2001) are having an impact on the health of wild animals, with resulting deleterious impacts on major human interests. In this paper, we review the relationship between climate change and wildlife health and argue that monitoring wildlife health provides an effective and sensitive indicator and predictor of climate-related emerging infectious diseases.

[7]Global Health Programs.

Effects of Climate Change on Wildlife Health

After a long period of neglect, pathogens have recently been suggested as important drivers of host population dynamics (Hudson et al., 1998; Tompkins and Begon, 1999). At the same time, disease has been implicated in major wildlife population declines (Pounds et al., 2006; Roelke-Parker et al., 1996). Wildlife health, therefore, is an important factor for population sustainability and system resilience and, hence, is drawing conservation attention. Because most human emerging infectious diseases originate from a wildlife reservoir (Jones et al., 2008), wildlife health is critically linked to public health. Wildlife species may serve as an early warning for some diseases, as is the case for howler monkeys and yellow fever in South America (Rawlins et al., 1990) or great apes and Ebola hemorrhagic fever in central Africa (Karesh and Reed, 2005). In addition, due to the flux of pathogens occurring at the wildlife-livestock interface, wildlife health is also important for domestic animal health.

Suggested Mechanisms

Climate change may affect wildlife health in several ways because the determinants of disease incidence are numerous and specific to the disease in question and climate change may influence each of these factors. Moreover, the direct and indirect impacts of climate change on host-pathogen interactions might favor some host species because they could release hosts from the population control exerted by pathogens by interfering with the precise conditions required for pathogen viability (either directly or indirectly; i.e., changes in vector abundance) and shift host population regulation to other factors, such as food or other resource availability. It is important to recognize that the components of a dynamically functioning ecosystem are interconnected; thus, the influence of climate change on animal distribution, abundance, or demography via infectious disease may be indirect due to shifts in relationships such as competitive advantages among conspecifics, predator-prey dynamics, and so forth.

In the interests of simplification, we propose that climate change may directly modify the patterns of infectious disease basically in two ways:

1. Favoring pathogens (increasing pathogen and/or vector proliferation, or pathogen and/or vector survivability)
2. Increasing the host's susceptibility to infection

Climate Change Favoring Pathogens and Their Vectors

Changes in climate shift the relationship among pathogen, host, and the environment. Focusing first on the pathogen, we know that climatic conditions play a significant role in the geographic and temporal distribution of pathogens and, as appropriate, their vectors. Environmental conditions also affect the viability and reproductive success of both pathogens and vectors, which can be thought of as the nonhost components of severity of infection.

Changes in distribution and seasonality It is known that, within limits, arthropod populations are favored by heat and moisture. Therefore, we anticipate that climate change will influence vector-borne diseases. In fact, a number of vector-borne human and domestic animal diseases have increased in incidence or geographic range in recent decades (e.g., malaria, African trypanosomiasis, tick-borne encephalitis, yellow fever, plague, dengue, African horse sickness, bluetongue) (Harvell et al., 2002). These changes have been or can be identified because the diseases are important for public health or domestic animals, and hence records to detect their occurrence currently exist in many cases. It is close to impossible to know if vector-borne diseases are changing in wildlife unless more widespread, systematic monitoring is put into place.

A large body of research has been conducted to attempt to predict the expansion of vector-borne diseases due to climate change. Largely, these efforts involve modeling. For example, a study addressed how climatic variables determined the abundance of *Ixodes scapularis* (the tick that transmits Lyme disease) and ehrlichiosis and babesiosis in eastern North America (Ogden et al., 2005). The authors used current knowledge of tick biology, originating from empirical and experimental data, to construct a theoretical model that predicted the abundance of ticks based on the climatic and seasonal variables they had measured. In a subsequent study, they used this model to simulate the expansion of tick distribution under two projected climate change scenarios, predicting a substantial northward movement of tick range (Ogden et al., 2006). These examples support the basic concept that climate change is expected to influence the geographic range of vectors and their transmitted diseases.

A similar situation may be observed for pathogens affecting ectothermic (cold-blooded) hosts or those that proliferate outside affected individuals, because the pathogens are more exposed to ambient temperature as opposed to those having a life cycle that is completed almost entirely inside a host that preserves a constant temperature (endothermic). A rise in average temperature may not only affect the proliferation of the pathogen, but also have the potential to modify the seasonality of the disease—which could occur earlier every year—and remain infective or active for a longer period of time (Harvell et al., 2002).

Harmful algal blooms (HABs), also known as "red tides," are events in which single-celled protists (dinoflagellates) proliferate rapidly and accumulate in the water column. These events are associated with wildlife mortalities because under certain circumstances these organisms can produce potent toxins. A systematic increase in seawater temperature may contribute to the occurrence of HABs (Juhl, 2005). For example, the Wildlife Conservation Society investigated an episode of mass bird die-off in the Malvinas-Falkland Islands in which high levels of toxins produced by these dinoflagellates were detected in sick or dead gentoo penguins as well as in the marine prey species found in the digestive tracts of affected animals (Uhart et al., 2004). This was the first report of paralytic shellfish poisoning affecting seabirds in the southwest Atlantic, which might suggest that climate change is aiding the expansion of this type of disease to more extreme latitudes.

Increased severity of disease An increased intensity of parasitism or severity of infection may also result from favorable conditions for pathogens. In the St. Kilda archipelago of Scotland, for example, feral populations of Soay sheep experience periodic mass mortalities (Coulson et al., 2001). Although the proximate cause of death has been determined to be protein-energy malnutrition, parasites have been implicated as a contributory factor (Grenfell et al., 1995). At first observation, the depth of the population crashes was critically dependent on the weather, and large numbers of trichostrongylid[8] nematodes were found in dead animals. An experimental study showed that the administration of antihelminthic therapy reduced mortality considerably (Gulland et al., 1993), which supported the link between parasites and death. Trychostrongylids have a life cycle that involves several stages outside their hosts, making them highly vulnerable to environmental conditions. In particular, larvae are very susceptible to desiccation, so humidity and precipitation regimes are crucial for their survival outside their hosts (Wharton, 1982). The emerging scenario would be that increased precipitation would allow for increased larval survival, which in turn would lead to higher parasite burdens, contributing to elevated mortality.

Climate Change Increasing Host's Susceptibility to Infection

As mentioned, changes in climate shift the relationship among pathogen, host, and the environment. Focusing on the hosts, we know that climatic conditions can affect their behavior and hence susceptibility to infectious organisms due to change in exposure or contact rates. While genetics provides a framework, host immunity or disease resistance is also dependent on physiologic mechanisms affected by environmental conditions. The impact of changes in climate can occur at a rate more rapid than a host's ability to adapt.

Increased exposure to pathogens One way in which climate change can result in an increased susceptibility to infection is by inducing changes in host behavior, which may determine increased exposure to pathogens. While some parts of the world are projected to become more moist, others are projected to become drier (IPCC, 2001). On the Patagonian Steppe, for example, water supplies are threatened (Barros et al., 2000). As a result, the concentration of individuals

[8]A common parasitic gastrointestinal nematode found in many ungulate species.

around water resources may grow, thereby increasing intraspecific interaction and indirect contact rates, and possibly shifting density-dependent infectious disease relationships. Since water supplies are frequently shared among wildlife and domestic animals, these alterations may increase the risk of pathogen exchange at the wildlife-livestock interface.

Another example of the way in which climate change may result in increased susceptibility to infection is its impact on feeding behavior. For instance, the reduction in sea ice is causing a change in the behavior and diet of walruses, which are becoming more pelagic and are preying more on ringed seals (a carnivore) and less on invertebrates. This, in turn, may increase the prevalence of trichinellosis in walruses (Rausch et al., 2007). Finally, climate change may determine that some vertebrate hosts expand their distribution, and thus expose immunologically naïve species to their pathogens and nonpathogenic commensal organisms, as well as expose themselves to new pathogens. For example, tropical deglaciation is causing an increase in the elevational limit of some anurans, which have taken with them the agent of chytridiomycosis (*Batrachochytrium dendrobatidis*) to unprecedented altitudes (Seimon et al., 2007).

Decreased host resistance The susceptibility of hosts may also increase if their intrinsic vulnerability to disease is affected. For many species, climate change will serve as an additional form of stress. The effect of stress on vertebrates is well known: a cascade of neuroendocrine mechanisms triggered by stress resulting in a reduction in immune function (Lochmiller and Dabbert, 1993). Since wild species usually live on tight energy budgets (Beldomenico et al., in press), a number of physiological functions compete for these limited resources. An increased demand by one system results in fewer resources for the rest. If climate change causes resources to become more scarce, or of poor quality, or if other physiological systems increase their demands (reproduction, molting, migration, etc.), then the share left for immunological investment will be reduced.

A recent study on rodents demonstrates that poor body condition predisposes individuals to a variety of infections, and these infections further decrease the condition of individuals, triggering a "vicious cycle" that eventually ends up in death and, therefore, population declines (Beldomenico et al., 2008). Thus, maintaining good body condition is important to reducing infection, and avoiding infection is essential to maintaining good condition. If climate change-induced food resource limitations or stress impoverish the condition of many individuals in a population, this type of infection-declining condition cycle may be triggered and the population will fail.

Amphibians are particularly sensitive to climate disturbance. In the last three decades, thousands of species have experienced population declines worldwide and more than 100 have disappeared (Stuart et al., 2004), many of them in seemingly undisturbed environments. The chytrid fungus *B. dendrobatidis* has been implicated in many of these amphibian population crashes (Berger et al., 1998;

Weldon et al., 2004); however, amphibians have also been declining in regions where the fungus is absent, and the fungus has been found in places with no affected frogs (Di Rosa et al., 2007). In declining populations where the chytrid fungus was not present, other pathogens were found at high prevalences, namely, *Saprolegnia ferax* (Kiesecker et al., 2001), *Amphibiocystidium ranae* (Di Rosa et al., 2007), ranaviral disease (Bollinger et al., 1999), and metazoan parasites (García et al., 2007).

Amphibian declines have been correlated with climatic change, and a hypothesis for climate-driven epidemics arising from climate favoring pathogens differentially over hosts has been proposed to explain amphibian declines (Pounds et al., 2006). However, the vicious cycle hypothesis (Beldomenico et al., 2008) may also be a satisfactory explanation. Reading (2007) presents evidence that environmental warming negatively affects the body condition of toads. Alford et al. (2007) observed that frog population declines are preceded by an increase in indicators of stress. The emerging vicious cycle hypothesis proposes that climate disturbance is affecting the condition of amphibians, which predisposes them to more frequent infections and/or infections of increased severity, which triggers a vicious cycle with the potential to cause amphibian population declines. Thus, while the synergy between poor condition and infection may be a proximate cause of these declines, the ultimate cause would be a condition impoverished by climatic changes.

In summary, there are multiple mechanisms by which climate change could affect wildlife health including, but not limited to, the following:

- Expansion in the geographic distribution of pathogens, vectors, or hosts
- Changes in the seasonality of some diseases
- Increased severity of disease
- Increased exposure to pathogens
- Decreased host immunity

All may result in a disruption of population and system health dynamics. Thus, independent of mechanism, monitoring the health of wildlife populations provides a sensitive and quantitative method to detect changes and serve as an early warning system.

One subject in need of further investigation is the relationship between evolution and an organism's ability to adapt to these rapid changes. It is possible that genetic shifts will modulate local effects of climate change. However, there is little evidence that evolution will mitigate negative effects of climate change at the species level (Parmesan, 2006). Moreover, it should be considered that the speed of evolution is different among different taxa. Bacteria and viruses, for example, have the capacity to evolve rapidly, adapting to environmental changes before their hosts. This might result in a differential adaptation favoring pathogens and inadvertently causing host population declines or extinctions.

Summary

Many studies have shown that climate change can influence the dynamics of wildlife diseases. The question is: Can this relationship be utilized to monitor the ecological effects of climate change and to predict and prevent the emergence of new diseases threatening wildlife, livestock, and human health? We argue that in the context of health and climate change, monitoring wildlife health is of direct relevance for several reasons. First, the emergence of human and livestock diseases is closely tied to wildlife health (Jones et al., 2008; Karesh and Cook, 2005). As a result, detection of climate-related emergence of disease in wildlife populations provides an early warning of system disturbance and, thus, potential human and domestic animal health concerns (early-warning capability being the useful feature of any indicator of ecological change [Carignan and Villard, 2002]). Second, the range of population turnover times in hosts and wildlife pathogens, from short generation times in bacterial and viral pathogens to relatively longer generation times in helminths and other parasites, and even decades in some hosts, provides an opportunity to evaluate change at a variety of temporal scales. Third, for some wildlife populations—particularly hunted or managed populations—good long-term baseline health data exist. Consequently, we know the range of what is normal and can more easily determine what is abnormal or different. Finally, and most relevant to this discussion, as "system integrators," the health of wild animals is tuned to a set range of natural variation and, therefore, provides a sensitive indicator of change.

USE OF CLIMATE VARIATION IN VECTOR-BORNE DISEASE DECISION SUPPORT SYSTEMS

William K. Reisen, Ph.D.[9]
University of California, Davis

Christopher M. Barker, M.S.[9]
University of California, Davis

Introduction

Vector-borne pathogen transmission cycles minimally consist of an arthropod vector, a vertebrate host, and a pathogen, but many are zoonotic and transmitted among a complex array of vectors and vertebrate hosts (e.g., West Nile virus; see Figure 3-2). For most zoonotic arboviruses, transmission to humans and to

[9]Center for Vectorborne Diseases, School of Veterinary Medicine, University of California, Old Davis Road, Davis, CA 95616, Reisen E-mail: arbo123@pacbell.net; Barker E-mail: cmbarker@ucdavis.edu.

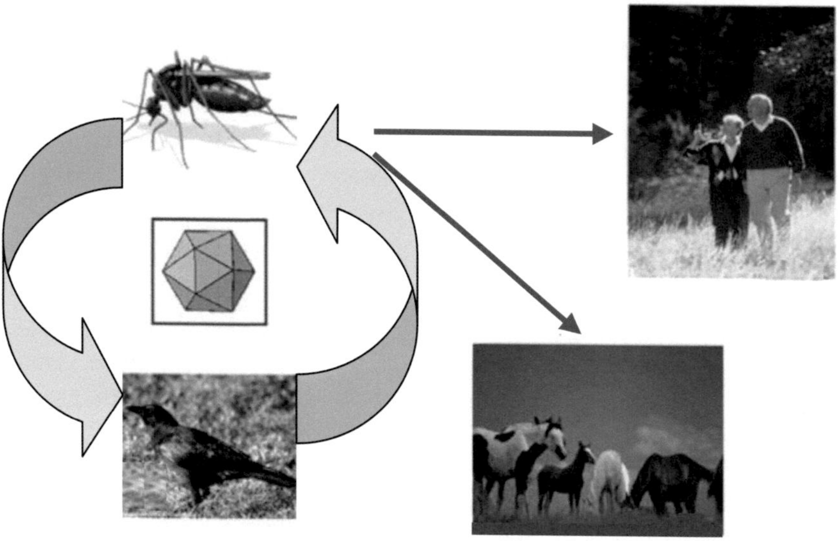

FIGURE 3-2 West Nile virus transmission cycle.
SOURCE: CDC (2005).

some extent domestic animals causes disease but is a "dead end" for the virus. Surveillance data on arthropod vectors or infection in reservoir hosts typically are skeletal, often leaving the passive detection of human or veterinary illness as the only consistent measure of pathogen activity. However, the diverse spectrum of clinical symptoms frequently makes syndromic surveillance difficult, and for many zoonoses, symptomatic individuals represent only a small proportion of the infected population, making them an insensitive measure of pathogen activity.

Regardless of the intensity of surveillance or transmission cycle complexity, pathogen dynamics are directly affected by climate at a variety of spatial and temporal scales. Long-term surveillance programs by control agencies provide one of the few measures of vector populations suitable for assessing the impact of climate variation on vector-pathogen-host systems. A detailed understanding of these climate-health relationships is the first step toward developing models and forecasting risk, which then can be assessed by measuring ecosystem and pathogen transmission dynamics. Risk forecasts are extremely useful in intervention programs charged with mitigating pathogen amplification and protecting the public using preventive methods, whereas measures of risk in real time form an integral part of decision support systems.

In this paper, we explore how climate variation impacts the transmission dynamics of vector-borne disease using California's mosquito-borne encephalitis virus surveillance and control program as an example. The California program provides an excellent model because (1) the state encompasses multiple biomes that vary markedly across north-south latitudinal and east-west elevational gradients; (2) an intensive surveillance program has been consistently monitoring mosquito abundance and encephalitis virus activity for more than 50 years; and (3) there is a statewide decision support system, including a response plan, that uses surveillance data to estimate risk and recommend appropriate levels of control.

Defining Climate Variation: Importance of Scale

Climate encompasses a variety of meteorological parameters, including temperature and wetness, which normally are averaged over a defined time period to delineate "average" conditions for a specific geographic region. Climate variation describes deviations about these long-term means that may be measured at a variety of scales from days to years, whereas climate change is directional and consists of long-term shifts in means over decades to centuries. Carefully monitoring climate variation and understanding its potential impact on ecosystem dynamics provides an important tool for forecasting vector-borne pathogen transmission. Models capturing several climate parameters have provided estimates of hydrologic conditions that were related to outbreaks of mosquito-borne encephalitis in Florida (Shaman et al., 2002, 2004). These models are less useful, however, when vectors exploit anthropogenic water sources in urban or agricultural ecosystems. Other indices measure biological parameters directly, such as the Normalized Difference Vegetation Index, which uses remotely sensed reflectance to estimate the vigor and density of live green vegetation (Tucker, 1979) as a surrogate for other biotic factors influencing vector populations. The value of raw data from satellite and ground sensors is enhanced through additional processing such as NASA's Terrestrial Observation and Prediction System (TOPS; see Figure 3-3), a modeling framework that integrates and preprocesses data so that land surface models can be run in near real time (Nemani et al., 2003). These models use ground and satellite instruments to measure various water (evaporation, transpiration, stream flows, and soil moisture), carbon (net photosynthesis, plant growth), and nutrient (uptake and mineralization) processes at a variety of spatial scales, from global net primary productivity (NPP) anomalies at 0.5 × 0.5-degree resolution to local estimates of ecosystem parameters at resolutions as fine as 250 m. At each spatial resolution, TOPS uses different sources of satellite data (Moderate Resolution Imaging Spectroradiometer [MODIS] to Ikonos) and meteorological data (single weather station to global atmospheric model outputs).

Once average climate conditions have been established, deviations or anomalies can be tracked at varying scales. Short-term changes (weather) can be forecast days to weeks in advance and predict events such as rainstorms or heat waves that may immediately affect vector-borne pathogen transmission. Interannual variation—driven by global cycles such as El Niño—may be used to forecast ecosystem change seasons in advance and therefore forecast changes in vector

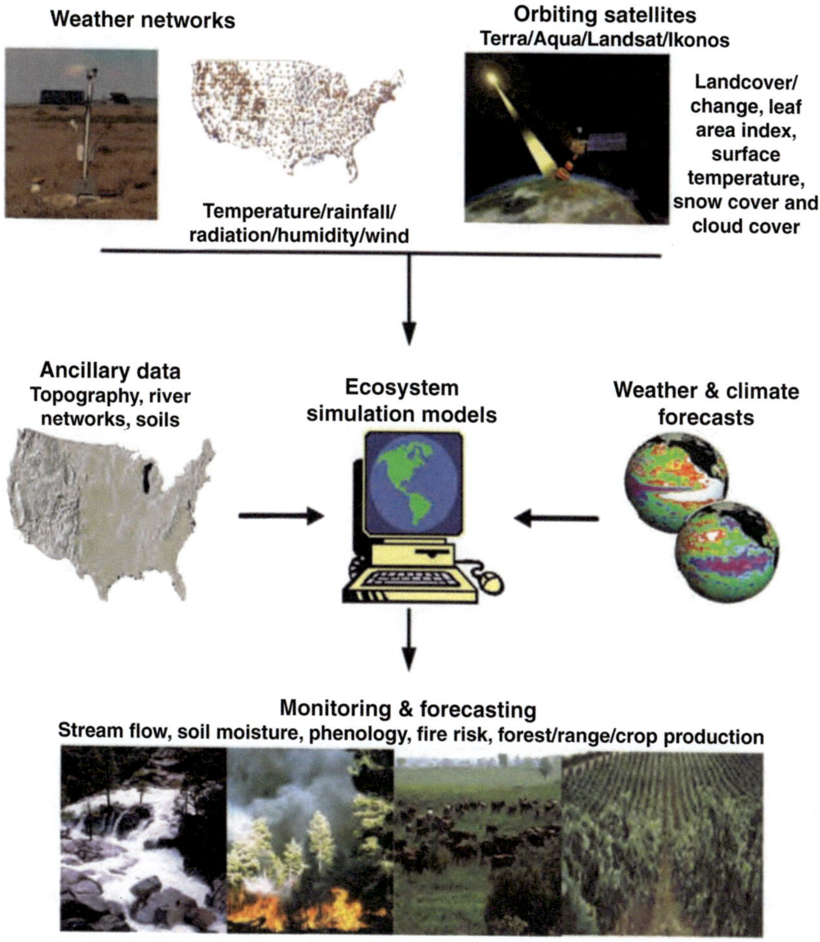

FIGURE 3-3 TOPS system brings ground and remote measures of climate into ecological models to monitor and forecast risk.
SOURCE: NASA (2007).

abundance (Reisen et al., 2008a) and outbreaks of arboviruses such as Rift Valley fever virus (Anyamba et al., 2002; Linthicum et al., 1991). Longer, interdecadal trends may indicate shifts in baselines (i.e., climate change), and these gradual changes may elongate transmission seasons and extend vector and pathogen distributions. Change has been most clearly detected at northern latitudes (Githeko et al., 2000) and in urban landscapes that present their own microcosms for climate change and variation (Kalnay and Cai, 2003). Certainly the consistently high incidence of WNV in the U.S. central plains and central Canada (data not shown) was unexpected and generally has tracked above-normal summer temperatures (see Figure 3-4), even though these relatively low mean temperatures were considered suboptimal for virus amplification within the primary mosquito vector *Culex tarsalis* (Reisen et al., 2006b).

Impact of Climate Variation on Vectors and Pathogens

Vectors and the pathogens they transmit are especially subject to climate variation because climate markedly affects arthropod population size and dynamics, and because pathogen replication rates are influenced directly by ambient temperatures during infection of the poikilothermic[10] arthropod vector. This is especially true for the mosquito-borne encephalitides at temperate latitudes where temperature delineates amplification and transmission season duration. Climate variation also indirectly determines the size and age structure of avian maintenance and amplifying host populations by impacting primary productivity and therefore the abundance and distribution of food sources.

In contrast, the impact of climate on mosquito populations is rapid and direct. Many vector mosquitoes utilize surface water accumulations for larval development that depend directly or indirectly on precipitation. For rural species such as *Culex tarsalis*, the timing and size of the adult population peaks depend directly on winter and spring rainfall and snowmelt runoff, and have been related to El Niño conditions and the associated wet winter and spring seasons (Reisen et al., 2008a). Conversely, urban species, such as those in the *Culex pipiens* complex, utilize peridomestic and underground drainage systems and may be favored by La Niña conditions of high temperature and low rainfall. In urban centers such as Los Angeles, high rainfall and associated runoff scour the underground larval habitats and actually reduce vector abundance (Su et al., 2003).

Warm temperatures increase the rate of mosquito population growth, reduce adult survival, and increase the frequency of blood feeding as well as the chances of virus acquisition and transmission (Reeves et al., 1994; Reisen, 1995). Temperature also is positively associated with encephalitis virus replication within the mosquito vector. The time from virus infection to transmission (the extrinsic incubation period [EIP]) is directly related to ambient conditions and can be described by degree-day models (Reisen et al., 1993, 2006b). These regimens frequently delineate episodic waves of transmission during outbreaks (Nielsen et al., 2008) as well as transmission seasons and the geographic distribution of virus amplification.

[10]Any organism whose body temperature varies with the temperature of its surroundings.

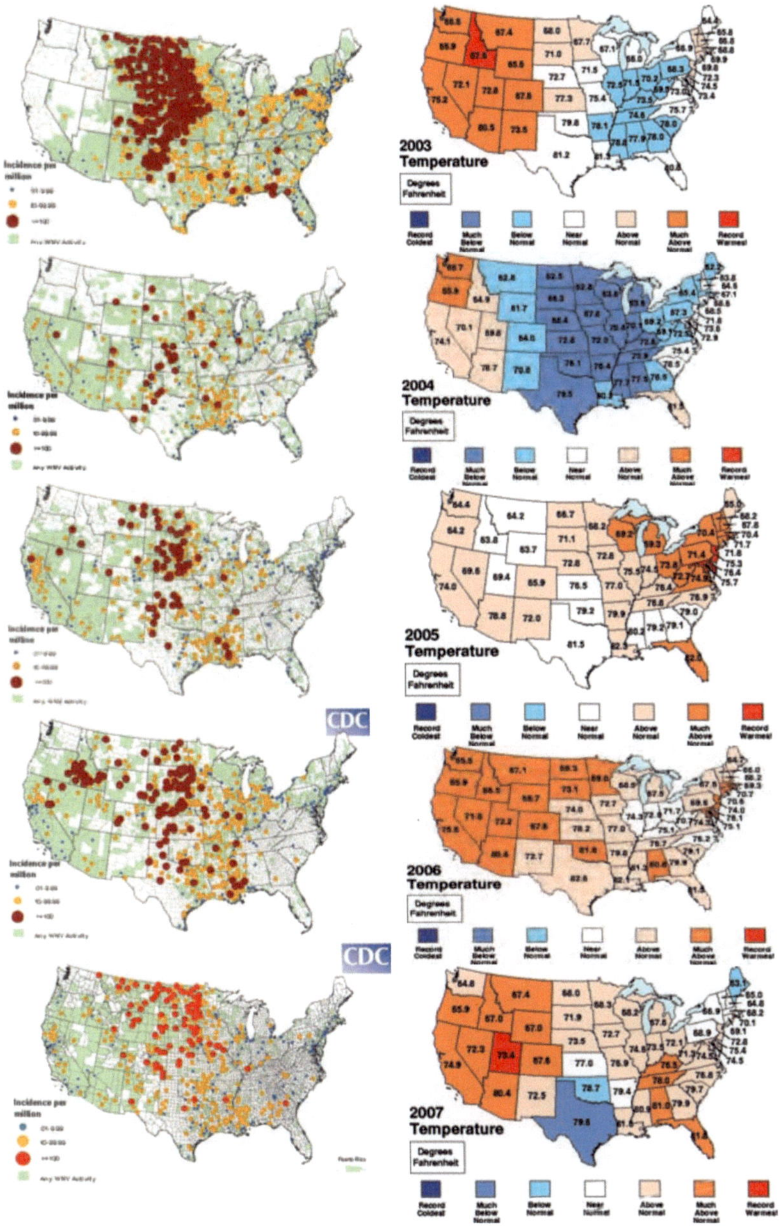

FIGURE 3-4 Incidence of human West Nile virus cases per million population and temperature anomalies for the United States, 2003-2007.
SOURCES: CDC (2007); NOAA (2008).

Components of a Vector-Borne Pathogen Surveillance Program

Comprehensive vector-borne disease surveillance programs include measures of early-season meteorological conditions that may be used to forecast risk for the coming virus transmission season. As temperatures increase and vectors become active, these programs begin monitoring vector abundance and virus activity within the primary enzootic cycle as a measure of the risk of pathogen transmission to humans and to validate earlier forecasts. These data can be combined in models that forecast or measure risk. For WNV in California, climate measures focus on temperature because the effects of precipitation vary among regions (Reisen et al., 2008a) and vector species (e.g., urban *Cx. pipiens* mosquitoes do well under hot, dry conditions, whereas rural *Cx. tarsalis* do well under wet conditions in many areas). *Culex* abundance is monitored by New Jersey or American light traps (Mulhern, 1985), the Centers for Disease Control and Prevention's (CDC's) dry ice-baited traps (Newhouse et al., 1966), and gravid female traps (Reiter, 1987) in urban situations. Mosquitoes and dead birds are tested rapidly for infection by detecting viral RNA using robotic extraction and real-time reverse transcriptase-polymerase chain reaction (RT-PCR) testing procedures (Shi et al., 2001). Avian infection indicating active transmission is measured by sequentially bleeding sentinel birds, such as chickens, for seroconversion[11] and by recording and testing dead birds reported by the public. Equine and human cases are diagnosed by healthcare providers and confirmed serologically at local laboratories. The temporal cascade of events and surveillance data along a typical WNV amplification curve is shown in Figure 3-5. The horizontal line shows the hypothetical level of amplification required before tangential transmission to humans is frequent.

The key to effective surveillance is the rapid dissemination of results and analysis to individuals responsible for intervention decisions. To expedite data exchange in California, a web-based management system called the Surveillance Gateway© has been implemented in which data are entered, stored, analyzed, and displayed (see Figure 3-6). When combined with rapid laboratory diagnostics, mosquitoes collected on Monday or Tuesday and immediately shipped to the laboratory are routinely tested, recorded, reported, and mapped online as early as Thursday or Friday of the same week. These data are then compared to historical records, and a risk score is assigned for each parameter (see Table 3-5). The scores for individual surveillance parameters are then averaged to obtain an overall risk score ranging from 1 to 5, where 1.0-2.5 denotes a "normal season," 2.6-4.0 represents increasing risk requiring "emergency planning," and 4.1-5.0 indicates "epidemic conditions."

Risk levels for WNV can be forecast and then measured based on average daily temperatures (see Figure 3-7). These risk levels, based on temperature,

[11]Conversion of a sentinel host from antibody negative to positive after infection.

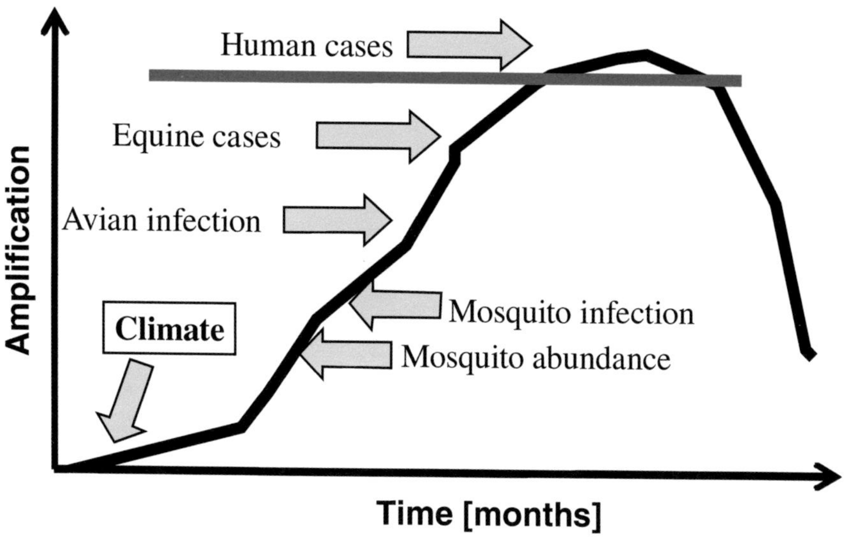

FIGURE 3-5 Sequence of surveillance data collected during seasonal virus amplification.

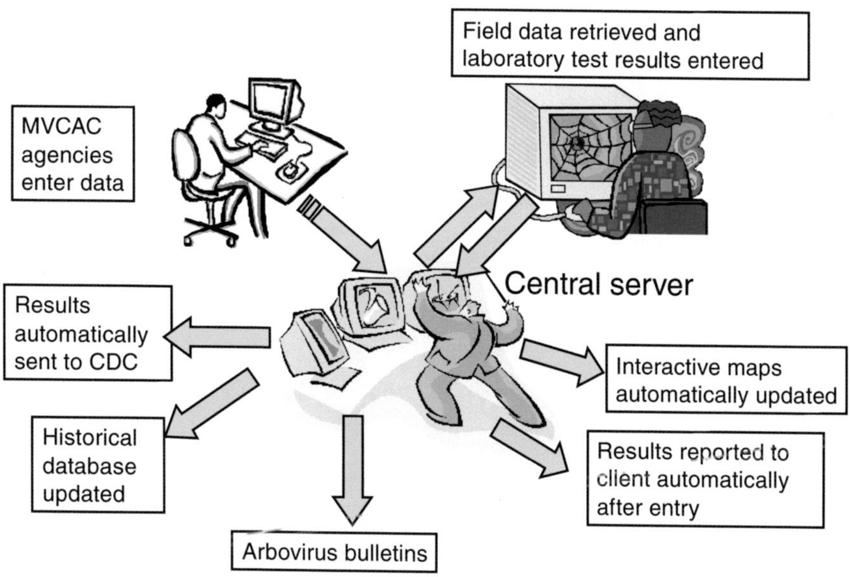

FIGURE 3-6 Data flow through the Surveillance Gateway© system.

TABLE 3-5 California Mosquito-Borne Virus Surveillance and Response Plan Model Scores for Each Surveillance Parameter

Risk Level	Average Daily Temperature	Adult Mosquito Abundance	Mosquito MIR/1,000	Chicken Seroconversions	Dead Bird Infections	Equine Cases	Human Cases
1	<13°C	<50% 5 year avg.	0	0 in region[a]	0 in region	0 in region	0 in region
2	13-18°C	50-90% 5 year avg.	0.1-1.0	≥1 in region,[a] 0 in agency	≥1 in region, 0 in agency		
3	18-22°C	91-150% 5 year avg.	1.1-2.0	1 flock in agency[b]	1 in agency	≥1 in region, 0 in agency	≥1 in region, 0 in agency
4	22-26°C	151-300% 5 year avg.	2.1-5.0	2 flocks in agency[b]	2-5 in agency	1-2 in agency	1 in agency
5	>26°C	>300% 5 year avg.	>5.0	>2 flocks in agency[b]	>5 in agency	>2 in agency	>1 in agency

[a]Region refers to areas encompassed by the formalized regions of the California Mosquito Control Association.
[b]Agency refers to the boundaries of the various mosquito districts.
SOURCE: Adapted from California Department of Public Health (2007).

relate directly to the impact of temperature on the duration of EIP estimates from the degree-day model (Reisen et al., 2006b). Cool temperatures, which require elongated periods before transmission is possible, are assigned low levels of risk; hot temperatures, which result in rapid and efficient transmission, are assigned elevated risk scores. Figure 3-7A shows risk scores based on temperature assigned to mosquito control districts in California during each month in 2007, with elevated risk shown in red. Above-normal temperatures in the southern San Joaquin Valley produced earlier-than-normal increases in urban *Cx. pipiens* complex mosquito abundance and virus infection rates, and epidemic risk levels by June were higher and earlier than in other areas of California (Figure 3-7B). Coincidentally, the 2007 WNV epicenter for California was Bakersfield in Kern County, and the number of human cases increased with risk, peaking in late July (Figure 3-7C).

Although temperature was predictive of risk and receptivity for amplification in the Central Valley, other epidemiological factors are clearly needed because there was little enzootic activity in the hot southeastern deserts during 2007. This area has had limited annual enzootic amplification and few human cases, even though temperatures are very high for most of the year (Reisen et al., 2008b). Vector mosquitoes are plentiful here, but virus amplification rarely has reached outbreak levels, because American crows and other corvids are uncommon in the desert biome of California (Reisen et al., 2006a). Lack of abundant corvid populations may thereby limit infection in peridomestic *Cx. pipiens* complex mosquitoes that are less competent hosts than *Cx. tarsalis* (Reisen et al., 2005).

Response to Surveillance Data

Integrated vector management programs are driven by decision support systems based on surveillance data (see Figure 3-8). The California Mosquito-borne Encephalitis Virus Surveillance and Response Plan describes the attributes of a comprehensive surveillance program, provides methods for calculating risk, and provides a series of response guidelines for each level of increased risk.[12] Climate variation and forecasts provide the only early-season surveillance information upon which to gauge the response program to pending risk. Early-season response activities include preventive methods, such as equine vaccination, source reduction, larval mosquito control, and public education. Surveillance activities, such as mosquito collection and bird testing, typically begin during spring, and the timing and sensitivity of these programs determine the limit of virus detection and therefore the time available for control operations (control window) to arrest amplification at levels where human risk is minimal (see Figure 3-8A, dashed line). If routine control methods fail to arrest virus amplification and risk increases to emergency planning or epidemic levels, emergency adult

[12]See http://westnile.ca.gov/resources.php.

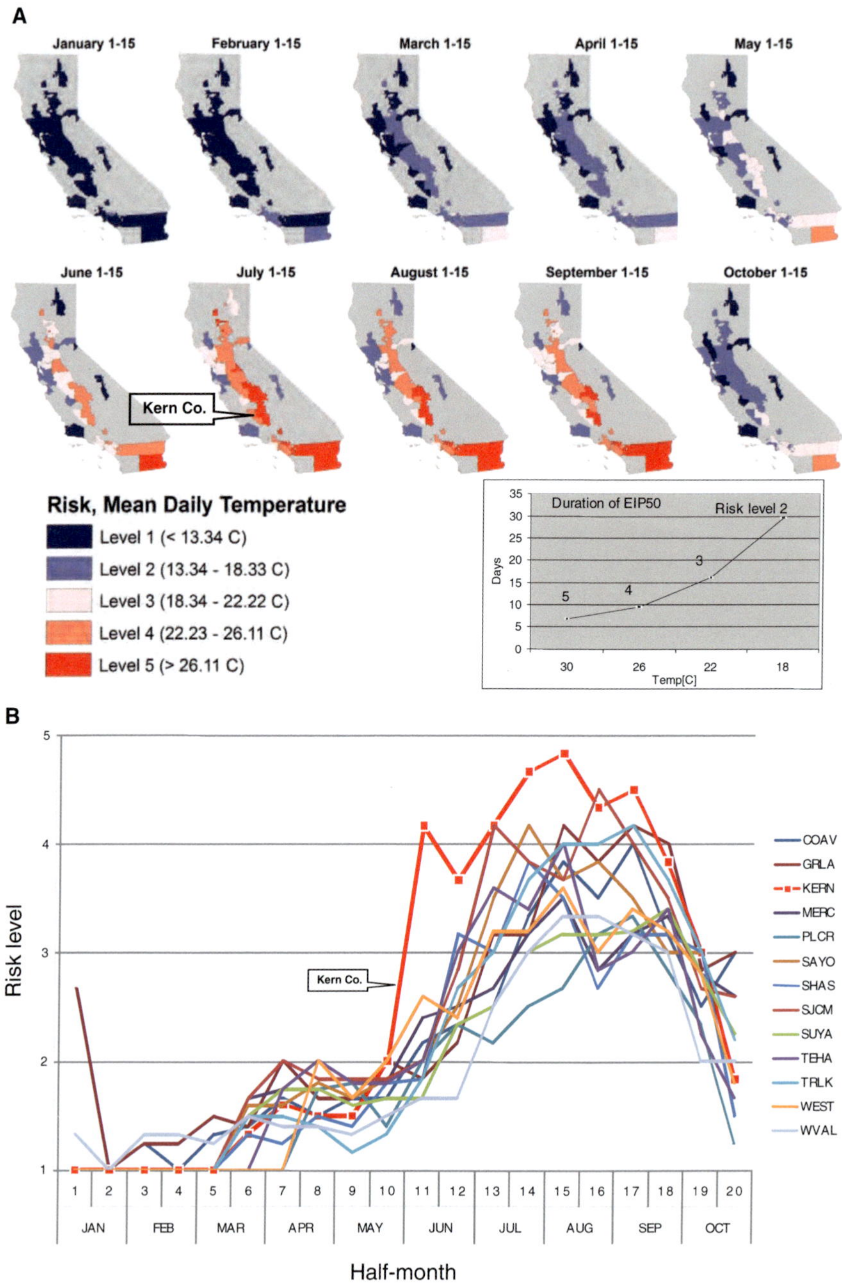
A
January 1-15
February 1-15
March 1-15
April 1-15
May 1-15
June 1-15
July 1-15
August 1-15
September 1-15
October 1-15
Kern Co.
Risk, Mean Daily Temperature
Level 1 (< 13.34 C)
Level 2 (13.34 - 18.33 C)
Level 3 (18.34 - 22.22 C)
Level 4 (22.23 - 26.11 C)
Level 5 (> 26.11 C)
Duration of EIP50
Risk level 2
Days
Temp[C]
B
Risk level
Half-month
Kern Co.
JAN
FEB
MAR
APR
MAY
JUN
JUL
AUG
SEP
OCT
COAV
GRLA
KERN
MERC
PLCR
SAYO
SHAS
SJCM
SUYA
TEHA
TRLK
WEST
WVAL

FIGURE 3-7

C

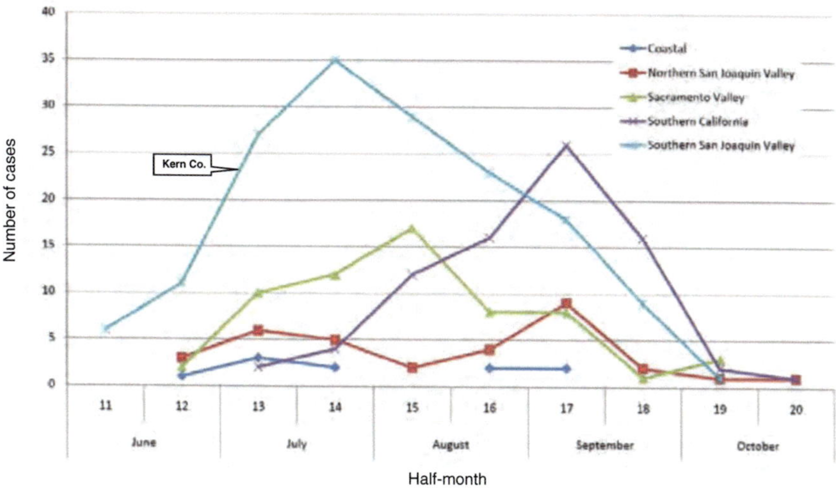

FIGURE 3-7 California mosquito district risk levels 1-5 for WNV transmission: estimated from (A) temperatures downloaded from the TOPS system (inset shows the increasing duration of extrinsic incubation period with decreasing temperature and associated risk), (B) the entire risk assessment model calculated bimonthly for selected mosquito control agencies, and (C) the number of human cases within each mosquito control region of California.

mosquito control and recommendations for human personal protection provide the only methods to interrupt transmission and protect the public.

Use of forecasts and nowcasts[13] for making control decisions have inherent problems, especially in California and other areas where the public (especially anti-insecticide advocates) frequently believes that the risk of applying insecticides, even in ultralow volumes (ULVs), exceeds the risk of illness or death from viral infection. Using forecasts and surveillance measures, it is possible to accurately determine that the risk of an epidemic is imminent and to apply large-scale aerial ULV adulticide treatments to immediately reduce vector abundance and transmission. In this scenario, cases are prevented, but the rationale for a large monetary expenditure and exposure of the human population appears unjustified to some because we cannot know whether the prevented cases would have occurred in the absence of vector control measures. The actions of vector control and public health officials are then questioned in the press, despite the fact that risks from pyrethrin insecticides and synergists are minimal (Peterson et al., 2005). If concurrent measures of risk are used instead of forecasts, the virus may have amplified to higher levels, some humans will have been infected before

[13]Forecasts or real-time measures of events in the immediate future.

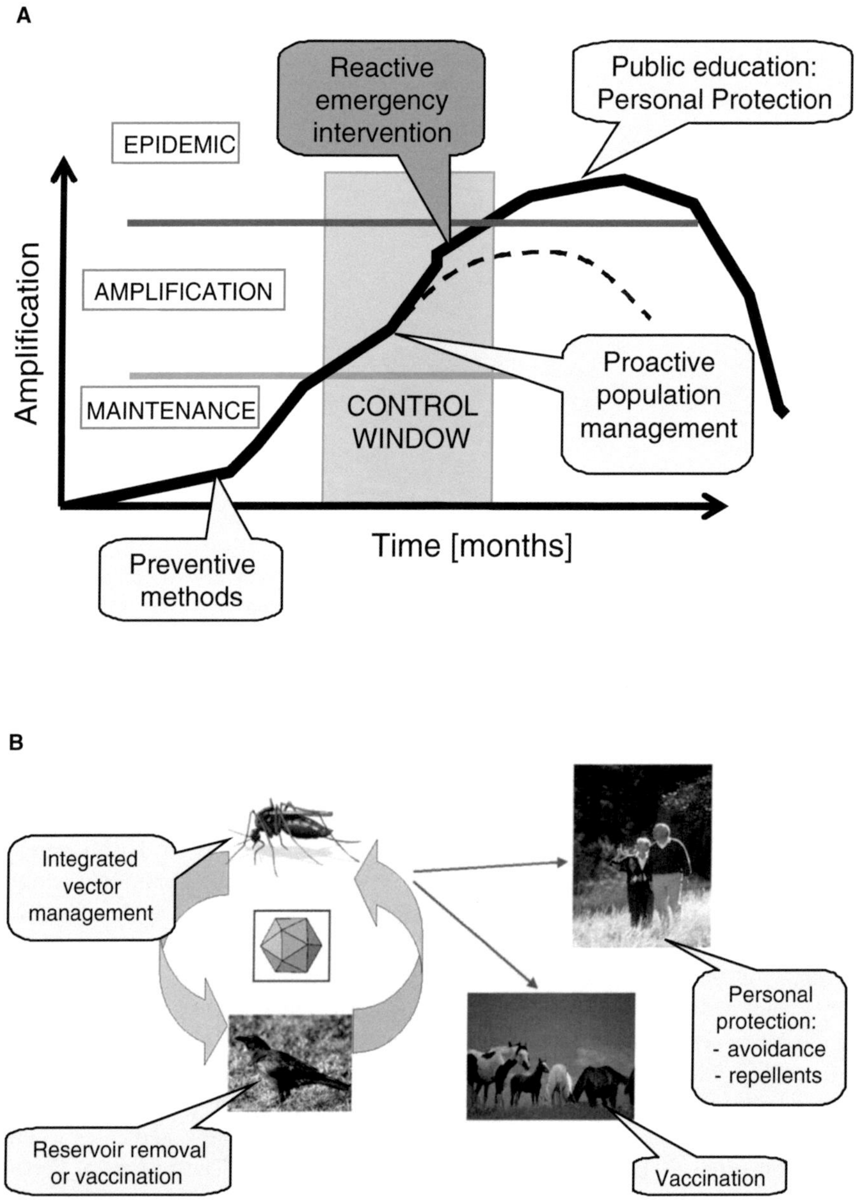

FIGURE 3-8 Intervention options for WNV shown in relation to (A) the amplification curve and (B) the transmission cycle.
SOURCE: Figure 3-8B is modified from CDC (2005).

adulticides are applied, and the public health consequences can be considerable. In this scenario, cases of severe illness and some deaths occur, the epidemic peak frequently is realized and recognized by the public, and the population is exposed to the same level of adulticides as during preventive sprays. In the latter case, both the antispray advocates and the families of affected cases are perplexed and the cost-benefit ratio is increased.

Surveillance in a Changing World

West Nile virus will not be the last invading zoonosis to reach North America. Rapid globalization of commerce and travel increases the probability of additional vector or pathogen introductions. For example, *Aedes albopictus* mosquitoes, which are vectors of dengue, chikungunya, and other viruses, were transported to California and distributed throughout Los Angeles with imported "Lucky Bamboo" (*Dracaena spp.*) from China (Linthicum et al., 2003; Madon et al., 2002). Fortunately, surveillance detected this introduction and the invading mosquitoes were eradicated. Vector-borne pathogens also frequently enter the United States. Travelers from India with chikungunya viremia levels sufficient to infect mosquitoes have been detected in the United States during the current epidemic (Lanciotti et al., 2007), but the risk for local transmission is unknown because the vector competence of the local mosquito populations is unknown. Climate change exacerbates these situations in time and space by elongating transmission seasons and increasing the geographic area receptive to pathogen introduction (Epstein, 1998; Patz et al., 2000). The WNV epidemic has provided a wake-up call for public health agencies and clearly delineated the inability of current control programs to contain an invading, mosquito-borne zoonosis (Holloway, 2000). Unfortunately, the waning WNV epidemic has resulted in the diversion of both research and surveillance funding to other health problems, and many of those programs still in place focus surveillance diagnostics only on WNV and will not detect the next invading pathogen.

Acknowledgments

This research was funded by the Climate Variability and Human Health and the California Applications Programs, Office of Global Programs, NOAA; Decision Support through Earth Science Results, NASA; Research Grant AI55607 from the National Institute of Allergy and Infectious Diseases, NIH. We are especially indebted to the corporate members of the Mosquito and Vector Control Association of California and the California Department of Public Health who granted permission for us to utilize their surveillance data for this project. Forrest Melton, NASA, and Bborie Park, UC Davis, assisted with data manipulations.

REFERENCES

Overview References

Leroy, E. M., B. Kumulungui, X. Pourrut, P. Rouquet, A. Hassanin, P. Yaba, A. Delicat, J. T. Paweska, J. P. Gonzalez, and R. Swanepoel. 2005. Fruit bats as reservoirs of Ebola virus. *Nature* 438(7068):575-576.

Linthicum, K. J., A. Anyamba, C. J. Tucker, P. W. Kelley, M. F. Myers, and C. J. Peters. 1999. Climate and satellite indicators to forecast Rift Valley fever epidemics in Kenya. *Science* 285(5426):397-400.

Pinzon, J. E., J. M. Wilson, C. J. Tucker, R. Arthur, P. B. Jahrling, and P. Formenty. 2004. Trigger events: enviroclimatic coupling of Ebola hemorrhagic fever outbreaks. *American Journal of Tropical Medicine and Hygiene* 71(5):664-674.

Reisen, W. K., and C. M. Barker. 2008. Use of climate variation in vector-borne disease decision support systems. In Institute of Medicine. *Global climate change and extreme weather events: understanding the contributions to the emergence, reemergence, and spread of infectious disease*. Washington, DC: The National Academies Press.

Reisen, W. K., D. Cayan, M. Tyree, C. M. Barker, B. F. Eldridge, and M. Dettinger. In press. Impact of climate variation on mosquito abundance in California. *Journal of the Society for Vector Ecology*.

Tucker, C. J., A. Anyamba, K. Linthicum, and J. Small. In press. *The 2000 Rift Valley hemorrhagic fever outbreak in southwestern Arabia: bio-terrorism or natural occurrence?* Paper submitted for publication.

Acuña-Soto et al. References

Acuña, R., ed. 1982. *Relaciones Geográficas*. Mexico City: Instituto de Investigaciones Antropológicas, Universidad Nacional Autónoma de México.

Acuña-Soto, R., L. Calderón-Romero, and J. H. Maguire. 2000. Large epidemics of hemorrhagic fevers in Mexico, 1545-1815. *American Journal of Tropical Medicine and Hygiene* 62(6):733-739.

Acuña-Soto, R., D. W. Stahle, M. K. Cleaveland, and M. D. Therrell. 2002. Megadrought and megadeath in 16th century Mexico. *Emerging Infectious Diseases* 8(4):360-362.

Acuña-Soto, R., D. W. Stahle, M. Therrell, D. Griffin, and M. K. Cleaveland. 2004. When half of the population died: the epidemic of hemorrhagic fevers of 1576 in Mexico. *FEMS Microbiology Letters* 240(1):1-5.

Acuña-Soto, R., D. W. Stahle, M. D. Therrell, S. Gomez Chavez, and M. Cleaveland. 2005. Drought, epidemic disease, and the fall of classic period cultures in Mesoamerica (AD 750-950). Hemorrhagic fevers as a cause of massive population loss. *Medical Hypotheses* 65(2):405-409.

Anonymous. 1978. Tira de Tepechpan. In *Biblioteca Enciclopédica del Estado de México*, edited and commentary by X. Noguez. Mexico City: Estado de México.

Anonymous. 1980a. *Anales de Tlaltelolco, unos anales históricos de la Nación Mexicana*, edited by Rafael Porrúa. Mexico City.

Anonymous. 1980b. *Códice Aubin (Códice de 1576)*, edited by A. Peñafiel (1902). Mexico City: Editorial Innovación.

Anonymous. 1981. *Anales de Tecamachalco, crónica local y colonial en idioma Nahuatl 1398-1590*. Edited by A. Peñafiel. Mexico City: Editorial Innovación.

Anonymous. 1995. Anales de Tula, Hidalgo. In *Fuentes y estudios sobre el México indígena*, Vol. 6. Mexico City: Instituto Nacional de Antropología e Historia, Universidad de las Américas.

Anonymous. 1998. *Histoire mexicaine depuis 1221 jusqu'en 1594: Manuscrito num. 40 del Fondo de Manuscritos Mexicanos, Biblioteca Nacional de Francia (Coleccion cientifica)*. Translation by X. M. González. Mexico City: Instituto Nacional de Antropología e Historia.

Camargo, D. M. 1999. *Relaciones geográficas de Tlaxcala*. Tlaxcala/San Luis Potosí, Mexico: Gobierno del Estado de Tlaxcala/El Colegio de San Luis.

Chimalpahin, D. 1998. *Las Ocho Relaciones y el Memorial de Colhuacan*, 2 vols., edited and translated by R. Tena. Mexico: Consejo Nacional para la Cultura y las Artes.

Chimalpahin, D. 2001. *Diario*. Mexico: Consejo Nacional para la Cultura y las Artes.

Clavijero, F. J. 1945. *Historia antigua de Mexico*, Vol. 4. Mexico City: Editorial Porrúa.

Cook, S. F., and F. Sherburne. 1985. El Hospital del Hambre de Guadalajara: un experimento de asistencia médica. In *Ensayos sobre la Historia de las Epidemias en México*, edited by C. E. Flores and E. Malvido. Mexico City: Instituto Mexicano del Seguro Social.

Cook, S. F., and L. B. Simpson. 1948. The population of central Mexico in the sixteenth century. *Ibero Americana* (31):43.

Cooper, D. B. 1965. *Epidemic disease in Mexico City, 1761-1813: an administrative, social, and medical study*. Austin, TX: University of Texas Press.

de Alva Ixtlixóchitl, F. 1975. *Obras Históricas*, 2 vols. Mexico City: Instituto de Investigaciones Históricas, Universidad Nacional Autónoma de México.

de Cárdenas, J. 1945. Problemas y secretos maravillosos de las Indias. In *Colección Incunables Americanos*, Vol. 9. Madrid, Spain: Ediciones de Cultura Hispanica.

de Grijalva, J. 1924. De las señales que en esta tierra se vieron por este tiempo, y de la gran peste que llamaron cocoliztli, Crónica de la Orden de San Agustín en las provincias de la Nueva España. In *Cuatro edades desde el año de 1533 hasta el de 1592*. Mexico City: Imprenta Victoria.

de Lorenzana, F. A. 1992. *Historia de la Nueva España*. Mexico City: Universidad de Castilla-La Mancha and Miguel Angel Porrúa.

de Obregón, B. 1988. *Historia de los descubrimientos antiguos y modernos de la Nueva España escrita por el Conquistador en el año de 1584*, edited by M. Cuevas. Mexico City: Editorial Porrúa.

de Ortega Montañez, J. 1955. *Instrucción reservada al conde de Moctezuma*. Mexico City: Editorial Jus.

de Torquemada, J. 1969. *Monarquia Indiana*, Vol. 5. Mexico City: Editorial Porrúa.

Farfan, A. 1592. Third book, Of fever and corrupt blood, Treatise of Medicine. *Libro* tercero. De la Calentura de Sangre corrompida y su Cura. *Tractado Breve de Medicina y de Todas las Enfermedades que a Cada Paso se Ofrecen. Hecho por el Padre fray Agustin Farfan, Doctor en Medicina y Religioso de la Orden de San Agustín en esta Nueva España. En la casa de Pedro Ocharte, Mexico City.*

Flores, C. E., and E. Malvido, eds. 1985. *Ensayos Sobre la Historia de las Epidemias en México*, 2 vols. Mexico City: Instituto Mexicano del Seguro Social.

Flores, T. A. 1888. Historia de la medicina en México. Mexico City: Oficina Tipográfica de la Oficina de Fomento.

Garcia, A. V., C. J. M. Pérez, and V. A. Molina. 2003. *Desastres agrícolas en México*. Catálogo histórico. Centro de Investigación y Estudios Superiores en Antropología Social. Fondo de Cultura Económica.

Gerhard, P. 1986. *Geografía histórica de la Nueva España 1519-1821*. Mexico City: Universidad Nacional Autónoma de México.

Gerhard, P. 1996. *La frontera norte de la Nueva España*. Mexico City: Universidad Nacional Autónoma de México.

Gibson, C. 1964. *The Aztecs under Spanish rule. A history of the Indians of the Valley of Mexico*. Stanford, CA: Stanford University Press.

Hodell, D. A., J. H. Curtis, and M. Brenner. 1995. Possible role of climate in the collapse of classic Maya civilization. *Nature* 375(6030):391-394.

Hugh, T. 1993. *Conquest, Montezuma, Cortés and the fall of old Mexico*. New York: Simon and Schuster.

Keber, E. Q. 1995. *Codex telleriano remensis*. Austin: University of Texas Press.

León, N. 1982. What were the Matlazahuatl and Cocoliztli in pre-Columbian and in Colonial Period. In *Ensayos sobre la historia de las epidemias en México*. Edited by C. E. Flores and E. Malvido. Mexico City: Instituto Mexicano del Seguro Social.

Lopez de Hinojoso, A. 1578. *Tratado VII. De Pestilencia. Summa y Recopilación de Cirugia con un Arte para Sangrar muy Util y Provechoso*. Mexico City: Imprenta de Antonio Ricardo.

Lopez de Hinojoso, A. 1595. *Libro Octavo. De tabardete y Cocoliztle. Summary Recopilación de Cirugia con un Arte para Sangrar Muy Util y Provechoso y Examen de Barberos*. Mexico City: Imprenta de Pedro Balli.

Marr, J. S., and J. B. Kiracofe. 2000. Was the huey cocoliztli a haemorrhagic fever? *Medical History* 44(3):341-362.

Mendieta, G. 1997. *Historia Eclesiástica Indiana*, 2 vols. Mexico City: Consejo Nacional para la Cultura y las Artes.

Metcalfe, S. E., and P. Hales. 1994. *Holocene diatoms from a Mexican crater*. Memoirs of the California Academy of Sciences. Pp. 501-515.

Millon, R. 1970. Teotihuacan: completion of map of giant ancient city in the Valley of Mexico. *Science* 270(3962):1070-1082.

Ocaranza, F. 1933. Las grandes epidemias del siglo XVI en la Nueva España. *Medicina* (Mex) 13:176-178.

Sahagún, B. 1997. *Primeros Memoriales: Paleography of Nahuatl text and English translation*. Translated by T. D. Sulllivan. Norman: University of Oklahoma Press.

Somolinos, A. G. 1956. Hallazgo del manuscrito sobre el cocoliztli (original del Dr. Francisco Hernández). *La Prensa médica mexicana* 21:115-123.

Stahle, D. W., E. R. Cook, M. K. Cleaveland, M. D. Therrell, D. M. Meko, H. D. Grissino-Mayer, E. Watson, and B. H. Luckman. 2000. Tree-ring data document 16th century megadrought over North America. *Eos, Transactions American Geophysical Union* 81(12):121, 125.

Tezozómoc, F. A. 1975. *Crónica Mexicayotl*. Mexico City: Instituto de Investigaciones Históricas, Universidad Nacional Autónoma de México.

Therrell, M. D., D. W. Stahle, and R. Acuña-Soto. 2004. Aztec drought and the curse of one rabbit. *Bulletin of the American Metereological Association* 85:1263-1272.

Yu-ping, C. N., and L. Heligman. 1994. Growth of the world's megalopolises. In *Mega-city growth and the future*, edited by R. J. Fuchs, E. Brennan, J. Chamie, L. Fu-Chen, and J. I. Uitto. Tokyo: United Nations University Press.

Beldomenico et al. References

Alford, R. A., K. S. Bradfield, and S. J. Richards. 2007. Ecology: global warming and amphibian losses. *Nature* 447(7144):E3-E4.

Barros, V., M. E. Castañeda, and M. Doyle. 2000. Recent precipitation trends in Southern America to the East of the Andes: an indication of a mode of climatic variability. In *Southern hemisphere paleo and neoclimates: key sites, methods, data and models*, edited by P. P. Smolka and W. Volkheimer. Berlin: Springer-Verlag.

Beldomenico, P. M., S. Telfer, S. Gebert, L. Lukomski, M. Bennett, and M. Begon. 2008. Poor condition and infection: a vicious circle in natural populations. *Proceedings of the Royal Society B: Biological Sciences*, epub ahead of print.

Beldomenico, P. M., S. Telfer, S. Gebert, L. Lukomski, M. Bennett, and M. Begon. In press. The dynamics of health of wild field vole (*Microtus agrestis*) populations: a haematological perspective. *Journal of Animal Ecology*.

Berger, L., R. Speare, P. Daszak, D. E. Green, A. A. Cunningham, C. L. Goggin, R. Slocombe, M. A. Ragan, A. D. Hyatt, K. R. McDonald, H. B. Hines, K. R. Lips, G. Marantelli, and H. Parkes. 1998. Chytridiomycosis causes amphibian mortality associated with population declines in the rain forests of Australia and Central America. *Proceedings National Academies of Science* 95(15):9031-9036.

Bollinger, T. K., J. Mao, D. Schock, R. M. Brigham, V. G. Chinchar. 1999. Pathology, isolation and preliminary molecular characterization of a novel iridovirus from tiger salamanders in Saskatchewan. *Journal of Wildlife Diseases* 35:413-429.

Carignan, V., and M.-A. Villard. 2002. Selecting indicator species to monitor ecological integrity: a review. *Environmental Monitoring and Assessment* 78(1):45-61.

Coulson, T., E. A. Catchpole, S. D. Albon, B. J. Morgan, J. M. Pemberton, T. H. Clutton-Brock, M. J. Crawley, and B. T. Grenfell. 2001. Age, sex, density, winter weather, and population crashes in Soay sheep. *Science* 292(5521):1528-1531.

Di Rosa, I., F. Simoncelli, A. Fagotti, and R. Pascolini. 2007. Ecology: the proximate cause of frog declines? *Nature* 447(7144):E4-E5.

García, I., R. Albornoz, and E. La Marca. 2007. Perturbaciones climáticas y disminución de *Atelopus oxyrhynchus* (Amphibia: Anura) en los Andes de Venezuela. *Herpetotropicos* 2:64-72.

Grenfell, B. T., K. Wilson, V. S. Isham, H. E. Boyd, and K. Dietz. 1995. Modelling patterns of parasite aggregation in natural populations: trichostrongylid nematode-ruminant interactions as a case study. *Parasitology* 111(Suppl):S135-S151.

Gulland, F. M., S. D. Albon, J. M. Pemberton, P. R. Moorcroft, and T. H. Clutton-Brock. 1993. Parasite-associated polymorphism in a cyclic ungulate population. *Proceedings Biological Sciences* 254(1339):7-13.

Harvell, C. D., C. E. Mitchell, J. R. Ward, S. Altizer, A. P. Dobson, R. S. Ostfeld, and M. D. Samuel. 2002. Climate warming and disease risks for terrestrial and marine biota. *Science* 296(5576):2158-2162.

Hudson, P. J., A. P. Dobson, and D. Newborn. 1998. Prevention of population cycles by parasite removal. *Science* 282(5397):2256-2258.

IPCC (Intergovernmental Panel on Climate Change). 2001. *Climate change 2001: the scientific basis.* Contribution of Working Group I to the third assessment report of the Intergovernmental Panel on Climate Change. Cambridge, UK: Cambridge University Press.

Jones, K. E., N. G. Patel, M. A. Levy, A. Storeygard, D. Balk, J. L. Gittleman, and P. Daszak. 2008. Global trends in emerging infectious diseases. *Nature* 451(7181):990-993.

Juhl, A. 2005. Growth rates and elemental composition of *Alexandrium monilatum*, a red-tide dinoflagellate. *Harmful Algae* 4(2):287-295.

Karesh, W. B., and R. A. Cook. 2005. The human-animal link. *Foreign Affairs* 84(4):38-50.

Karesh, W. B., and P. Reed. 2005. Ebola and great apes in central Africa: current status and future needs. *Bulletin de la Société de Pathologie Exotique* 98(3):237-238.

Kiesecker, J. M., A. R. Blaustein, and L. K. Belden. 2001. Complex causes of amphibian population declines. *Nature* 410(6829):681-684.

Lochmiller, R. L., and C. Dabbert. 1993. Immunocompetence, environmental stress, and the regulation of animal populations. *Trends Comparative Biochemical Physiology* 1:823-855.

Ogden, N. H., M. Bigras-Poulin, C. J. O'Callaghan, I. K. Barker, L. R. Lindsay, A. Maarouf, K. E. Smoyer-Tomic, D. Waltner-Toews, and D. Charron. 2005. A dynamic population model to investigate effects of climate on geographic range and seasonality of the tick *Ixodes scapularis*. *International Journal of Parasitology* 35(4):375-389.

Ogden, N. H., A. Maarouf, I. K. Barker, M. Bigras-Poulin, L. R. Lindsay, M. G. Morshed, C. J. O'Callaghan, F. Ramay, D. Waltner-Toews, and D. F. Charron. 2006. Climate change and the potential for range expansion of the Lyme disease vector *Ixodes scapularis* in Canada. *International Journal of Parasitology* 36(1):63-70.

Parmesan, C. 2006. Ecological and evolutionary responses to recent climate change. *Annual Review of Ecology, Evolution, and Systematics* 37:637-639.

Pounds, J. A., M. R. Bustamante, L. A. Coloma, J. A. Consuegra, M. P. Fogden, P. N. Foster, E. La Marca, K. L. Masters, A. Merino-Viteri, R. Puschendorf, S. R. Ron, G. A. Sanchez-Azofeifa, C. J. Still, and B. E. Young. 2006. Widespread amphibian extinctions from epidemic disease driven by global warming. *Nature* 439(7073):161-167.

Rausch, R. L., J. C. George, and H. K. Brower. 2007. Effect of climatic warming on the Pacific walrus, and potential modification of its helminth fauna. *Journal of Parasitology* 93(5):1247-1251.

Rawlins, S. C., B. Hull, D. D. Chadee, R. Martinez, A. LeMaitre, F. James, and L. Webb. 1990. Sylvatic yellow fever activity in Trinidad, 1988-1989. *Transactions of the Royal Society of Tropical Medicine and Hygiene* 84(1):142-143.

Reading, C. J. 2007. Linking global warming to amphibian declines through its effects on female body condition and survivorship. *Oecologia* 151(1):125-131.

Roelke-Parker, M. E., L. Munson, C. Parker, R. Cock, S. Cleaveland, M. Carpenter, S. J. O'Brien, A. Pospochil, R. Hofmann-Lehmann, H. Lutz, G. L. M. Mwamengele, M. N. Mgasa, G. A. Machange, B. A. Summers, and M. J. G. Appel. 1996. A canine distemper virus epidemic in Serengeti lions (*Panthera leo*). *Nature* 379(6564):441-445.

Seimon, T. A., A. Seimon, P. Daszak, S. R. P. Halloy, L. M. Schloegel, C. A. Aguilar, P. Sowell, A. D. Hyatt, B. Konecky, and J. E. Simmons. 2007. Upward range extension of Andean anurans and chytridiomycosis to extreme elevations in response to tropical deglaciation. *Global Change Biology* 13:288-299.

Stuart, S. N., J. S. Chanson, N. A. Cox, B. E. Young, A. S. Rodrigues, D. L. Fischman, and R. W. Waller. 2004. Status and trends of amphibian declines and extinctions worldwide. *Science* 306(5702):1783-1786.

Tompkins, D. M., and M. Begon. 1999. Parasites can regulate wildlife populations. *Parasitology Today* 15(8):311-313.

Uhart, M., W. Karesh, R. Cook, N. Huin, K. Lawrence, L. Guzman, H. Pacheco, G. Pizarro, R. Mattsson, and T. Mörner. 2004. *Paralytic shellfish poisoning in gentoo penguins* (Pygoscelis papua) *from the Falkland* (Malvinas) *Islands*. Proceedings AAZV/AAWV/WDA Joint Conference, San Diego, California, August 28-September 3, 2004. Pp. 481-486.

Weldon, C., L. H. du Preez, A. D. Hyatt, R. Muller, and R. Spears. 2004. Origin of the amphibian chytrid fungus. *Emerging Infectious Diseases* 10(12):2100-2105.

Wharton, D. A. 1982. The survival of desiccation by the free-living stages of *Trichostrongylus colubriformis* (Nematoda: *Trichostrongylidae*). *Parasitology* 84(Pt 3):455-462.

Reisen and Barker References

Anyamba, A., K. J. Linthicum, R. Mahoney, C. J. Tucker, and P. W. Kelley. 2002. Mapping potential risk of Rift Valley fever outbreaks in African savannas using vegetative index time series data. *Photogrammetric Engineering and Remote Sensing* 68:137-145.

California Department of Public Health. 2007. *California mosquito-borne virus surveillance and response plan*, http://westnile.ca.gov/downloads.php?download_id=324&filename=2007%20CA%20Mosq%20Surv%20and%20Resp%20Plan%20JULY.pdf (accessed February 25, 2008).

CDC (Centers for Disease Control and Prevention). 2005. *West Nile virus transmission cycle*, http://www.cdc.gov/ncidod/dvbid/westnile/cycle.htm (accessed February 25, 2008).

CDC. 2007. *Human incidence maps*, http://www.cdc.gov/ncidod/dvbid/westnile/surv&control.htm (accessed February 6, 2008).

Epstein, P. R. 1998. Global warming and vector-borne disease. *Lancet* 351(9117):1737.

Githeko, A. K., S. W. Lindsay, U. E. Confalonieri, and J. A. Patz. 2000. Climate change and vector-borne diseases: a regional analysis. *Bulletin of the World Health Organization* 78(9):1136-1147.

Holloway, M. 2000. Outbreak not contained. West Nile virus triggers a reevaluation of public health surveillance. *Scientific American* 282(4):20-22.

Kalnay, E., and M. Cai. 2003. Impact of urbanization and land-use change on climate. *Nature* 423(6939):528-531.

Lanciotti, R. S., O. L. Kosoy, J. J. Laven, A. J. Panella, J. O. Velez, A. J. Lambert, and G. L. Campbell. 2007. Chikungunya virus in U.S. travelers returning from India, 2006. *Emerging Infectious Diseases* 13(5):764-767.

Linthicum, K. J., C. L. Bailey, C. J. Tucker, D. R. Angleberger, T. Cannon, T. M. Logan, P. H. Gibbs, and J. Nickeson. 1991. Towards real-time prediction of Rift Valley fever epidemics in Africa. *Preventive Veterinary Medicine* 11:325-334.

Linthicum, K. J., V. L. Kramer, M. B. Madon, K. Fujioka, and the Surveillance-Control Team. 2003. Introduction and potential establishment of *Aedes albopictus* in California in 2001. *Journal of the American Mosquito Control Association* 19(4):301-308.

Madon, M. B., M. S. Mulla, M. W. Shaw, S. Kluh, and J. E. Hazelrigg. 2002. Introduction of *Aedes albopictus* (Skuse) in southern California and potential for its establishment. *Journal of Vector Ecology* 27(1):149-154.

Mulhern, T. D. 1985. The New Jersey mechanical trap for mosquito surveys. 1942. *Journal of the American Mosquito Control Association* 1(4):411-418.

NASA (National Aeronautics and Space Administration). 2007. *Terrestrial Observation and Prediction System*, http://ecocast.arc.nasa.gov/content/view/76/133/ (accessed February 6, 2008).

Nemani, R. R., C. D. Keeling, H. Hashimoto, W. M. Jolly, S. C. Piper, C. J. Tucker, R. B. Myneni, and S. W. Running. 2003. Climate-driven increases in global terrestrial net primary production from 1982 to 1999. *Science* 300(5625):1560-1563.

Newhouse, V. F., R. W. Chamberlain, J. G. Johnston, Jr., and W. D. Sudia. 1966. Use of dry ice to increase mosquito catches of the CDC miniature light trap. *Mosquito News* 26:30-35.

Nielsen, C. F., V. Armijos, S. S. Wheeler, T. E. Carpenter, K. Kelley, W. M. Boyce, D. A. Brown, and W. K. Reisen. 2008. Risk factors associated with the 2006 West Nile virus outbreak in Davis, a residential community in Northern California. *American Journal of Tropical Medicine and Hygiene* 78(1):53-62.

NOAA (National Oceanic and Atmospheric Administration). 2008. *U.S. climate at a glance*, http://www.ncdc.noaa.gov/oa/climate/research/cag3/cag3.html (accessed February 25, 2008).

Patz, J. A., D. Engelberg, and J. Last. 2000. The effects of changing weather on public health. *Annual Review of Public Health* 21:271-307.

Peterson, R. K. D., P. A. Macedo, and R. S. Davis. 2005. A human-health risk assessment for West Nile virus and insecticides used in mosquito management. *Environmental Health Perspectives* 114(3):366-372.

Reeves, W. C., J. L. Hardy, W. K. Reisen, and M. M. Milby. 1994. Potential effect of global warming on mosquito-borne arboviruses. *Journal of Medical Entomology* 31(3):323-332.

Reisen, W. K. 1995. Effect of temperature on *Culex tarsalis* (Diptera: *Culicidae*) from the Coachella and San Joaquin Valleys of California. *Journal of Medical Entomology* 32(5):636-645.

Reisen, W. K., R. P. Meyer, S. B. Presser, and J. L. Hardy. 1993. Effect of temperature on the transmission of western equine encephalomyelitis and St. Louis encephalitis viruses by *Culex tarsalis* (Diptera: *Culicidae*). *Journal of Medical Entomology* 30(1):151-160.

Reisen, W. K., Y. Fang, and V. M. Martinez. 2005. Avian host and mosquito (Diptera: *Culicidae*) vector competence determine the efficiency of West Nile and St. Louis encephalitis virus transmission. *Journal of Medical Entomology* 42(3):367-375.

Reisen, W. K., C. M. Barker, R. Carney, H. D. Lothrop, S. S. Wheeler, J. L. Wilson, M. B. Madon, R. Takahashi, B. Carroll, S. Garcia, Y. Fang, M. Shafii, N. Kahl, S. Ashtari, V. Kramer, C. Glaser, and C. Jean. 2006a. Role of corvids in epidemiology of West Nile virus in southern California. *Journal of Medical Entomology* 43(2):356-367.

Reisen, W. K., Y. Fang, and V. M. Martinez. 2006b. Effects of temperature on the transmission of West Nile virus by *Culex tarsalis* (Diptera: *Culicidae*). *Journal of Medical Entomology* 43(2):309-317.

Reisen, W. K., D. Cayan, M. Tyree, C. M. Barker, B. F. Eldridge, and M. Dettinger. 2008a. Impact of climate variation on mosquito abundance in California. *Journal of Vector Ecology* 33(2):89-98.

Reisen, W. K., H. D. Lothrop, S. S. Wheeler, M. Kensington, A. Gutierrez, Y. Fang, S. Garcia, and B. Lothrop. 2008b. Persistent West Nile virus transmission and the displacement St. Louis encephalitis virus in southeastern California, 2003-2006. *Journal of Medical Entomology* 45(3):494-508.

Reiter, P. 1987. A revised version of the CDC gravid mosquito trap. *Journal of the American Mosquito Control Association* 3(3):325-327.

Shaman, J., M. Stieglitz, C. Stark, S. Le Blancq, and M. Cane. 2002. Using a dynamic hydrology model to predict mosquito abundances in flood and swamp water. *Emerging Infectious Diseases* 8(1):6-13.

Shaman, J., J. F. Day, M. Stieglitz, S. Zebiak, and M. Cane. 2004. Seasonal forecast of St. Louis encephalitis virus transmission, Florida. *Emerging Infectious Diseases* 10(5):802-809.

Shi, P. Y., E. B. Kauffman, P. Ren, A. Felton, J. H. Tai, A. P. DuPuis, S. A. Jones, K. A. Ngo, D. C. Nicholas, J. Maffei, G. D. Ebel, K. A. Bernard, and L. D. Kramer. 2001. High-throughput detection of West Nile virus RNA. *Journal of Clinical Microbiology* 39(4):1264-1271.

Su, T., J. P. Webb, R. P. Meyer, and M. S. Mulla. 2003. Spatial and temporal distribution of mosquitoes in underground storm drain systems in Orange County, California. *Journal of Vector Ecology* 28(1):79-89.

Tucker, C. J. 1979. Red and photographic infrared linear combinations for monitoring vegetation. *Remote Sensing of Environment* 8(2):127-150.

4

Policy Implications of the Health Effects of Climate Change and Extreme Weather Events

OVERVIEW

The urgent question of how to act on what is known—or even suspected—about the potential health consequences of climate change, which underscored discussion throughout the workshop, is taken up explicitly in this chapter. The first paper, coauthored by speaker Douglas MacPherson of McMaster University and Migration Health Consultants, Incorporated, focuses on the complex, two-way association between climate change and human mobility and its role in infectious disease emergence. The authors discuss the burgeoning influence of human mobility and migration on infectious disease emergence and describe a variety of ways in which interactions between climate and human behavior shape infectious disease dynamics, from the cataclysmic repercussions of extreme weather events to the slowly evolving impacts of increasing temperatures, sea level rise, and decreasing freshwater availability. If, as expected, climate change drives the simultaneous emergence of multiple infectious diseases along many possible pathways, the authors observe, preventing the global spread of individual pathogens is unlikely to be feasible. Therefore, they argue, "mitigation efforts and the policies that guide them will need to become more process-related and anticipatory rather than outcome-based and reactive."

The development of such policies is fraught with challenges, as described in the chapter's second paper by Diarmid Campbell-Lendrum of the World Health Organization (WHO). He notes that some of the obstacles to creating international public health policies to address the implications of climate change are intrinsic to the health sector and its long-standing focus on single-disease threats, rather than "systemic and long-term stresses" that produce a broad range of health effects.

Within this spectrum of potential impacts, Campbell-Lendrum emphasizes "routine" health threats such as diarrhea, malnutrition, and malaria that are known to be climate sensitive, and the value of basic public health interventions (e.g., providing clean water and sanitation services, improving hygiene) as a means to reduce overall disease rates and to moderate the potential effects of climate change. This is the position taken by WHO, which has evolved from assessing the health risks associated with climate change to an increasingly operational role in addressing these risks; Campbell-Lendrum discusses the organization's primary objectives in this area, which include raising public and political awareness of climate change, promoting health through climate change mitigation, and strengthening health systems to manage the additional burden imposed by climate change.

Due in large part to its predicted far-reaching effects on health, climate change is viewed as a potentially powerful agent of geopolitical upheaval. At its summit in March 2008, a paper presented to the European Union included a grim catalog of threats to international security posed by climate change: conflicts over water, energy, and other increasingly scarce resources; loss of infrastructure and territory; border disputes; environmentally-induced migration; and political tension at all levels of governance (European Commission, 2008). These themes were taken up by Major General Richard Engel (U.S. Air Force, retired), National Intelligence Council (NIC) deputy national intelligence officer for science and technology, in his workshop presentation, which described efforts under way to conduct a National Intelligence Assessment (NIA) concerning the challenges posed to national security by global climate change over the next two decades. Although NIAs inform decision making at the highest levels of the U.S. government, Engel noted that the NIC intends to prepare this assessment as an unclassified document—a goal that he characterized as "increasingly challenging" to meet. The classified NIA entitled *National Security Implications of Global Climate Change Through 2030* was delivered to Congressional requestors in mid-June 2008.

The NIC has chosen to evaluate the potential impacts of climate change on the four classical elements that comprise national power: geopolitical power, military power, economic power, and social cohesion. To date, the NIC has received considerable nongovernmental expert opinion on this issue from a variety of sources including the Joint Global Change Research Institute, a partnership between the University of Maryland and Pacific Northwest National Laboratories; the U.S. Climate Change Research Program, an interagency group within the U.S. government; the Center for Naval Analysis; the Center for International Earth Science Information Network at Columbia University; Arizona State University; the RAND Corporation; the Global Business Network; and the Center for Strategic and International Studies. Engel emphasized that he had crafted his remarks from these "outside views" as they have been received—but not fully evaluated—by the NIC. The NIC and the intelligence community have yet to complete their own analysis and interpretation of these contributions.

Inherent uncertainties in predicting the course of climate change prompted the NIC to consider a "system vulnerability approach" for this assessment, which identifies existing internal vulnerability of states or regions of interest to U.S. security, then examines how the added stress of climate change could affect these states or regions. This process, with input from outside experts, has identified Africa as particularly vulnerable to the effects of climate change, Engel noted. Most immediately, the region is ill-equipped to deal with water shortages due to persistent poverty and to the number of agriculture-dependent states with weak governments. Experts advising the NIC have predicted that climate change will produce water issues and damaging weather extremes, such as the flooding of major coastal cities, in North America and Asia; stress on marine ecosystems in Asia and Australia; flooding and heat waves in Europe; both flooding and drought in South America; intensification of drought in Australia; and increasingly extreme environmental changes at the world's poles.

In addition to these climate change scenarios, which Engel characterized as "middle of the road," the NIC is considering the implications of unlikely but disastrous "threshold" events, such as the collapse of major ice sheets in Antarctica and Greenland or the stopping, slowing, or reversal of patterns of circulation in the North Atlantic current of the Gulf Stream, which provides warmth to Europe. There may also be yet-unidentified feedbacks within the climate system that could amplify the changes caused by greenhouse gasses alone. On the other hand, Engel noted, experts have identified a few sources of potential overestimates of the effects of climate change, including assumptions that interventions will not be introduced to mitigate the process or its consequences. He concluded that while the best estimates of future warming represent an average, the potential for greater-than-expected warming has not been well defined.

Climate change has the potential to create geopolitical divisions, several of which have already been reported in the global press. "Developed countries want the developing countries to participate so they don't bear the full burden [of the cost of addressing climate change], and the developing countries want the developed countries to pay for it," Engel observed. Experts have reported to the NIC that a north-south division even exists within Europe, resulting from the varied effects of climate change along this axis. The differential effects of climate change in regions of Asia—where some areas may suffer droughts, while others flood—may also prove a source of tension, particularly where water is concerned. Such situations, although removed from the United States, have the potential to compromise national security due to their influence on economic partners or security allies or by creating chaos that requires a U.S. response, thereby consuming national resources. The NIC's expert consultants agree that actions taken by the United States will profoundly influence the fate of a global consensus on climate change and, thereby, the nature and location of geopolitical fault lines in the coming decades.

INFLUENCES OF MIGRATION AND POPULATION MOBILITY

Douglas W. MacPherson, M.D., M.Sc. (CTM), F.R.C.P.C.[1,2]
McMaster University and Migration Health Consultants, Inc.

Brian D. Gushulak, M.D.[2]
Migration Health Consultants, Inc.

Introduction

The slow and evolutionary migration of the human race out of Africa (Mayell, 2003; Read, 2003) was associated with a parallel movement of hominoid diseases, including those associated with or produced by human interaction with microbial life. The infectious diseases we carried and the new microbial agents that were encountered during this primordial migration have very much contributed to the evolutionary development of the human race. The slow and evolutionary process of population mobility driven by expedient "push" forces of available food and water supply, environmental events, physical hazards, and predation, were combined with corresponding "pull" factors of exploration, conquest, environmental suitability, and human curiosity.

During the last 50 years the greater availability of inexpensive local and international transportation has taken place in an environment of increased trade, expanding global economies, and greater use of telecommunications systems. Together they have heralded a new era of population mobility. The modern era of mobility is associated with several factors that affect the international dispersion of human disease events, the diversity of populations, and disparities across the determinants of health. These factors include the magnitude of populations, associated goods, and conveyances on the move; the range of the population demographics and biometric characteristics of those moving (often across significant historical boundaries of disease prevalence); and the continuity (from local to international) of processes associated with these movements (MacPherson and Gushulak, 2001). The categorization and quantification of modern mobile populations provides insight regarding the significance of human movements across the sociological aspects of health and disease (see Table 4-1; DoJ, 2005; Hinrichsen, 1999; ILO, 2006; UNHCR, 2007).

---

[1]Department of Pathology and Molecular Medicine, Faculty of Health Sciences, Hamilton, Ontario, Canada.

[2]Cheltenham, Ontario, Canada/Singapore.

TABLE 4-1 Mobile Population Characteristics and Estimated Annual Magnitudes

Place	Process	Population	Magnitude (annual estimates)
Local to regional	Temporary	Commuters	Billions Bombay: 6.5 million commuters daily
		Internally displaced persons	24.5 million
	Permanent	Rural to urban	50% (developing nations) to 70% (developed nations) world population
		Inland to coastal	3.2 billion (<200 km) 4.0 billion (<400 km)
Regional to international	Temporary	Tourists, VFR	802 million
		Business travel	(Included in 802 million)
		Workers	86 million globally 49% international workers are women
	Permanent—regular	Immigrants	191 million
		Refugees	9.9 million
		Humanitarian	100,000 (est.)
	Permanent—irregular	Refugee claimants or asylum seekers	596,000 (plus 740,000 pending determination)
		Trafficked	900,000 (est.)
		Smuggled	Unknown
		Stateless people	5.8 million

NOTE: VFR = visiting family and friends.

The association of human population mobility with global climate change and adaptation can be viewed through two lenses. The first is related to the total population number and spatial density of human beings on the planet and the associated energy requirements necessary for individual-to-communal growth and development. This association is supported by the Third Assessment Report (2001) of the UN's Intergovernmental Panel on Climate Change (IPCC), which noted, "There is new and stronger evidence that most of the warming observed over the last 50 years is attributable to human activities" (IPCC, 2001). Atmospheric concentrations of energy-trapping gases are an essential part of the natural greenhouse effect that makes the Earth habitable. Global warming is arguably a consequence of increasing accumulation of human-generated atmospheric greenhouse gases (GHGs). The excess amounts of GHG observed today consist principally of carbon dioxide generated from fossil fuel combustion and the burning of forests. There are other heat-trapping gases associated with human activities, including methane associated with irrigated agriculture, animal husbandry, and oil extraction; nitrous oxide; and human-made halocarbons. Sustained population growth and development are accompanied by parallel growth in several of these activities, and increased population mobility supports their concomitant movement to new geographic areas.

The magnitude of "what" the world's human population is doing is complicated by "how" we are doing it. The linkages between and among human energy requirements and the global census accompanied by the push-pull factors behind human movement reflect the development of modern transportation and transportation systems. As surrogate markers of the impact of human activities on global climate change, the rate of change of population movements, the changing nature of human activities during the last 50 years, mobility, and transportation can all be linked to the creation of GHGs that contribute to global climate change. At the same time, those three factors also generate secondary influences on the distribution and epidemiology of infections through ecological change, contiguous extension of microbial and vector geographic patterns, and the incidental introduction of infectious diseases as a consequence of the movement of goods, conveyances, and people.

The second lens through which to view global climate change and population mobility is provided by examination of the changing physical environment and the potential for extreme weather events to provoke or generate secondary adaptive population movements. It is through this latter lens that this paper outlines the links between population mobility and the nature of emerging and reemerging infectious diseases that are associated with or result from climate change or extreme weather events.

This dynamic interaction between population mobility, infectious diseases, and climate change or extreme weather events is approached by reviewing the factors and forces in the following framework:

1. The establishment of new, noncontiguous prevalence zones of infectious diseases, or the "skip" pattern of diseases, introduced by population mobility
2. The temporary movement of populations from low to higher infectious disease prevalence zones for humanitarian, business, or recreational purposes
3. The slowly evolving ecological impacts of environmental change that either force or attract population movement and the classical determinants of health
4. The precipitous and forced movement of people due to extreme and sudden weather events that result in new ecological niches that support newly arrived populations with diverse infectious disease conditions

Evidence-Based Approach to the Adaptive Features of Population Mobility and Changes in Prevalence of Infectious Diseases Due to Climate Change or Extreme Weather Events

It has been suggested that there are three categories of research into the association between climatic conditions and infectious disease transmission. The first link examines evidence from the recent past of associations between climate variability and infectious disease occurrence. The next looks at early indicators of already-emerging infectious disease impacts of long-term climate change. The third link uses the above evidence to create predictive models to estimate the future burden of infectious disease under projected climate change scenarios (WHO, 2003a).

A fourth perspective on research in this area involves the linkage between human determinants of health (socioeconomic status, biology and genetics, environment, and behavior) and considers their contribution to global climate change and the impact on infectious diseases. Approaching the issues in this context allows human mobility and migration to be considered as a determinant of health that is directly linked to the globalization of microbial disease threats and risks.

Transportation and transportation systems have been identified in the *Fourth Assessment Report* of the IPCC, released on November 16, 2007, as an important contributor to global climate change (IPCC, 2007). Frameworks similar to those that frame the processes and associated policies on transportation—for example, agreements, regulations, inspections, or data sharing between regional authorities—are mirrored in the modern approaches to international trade, economics, environment, and health. The modern relationships between global transportation, trade, travel, and migration also provide opportunities for the introduction of transmissible disease risks between previously separated prevalence zones. Disease prevalence zones can be defined as regions of disparate epidemiological prevalence demarcated by geophysical, sociopolitical, environmental, biological or genetic, or behavioral characteristics, either alone or in combination.

Relationships Between Population Mobility and Global Infectious Disease Epidemiology

Human population mobility can affect the global distribution of infections through either chronic, sustained, or acute short-term processes. Either of these situations allows for or supports the establishment of new, noncontiguous prevalence zones of infectious diseases—what in this paper are called "skip" patterns of disease extension. This is often associated with what can be conceived as sustained, chronic mobility. The association between the international movement of infectious diseases and the risk of transmission or introduction of novel infectious diseases in nations or global regions where the diseases have been non-endemic or of very low prevalence is an issue of increasing importance in both public health and clinical care settings (Gushulak and MacPherson, 2000). The contribution of global warming and climate change to these events may be more indirect than in the past, but the relationship between population movement and disease introduction is an important aspect of human history. Examples originate with the beginning of historical records and include descriptions of the Athenian plague of 430 B.C.E. and the Justinian plague of 542 A.D. (Smith, 1996). The threat of newly introduced infections and the risks associated with those introductions provided the foundation for the traditional, quarantine-based approach to international travel and public health associated with human development until the last century (Yew, 1980).

From an international health policy perspective (MacPherson et al., 2007), there is continued attention to the risks posed by the acute arrival of some regulated infectious diseases, in a manner similar to public health control programs of the mid-nineteenth century (WHA, 2005; see next section). Additionally, greater attention is paid to the broader global epidemiological implications of population mobility and infectious diseases of importance to the health of the public. Several health authorities in regions where diseases have been of low or very low incidence report increasing burdens of infectious diseases related to foreign-born or foreign residents. Examples include tuberculosis (Health Protection Agency, 2007), syphilis (MacPherson and Gushulak, 2008), HIV/AIDS (MacPherson et al., 2006; Zencovich et al., 2006), and multidrug-resistant organisms (Maurer and Sneider, 1969; WHO, 2006).

The Temporary Movement of Populations from Low-Prevalence to Higher Infectious Disease Prevalence Zones for Humanitarian, Business, or Recreational Purposes Can Acutely Affect Infectious Disease Epidemiology

Examples, including cholera in Peru arriving in the United States (Eberhart-Phillips et al., 1996), plague in India (Mittal et al., 2006), falciparum malaria in India (Das et al., 2007), severe acute respiratory syndrome (SARS) in Asia and Canada (WHO, 2003b), and avian influenza spreading from Asia to Europe and Africa (WHO, 2008), link the risks of regional events or outbreaks to the potential for international spread via the movement of people, goods, and conveyances. For the majority of these examples, the focus is on the microbial agent rather than the human factors.

For each of the above noted situations through which population mobility can affect the global spread of infectious diseases, the role of environmental factors in relation to the events has often not been examined explicitly. More recently published explorations discussing the globalization of infectious diseases have made the components of international movement of infectious diseases part of the study design (Angell and Behrens, 2005; Leder et al., 2006), but there is still limited focus on the relationship between the environment and human mobility in this context. This relationship is an issue that requires study because the local environment, itself a product of climatic features, is a contributing factor influencing both the mobility of populations and the infectious diseases that afflict them.

The Relationship Between Slowly Evolving Ecological Impacts of Environmental Change and the Forces That Create the Push or Pull of Population Movements

The ends of the population distribution, representing economically advantaged populations and highly vulnerable populations, are often most directly affected by local environmental impacts related to climate change. As noted above, tourism and other voluntary and usually temporary population movements, all associated with economic capacity, have the potential to bring individuals from low-risk regions into diverse environments that may also represent greatly enhanced risks for acquired infectious diseases (Fenner et al., 2007; Pistone et al., 2007; Sejvar et al., 2003). Climate change has the potential to significantly affect travel and mobility of this type.

At the other end of the population distribution are the most vulnerable populations, where the concept of being an environmental refugee is emerging as a driver for displacement (Meyers, 2002). Most extremely, desertification and coastal flooding are examples of climate change impacts on local environments that will affect the most disparate and at-risk populations of the world (Black and Sessay, 1998; Rashid et al., 2007). The choices for local populations under these circumstances are very limited and are associated with internal displacement, international displacement as refugees, or refugee claimants/asylum seekers; complex humanitarian emergency evacuations (Coninx, 2007); or nefariously trafficked or smuggled persons.

As the global distribution of populations moves from rural to urban settings, where the majority of the world's largest cities are found in lowland coastal environments, the potential for environmental shifts—whether slowly evolving or as a result of disastrous events—is likely to result in the displacement of large numbers of people.

Disease risks resulting from environmentally triggered or mediated movements can be either acute or chronic. Acute consequences of infectious disease exposure and the clinical presentation of short-incubation diseases in areas where they are of low prevalence are often the subject of study, given the proximity of infection to the time of travel. Examples include malaria, influenza, cholera, and yellow fever, among others. Additionally, diseases with long incubation or latency periods are imported by mobile populations into low-incidence locations. Due to the natural history of these infections, they may have been acquired long before the time of the environmental event related to population movement. Examples include tuberculosis, HIV, and the nontransmissible sequelae of chronic infections, such as human papilloma virus (HPV) associated cervical and esophageal cancer.

The Precipitous and Forced Movement of People Due to Extreme and Sudden Weather Events

The link between global climate change and extreme weather events, such as high winds, extremes of temperature or precipitation, or such events occurring in combination, may be questioned, but the consequences of extreme weather events on human health can be measured.

It has been common to measure the health consequences of severe environmental events using the residual health infrastructure that has survived the event. This is frequently the case for the public health surveillance of diseases. As a consequence, the situations producing the most information are those that have resulted in areas where the infrastructure is most robust. This explains in part why the health impacts of Hurricanes Katrina and Rita on the southeastern United States (CDC, 2006a,b; Elledge et al., 2007) cannot be compared to Hurricane Georges's (O'Leary et al., 2002; Sanders et al., 1999) effects in Central America and the Caribbean. Almost certainly because the direct physical consequences of Hurricane Katrina affected a major urban center in the United States, the existence of a highly sophisticated social environment with considerable robustness and response capacity during emergencies was more permissive of several measurements of infectious disease outcomes being sought in both the displaced persons and the responders (CDC, 2005, 2006c; Corbin et al., 2007; Rao et al., 2007; Sevbold et al., 2007; Yee et al., 2007) compared to other weather-related catastrophes in the region.

The long-term health consequences of extreme weather events, particularly the infectious disease outcomes in affected or dispersed persons, are much more difficult to quantify. Tracking an individual over time and place, recording the changing administrative status from displaced person to returnee or from refugee to permanent resident, poses significant challenges to the way surveillance data are currently managed.

As rural to urban migration increases globally and as many of the megacities of the world are located at or near coastal regions, the impact on population placement due to global climate change and extreme weather events can only be expected to be of greater magnitude and associated with a higher risk of disease outcomes.

Policy Implications Related to Population Mobility, Climate Change, and Infectious Diseases

Recognizing the relevant linkages between population mobility, climate change, and infectious diseases, and being able to address them at a programmatic level, require policy responses that are informed, robust, and feasible and extend across a spectrum from local to international scope.

Existing methodologies used by some nations, such as the medical screening and health assessment of immigrants and refugees for transmissible infectious diseases of public health significance, are feasible in some situations. They may not be transferable between health jurisdictions or applicable in situations involving other types of mobile populations. For example, climatic events precipitating large population movements may themselves be of magnitudes that exceed the capacity to mitigate or prevent the introduction of infectious diseases. Very large movements, such as seen in the Americas during school winter breaks, when thousands of people move to warmer regions for vacations, may overwhelm the existing local public health detection systems originally designed for the host population. This can be particularly true for diseases that are not part of routine or anticipated surveillance designs, such as mumps in what is believed to be an immunized population (Peltola et al., 2007). Lack of anticipatory capacity is much greater with other types of population mobility. Humanitarian movements by nature include a balance between the expected benefit and potential risk resulting from the movement. Depending on the characteristics of the population itself, the local environment, and the nature of the physical journey, the potential for novel disease introductions related to the movement may fall into the "unknown unknowns" category of impossible planning challenges. Further complicating policies to mitigate the infectious disease risk associated with mobile populations are the indirect associations with other disease-related processes. Food production environments, the processes of food management and safety, may all be subject to local influences that can affect the risk of disease. The increasing dispersion both regionally and internationally of nutrition also has disease implications (FDA, 2007a,b).

Bio-adaptive responses to climate change, particularly those that involve the movement of conveyances, goods, and people, have implications for infectious disease management, including training, education, maintenance of competence, surveillance, notification, reporting systems, and consideration of multiple layers of service providers from border inspectors to traditional and nontraditional health services providers such as doctors, nurses, pharmacists, and linguistically or culturally aligned healers.

The resulting challenges will be particularly daunting for policy and decision makers. Increasing global mobility across all sectors involving people, goods, and the conveyances themselves is increasingly associated with movement across gradients of individual, population, and public health determinants. These determinants, represented by biological, genetic, environmental, socioeconomic, and behavioral characteristics of the populations on the move, are themselves dynamic and interactive.

While many health prevention and mitigation strategies recognize that there are disease risks associated with travel across these gradients, they tend to consider the mitigation of these risks through historically conceived disease control and prevention policies. Given the scope, diversity, and growing integration of current global mobility, preventing disease introduction and spread may be less manageable. Mitigation efforts and the policies that guide them will need to become more process-related and anticipatory rather than outcome based and reactive.

Not all of the climatic implications for population mobility will be negative in context. Some positive consequences of climate change already measured include longer growing seasons and increased rains that boost yields in many temperate regions (CCSP and the Subcommittee on Global Change Research, 2008). Records show that the crop-growing season has already lengthened in the United Kingdom, Scandinavia, Europe, and North America (UNFCCC, 2008). Young forest growth is also expected to occur more productively in these regions. There is expected to be a reduction in cold weather-associated mortality that will benefit the most vulnerable populations such as the elderly, under-housed/homeless, and displaced persons (Patz et al., 2000). In addition, economic benefits are expected from increasingly accessible shipping lanes in the Northwest Passage due to longer ice-free seasons. This will reduce the human and economic costs of international shipping and trade (Environment Canada, 1999; Nickerson, 2000). These regional environmental outcomes will also be associated with increased mobility of human populations, goods, and conveyances that move to take advantage of new opportunities in food security, labor markets, and other forms of trade.

Global climate change, extreme weather events, and infectious diseases are a complicated and complex interface involving several factors. Humans contribute to the issue, experience both the benefits and the deleterious consequences, and are the potential stewards of this dynamic interaction. How the management of global environmental systems proceeds will influence migration and population mobility and can be expected to raise new challenges to human endeavors on this planet (UNFCCC, 2007).

CLIMATE CHANGE, INFECTIOUS DISEASE, AND INTERNATIONAL PUBLIC HEALTH POLICY[3]

Diarmid Campbell-Lendrum, D.Phil.[4]
World Health Organization

Climate Change as a Challenge to Public Health Policy

Health concerns are increasingly evident in policy discussions and media reporting on climate change. In recent years, there has also been a corresponding

[3]The views expressed in this article are those of the author and do not necessarily reflect the position of the World Health Organization.

[4]Department of Public Health and Environment, CH-1211 Geneva 27, Switzerland.

increase in the engagement of national and international health agencies. For all of this emerging interest, the issue combines a range of characteristics that render policy making particularly difficult.

Some of these challenges arise within the health sector. As a generalization, the traditional assessment methods used in the health sciences are designed to deal with direct, compartmentalized, cause-effect relationships resulting in individual diseases, rather than systemic and long-term stresses leading to a range of health effects. This makes it difficult both to assess and to communicate the evidence for either health effects that can already be attributed to climate change or the risks of health impacts that may occur as the process continues. For example, it is often difficult to make clear that for well-described methodological reasons (Kovats et al., 2001; Wilkinson et al., 2003), the empirical evidence that gradual climate change is already affecting health is limited to a few specific examples,[5] but at the same time, the evidence that important diseases are climate sensitive, and therefore likely to be affected either now or in the future, is overwhelmingly strong.[6] Related to this, much of the effort of the health sector is applied to addressing individual diseases with specific, often curative, interventions. This tends to lead to a focus on identifying single technological tools to address new risks, with a relative hesitancy to work across sectors to address the root causes of multiple health exposures, including environmental and social determinants (Commission on Social Determinants of Health, 2007; McMichael et al., 2008).

Challenges in the health sector are further amplified because the health effects of climate change lie within a broader and very complex policy debate. As an international policy issue, climate change touches on many highly contentious fields, including the speed and nature of economic development, energy production and security, division of obligations between rich and poor countries, population growth, and individual lifestyle choices, to name but a few. Within this policy discussion, risks to health tend to attract particular attention because they present a tangible "human dimension" of the climate change debate. Evidence for or against health risks from climate change is therefore often used selectively by parties to promote political agendas that have little to do with the health risk itself.

Related to this, the public perception of health risks from climate change is filtered through media that can often tend to sensationalize stories and amplify disagreements to present polarized positions (Boykoff and Boykoff, 2004). For an example of the former, take the following headline from a recent edition of the *Daily Telegraph* newspaper following the release of a U.K. government report on health risks arising from climate change in Britain (Prince, 2008):

---

[5]See Bouma et al. (1996); Pascual et al. (2006, 2008); Stott et al. (2004).

[6]See reviews such as Confalonieri et al. (2007); McMichael et al. (2003).

> Malaria warning as UK becomes warmer: The UK is to be hit by regular malaria outbreaks, fatal heatwaves and contaminated drinking water within five years because of global warming, the Government has warned the NHS.

The cited U.K. report, however, actually stated (Department of Health/Health Protection Agency, 2008):

> We would therefore not expect more than a very few cases of autochthonous malaria in the UK over the next 50 years. Indeed one is more likely to be struck by lightning than to get malaria from an English mosquito.

In this environment, the production and clear communication of impartial, evidence-based, policy-relevant assessments becomes particularly important.

Placing Climatic Influences on Infectious Diseases in Context

The focus of this particular meeting of the Forum on Microbial Threats is on the potential effect of climate change on the emergence, reemergence, and spread of infectious disease. Such processes, bringing infections to new populations, have important public health implications. These include exposure of nonimmune populations that may suffer particularly high morbidity and mortality rates, as well as diversion of resources from routine endemic disease problems and, in some cases, more indirect effects, including disruption of important economic sectors such as tourism.

However, it is important to place these "novel" disease threats within the context of a range of public health implications of climate change. Many diseases that are well known and not necessarily spreading rapidly are also highly sensitive to climate conditions. For example, time-series and cross-sectional studies in Lima, Peru, and in the South Pacific Islands have shown that rates of diarrhea morbidity are influenced by both temperature and precipitation, higher rates being associated with higher temperatures in both studies and with either particularly high or low rainfall in the South Pacific (Checkley et al., 2000; Singh et al., 2001). Even small variations in meteorological conditions can have a marked effect on such diseases. In Lima, rates of admission for acute diarrhea at a pediatric clinic were 3 to 4 times higher in the summer than in the winter and increased by 8 percent for every 1°C increase in temperature, with an additional increase during the sustained high temperatures associated with the 1997-1998 El Niño event.

Such "routine" climate-sensitive diseases may attract less attention than emerging, reemerging, and spreading infections, but they are nonetheless of major importance, particularly in the developing world. For example, diarrhea kills more than 1.8 million people annually; malaria kills almost 1 million; and undernutrition (partly compounded by infections) kills approximately 3.5 million (Black et al., 2008; WHO, 2006a). These burdens are overwhelmingly concentrated on children in developing countries. The current trend toward a warmer and more variable climate, with reduced agricultural production at low latitudes, is expected to increase the difficulties of combating these and other persistent health problems (WHO, 2008a).

It is also essential to place the effect of climate change and climate variability in the context of other health determinants. While the evidence for meteorological effects on diarrhea is clear, other factors—most importantly, water, sanitation, and hygiene—have a very strong effect on diarrhea mortality and morbidity (Prüss-Üstün et al., 2004). Such factors not only influence the overall rate of disease, but can also moderate climate effects. There is evidence that improvements in water, sanitation, and hygiene services have a relatively greater effect on reducing infections with bacterial and protozoan pathogens that tend to be favored by high temperatures, than with viral pathogens that often survive better in colder conditions. This is a likely explanation for the occurrence of winter rather than summer peaks in diarrhea incidence in more developed countries with improved infrastructure.

For these reasons, separating the effects of climate variability and change from the context of other determinants, or assessing the influence of climate versus other factors as a mutually exclusive "either/or" debate, is unproductive. Instead, considering climate as one important determinant of health risks, mediated by other contextual determinants, is more likely to lead to sound health policy.

International Developments in Health and Climate Policy

Since climate change began to emerge as a major issue at the end of the 1980s, research output has steadily increased (see Figure 4-1). The WHO has been involved throughout this process. Beginning in 1990, WHO released a series of reports on the health implications of climate change regionally and globally; published methods for qualitative and quantitative assessment of health vulnerability; and participated in major review processes, such as the production of the assessment reports of the IPCC (WHO, 2008b).

In recent months, there has been rapid change in the scale and nature of the engagement of health agencies in climate change. This has included strong statements and dedicated policy speeches by the WHO director-general (Chan, 2007) and the selection of "Health Protection from Climate Change" as the theme for World Health Day 2008 (WHO, 2008a). This has been mirrored by similar highlighting of the issue by major U.S. organizations, such as the American Public Health Association, the American Institute of Biological Sciences, and the U.S. Centers for Disease Control and Prevention, as well as many others internationally. The Member States of the Executive Board of WHO have also proposed a resolution on climate change and health to be discussed at the 2008 World Health Assembly—the first on this subject for 10 years (Box 4-1). Climate change has therefore moved firmly to the center of the international public health agenda.

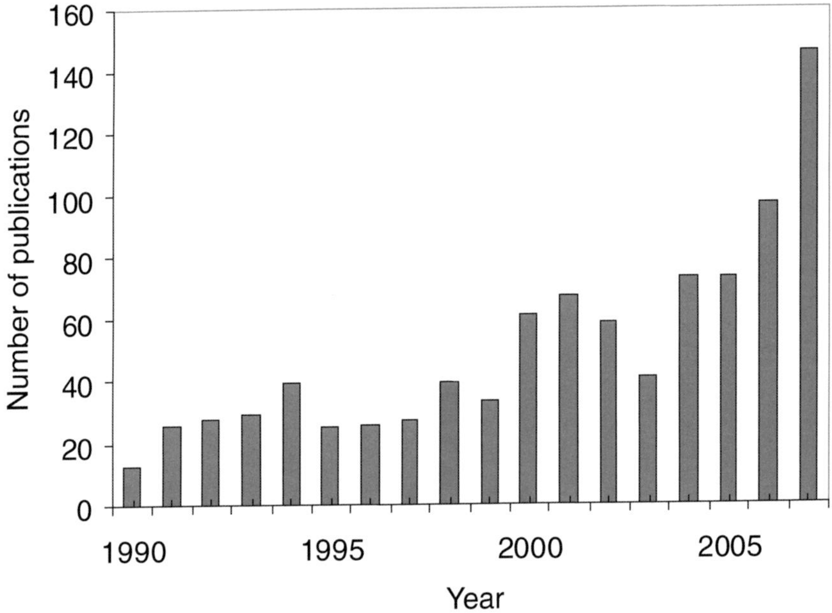

FIGURE 4-1 Number of publications in PubMed referring to "health" and either "climate change" or "global warming" from 1990 to 2007.

Putting Together a Coherent Response to Health Risks from Climate Change

This has called for a development of the field away from only describing health risks from climate change to a more serious consideration of how best to support public health policy and disease control programs to cope with this emerging threat. In line with the Executive Board resolution, WHO is reframing its support to Member States on health and climate change around three main objectives: (1) raising awareness, (2) promoting health through climate change mitigation, and (3) strengthening health systems to address the additional health risks from climate change.

In terms of raising awareness, there is growing appreciation that climate change can no longer be considered simply an environmental or a developmental issue. More importantly, it puts at risk the protection and improvement of human health and well-being. A greater appreciation of the human health dimensions of climate change is necessary for both the development of effective policy and the mobilization of public engagement. A range of health agencies are now making clear that the ultimate aim of mitigation and adaptation, and related development

BOX 4-1
The International Mandate for Stronger Action on Health and Climate Change

In January 2008, the 34 Member States of the Executive Board of WHO drafted a resolution on "Climate change and health," for consideration by the World Health Assembly, comprising the 193 WHO Member States. The draft resolution places on record the shared concern of the Member States regarding the strengthening evidence of the effect of atmospheric greenhouse gases, and the potential negative consequences for human health, including risks to the achievement of the Millennium Development Goals and undermining of efforts to improve public health and reduce health inequalities globally.

The resolution recognizes the joint responsibility of all Member States to support solutions to address the health impacts of climate change. It calls on WHO to work with its partners to strengthen support to Member States, through four main areas of work. These can be summarized as follows:

1. Drawing attention to the serious risk of climate change to global health security and to the achievement of the health-related Millennium Development Goals, to ensure that health concerns are taken into account in national and international responses.
2. Engaging in the UN cross-sectoral program on adaptation to climate change, to ensure its relevance to health and to facilitate participation by Member States.
3. Developing capacity to assess the risks from climate change for human health and to implement effective response measures, through further research and pilot projects, including on

- the scale and nature of health vulnerability to climate change;
- assessment of protection strategies and measures;
- the health impacts of potential adaptation and mitigation measures in other sectors such as water resources, land use, and transport;
- decision support and other tools, such as surveillance and monitoring, for assessing vulnerability and targeting protection measures appropriately; and
- assessment of the likely financial costs and other resources necessary for health protection from climate change.

4. Consultation with Member States on scaling up WHO's technical support for assessing and addressing the implications of climate change for health and health systems.

SOURCE: The full text of the resolution and background documentation can be found in WHO (2008c).

decisions, should be the protection and improvement of human well-being (i.e., health should be at the heart of climate policy).

On promoting health through climate change mitigation, it is clear that in the long term, sustainable development and protection of ecosystem services are fundamentally necessary for human health (Millennium Ecosystem Assessment, 2005). However, even in the short term, development choices that protect the climate have the potential for important public health benefits. The intersection of energy, transport, climate, and health provides a clear example. It is widely appreciated that both modes of transport (i.e., relative use of active, public, and private modes) and choice of fuels have an important influence on greenhouse gas emissions. It is less widely appreciated that these decisions also have an effect on the 800,000 annual global deaths from outdoor air pollution; the 1.2 million annual deaths from traffic accidents; and the 1.9 million annual deaths from physical inactivity (Ezzati et al., 2004; WHO, 2002). Providing the poorest communities with access to cleaner domestic energy technologies could reduce the 1.5 million annual indoor air pollution deaths, as well as slow the growth in greenhouse gas emissions (Bailis et al., 2005; Smith et al., 2004; WHO, 2006b). There is, therefore, the potential for important public health improvements through a closer engagement with relevant environmental and economic sectors.

The most obvious and immediate area of engagement, however, is in strengthening public health systems to protect populations against emerging health risks. In climate change policy, this corresponds to "adaptation" (i.e., minimizing the damages caused by climate change that is now inevitable). To some extent, planning of interventions may be less complex than assessing and communicating the health risks from climate change. While there is vocal disagreement over some aspects of the evidence for effects of climate change on health, many of these disagreements are of little relevance to identifying effective interventions.

The fundamental principle put forward by WHO and others is that protection from climate change is part of a basic, preventive approach to public health, not a separate or competing demand (Campbell-Lendrum et al., 2007; WHO, 2008a). Many of the most important actions are public health interventions of proven effectiveness—from controlling vector-borne disease, to providing clean water and sanitation. All would improve health now, as well as reducing vulnerability to climate change in the future. This can be illustrated by examples of two broad areas: preventive environmental health interventions and infectious disease surveillance and response.

Preventive Environmental Health Interventions

One example of an emerging health threat associated with climate change is the decline in global freshwater resources, caused mainly by increasing rates of water extraction and contamination. A warmer and more variable climate is expected to further worsen the decline in water quantity and quality in many regions through the retreat of glaciers that supply freshwater in summer to much of the global population, greater surface evaporation, and shifting and more extreme precipitation (IPCC, 2007). Scaling-up water and sanitation services and providing point-of-use disinfection are highly cost-effective interventions (Hutton and Haller, 2004) that should both reduce the current burden of disease and ameliorate the health impacts of decreasing water supplies. As water stresses intensify, governments could protect health by strengthening and enforcing their regulatory frameworks to ensure that increasing use of new water sources, such as wastewater, excreta, and graywater in agriculture and aquaculture, does not bring increases in infectious or chemical risks (WHO, 2006c).

Infectious Disease Surveillance and Response

Effective surveillance and response systems are essential in managing any infectious disease. However, they become even more important under conditions of rapid change, including climatic shifts as well as increasing rates of movement of and contact between humans, pathogens, and reservoirs. This requires improved human health surveillance integrated with monitoring of climate and other environmental conditions that favor disease outbreaks, including disease in wildlife and agricultural animals. Climate change also strengthens the case for reinforcing response systems for infectious disease outbreaks, including predefined action plans and maintenance of the control resources and personnel capacity necessary to implement disease control. There is already a great deal of institutional infrastructure in place that can help to control these risks. Most importantly at the global level, the newly revised International Health Regulations define operating procedures for detection, notification, and control of disease risks, including but not restricted to preventing the spread of infections across international borders (WHO, 2007). New investments to meet the additional risks of climate change should build on, rather than replicate, these existing mechanisms.

A great deal of the scientific discussion in this field relates to the development of early warning systems, often based on new technologies, such as satellite-based remote sensing. Such tools have the potential to improve lead times and, therefore, improve control. However, one should not assume that advances in this area will by themselves have a major impact on cutting disease rates, particularly in developing countries. It is essential that future work in this area takes note of the warning that "forecasts based entirely on scientific objectives have little impact on policy because there is no stakeholder" (Clark et al., 2001) and follows the recommendation provided by a previous National Research Council (NRC) report that "development of early warning systems should involve active participation of the system's end users" (NRC, 2001). In many cases, the weak link is not the lead time of a warning but the capacity to respond effectively. A recent WHO literature review identified numerous examples of scientific studies linking climate variability and infectious disease, but could not locate any full descriptions of climate-based early warning systems being used to influence control decisions (WHO, 2005).

Conclusions

Health threats arising from climate change understandably attract widespread policy and public attention. Climate change does not act in isolation, but is likely to interact with other rapid changes to strain existing weak points in public health systems. The most effective responses are likely to be strengthening of key functions, such as environmental management, and surveillance and response to safeguard changes in infectious disease patterns and other hazards. The NRC (2001) reported that

> there will always be some element of unpredictability in climate variations and infectious disease outbreaks. Therefore, a prudent strategy is to set a high priority on reducing people's overall vulnerability to infectious disease through strong public health measures such as vector control efforts, water treatment systems, and vaccination programs.

As climate change continues to rise up the scientific and political agenda, this guidance remains as relevant as ever.

REFERENCES

Overview Reference

European Commission. 2008. *Climate change and international security*. Paper presented at European Union Summit, Brussels.

MacPherson and Gushulak References

Angell, S. Y., and R. H. Behrens. 2005. Risk assessment and disease prevention in travelers visiting friends and relatives. *Infectious Disease Clinics of North America* 19(1):49-65.

Black, R., and M. Sessay. 1998. Forced migration, natural resource use and environmental change: the case of the Senegal River Valley. *International Journal of Popular Geography* 4(1):31-47.

CCSP (U.S. Climate Change Science Program) and the Subcommittee on Global Change Research. 2008. *The effects of climate change on agriculture, land resources, water resources, and biodiversity in the United States*. Washington, DC: U.S. Environmental Protection Agency.

CDC (Centers for Disease Control and Prevention). 2005. Infectious disease and dermatologic conditions in evacuees and rescue workers after Hurricane Katrina—multiple states, August-September, 2005. *Morbidity and Mortality Weekly Report* 54(38):961-964.

———. 2006a. Surveillance for illness and injury after Hurricane Katrina—three counties, Mississippi, September 5-October 11, 2005. *Morbidity and Mortality Weekly Report* 55(9):231-234.

———. 2006b. Illness surveillance and rapid needs assessment among Hurricane Katrina evacuees—Colorado, September 1-23, 2005. *Morbidity and Mortality Weekly Report* 55(9):244-247.

———. 2006c. Two cases of toxigenic *Vibrio cholerae* O1 infection after Hurricanes Katrina and Rita—Louisiana, October 2005. *Morbidity and Mortality Weekly Report* 55(2):31-32.

Coninx, R. 2007. Tuberculosis in complex emergencies. *Bulletin of the World Health Organization* 85(8):637-640.

Corbin, A., C. Delatte, S. Besson, A. Guidry, A. H. Hoffmann III, P. Monier, and R. Nathaniel. 2007. *Budvicia aquatica* sepsis in an immunocompromised patient following exposure to the aftermath of Hurricane Katrina. *Journal of Medical Microbiology* 56(Pt 8):1124-1125.

Das, N. G., P. K. Talukdar, J. Kalita, I. Baruah, and R. B. Sribastave. 2007. Malaria situation in forest-fringed villages of Sonitpur district (Assam), India bordering Arunachal Pradesh during an outbreak. *Journal of Vector-Borne Diseases* 44(3):213-218.

DoJ (Department of Justice). 2005. *The human smuggling and trafficking centre. Fact sheet: distinctions between human smuggling and human trafficking*, http://www.usdoj.gov/crt/crim/smuggling_trafficking_facts.pdf (accessed December 29, 2007).

Eberhart-Phillips, J., R. E. Besser, M. P. Tormey, D. Koo, D. Feikin, M. R. Araneta, J. Wells, I. Kilman, G. W. Rutherford, P. M. Griffin, R. Baron, and L. Mascola. 1996. An outbreak of cholera from food served on an international aircraft. *Epidemiology and Infection* 116(1):9-13.

Elledge, B. L., D. T. Boatright, P. Woodson, R. E. Clinkenbeard, and M. W. Brand. 2007. Learning from Katrina: environmental health observations from the SWCPHP response team in Houston. *Journal of Environmental Health* 70(2):22-26.

Environment Canada. 1999. Responding to global climate change in the Arctic. In *The Canada Country Study*. Toronto, Ontario: Environmental Adaptation Research Group.

FDA (Food and Drug Administration). 2007a. *Pet food recall (melamine) tainted animal feed*, http://www.fda.gov/oc/opacom/hottopics/petfood.html (accessed December 31, 2007).

———. 2007b. *FDA warns consumers not to eat certain jars of Peter Pan peanut butter and Great Value peanut butter,* http://www.fda.gov/bbs/topics/NEWS/2007/NEW01563.html (accessed December 31, 2007).

Fenner, L., R. Weber, R. Steffen, and P. Schlagenhauf. 2007. Imported infectious disease and purpose of travel, Switzerland. *Emerging Infectious Diseases* 13(2):217-222.

Gushulak, B. D., and D. W. MacPherson. 2000. Population mobility and infectious diseases: the diminishing impact of classical infectious diseases and new approaches for the 21st century. *Clinical Infectious Diseases* 31(3):776-780.

Health Protection Agency. 2007. *Epidemiology data: tables and figures. Case reports (enhanced TB surveillance and national TB surveys)*, http://www.hpa.org.uk/infections/topics_az/tb/epidemiology/tables.htm#ets (accessed December 31, 2007).

Hinrichsen, D. 1999. The coastal population explosion. In *Trends and future challenges for U.S. national ocean and coastal policy*, edited by B. Cicin-Sain, R.W. Knecht, and H. Foster. Washington, DC: NOAA. Pp. 27-30. http://oceanservice.noaa.gov/websites/retiredsites/natdia_pdf/3hinrichsen.pdf (accessed December 29, 2007).

ILO (International Labour Organization). 2006. *Facts on labour migration*, http://www.ilo.org/public/english/bureau/inf/download/ecosoc/migration.pdf (accessed August 19, 2008).

IPCC (Intergovernmental Panel on Climate Change). 2001. *Climate change 2001: third assessment report (vol. I)*. Cambridge, UK: Cambridge University Press.

———. 2007. *Summary for policymakers of the synthesis of the IPCC fourth assessment report*. Cambridge, UK: Cambridge University Press.

Leder, K., S. Tong, I. Weld, K. C. Kain, A. Wilder-Smith, F. von Sonnenburg, J. Black, V. Brown, J. Torresi, and GeoSentinel Surveillance Network. 2006. Illness in travelers visiting friends and relatives: a review of the GeoSentinel Surveillance Network. *Clinical Infectious Diseases* 43(9):1185-1193.

MacPherson, D. W., and B. D. Gushulak. 2001. Human mobility and population health. New approaches in a globalizing world. *Perspectives in Biology and Medicine* 44(3):390-401.

———. 2008. Syphilis in immigrants, public health, and the Canadian immigration medical examination. *Journal of Immigrant and Minority Health* 10(1):1-6.

MacPherson, D. W., M. Zencovich, and B. D. Gushulak. 2006. Emerging pediatric HIV epidemic related to migration. *Emerging Infectious Diseases* 12(4):612-617.

MacPherson, D. W., B. D. Gushulak, and L. Macdonald. 2007. Health and foreign policy: influences of migration and population mobility. *Bulletin of the World Health Organization* 85(3):200-206.

Maurer, L. H., and T. J. Sneider. 1969. Gonococcal urethritis in males in Vietnam: three penicillin regimens and one tetracycline regimen. *Journal of the American Medical Association* 207(5):946-948.

Mayell, H. 2003. Documentary redraws humans' family tree. *National Geographic News*, http://news.nationalgeographic.com/news/2002/12/1212_021213_journeyofman.html (accessed February 26, 2008).

Meyers, N. 2002. Environmental refugees: a growing phenomenon of the 21st century. *Philosophical Transactions of the Royal Society of London. Series B, Biological Sciences* 357(1420): 609-613.

Mittal, V., D. Bhattacharya, U. V. Rana, A. Rai, S. T. Pasha, A. Kumar, A. K. Hart, R. L. Ishpujani, U. K. Baveia, S. Lal, and S. P. Agarwal. 2006. Prompt laboratory diagnosis in timely containment of a plague outbreak in India. *Journal of Communicable Diseases* 38(4):317-324.

Nickerson, C. 2000 (March 21). Girding for a sea change: with ice thinning, Canada claims a Northwest Passage. *Boston Globe*.

O'Leary, D. R., J. G. Rigau-Pérez, E. B. Hayes, A. V. Vonrdam, G. G. Clark, and D. J. Gubler. 2002. Assessment of dengue risk in relief workers in Puerto Rico after Hurricane Georges, 1998. *American Journal of Tropical Medicine and Hygiene* 66(1):35-39.

Patz, J. A., M. A. McGeehin, S. M. Bernard, K. L. Ebi, P. R. Epstein, A. Grambsch, D. J. Gubler, P. Reither, I. Romieu, J. B. Rose, J. M. Samet, and J. Trtanj. 2000. The potential health impacts of climate variability and change for the United States: executive summary of the report of the health sector of the U.S. National Assessment. *Environmental Health Perspectives* 108(4):367-376.

Peltola, H., P. S. Kulkarni, S. V. Kapre, M. Paunio, S. S. Jadhav, and R. M. Dhere. 2007. Mumps outbreaks in Canada and the United States: time for new thinking on mumps vaccines. *Clinical Infectious Diseases* 45(4):459-466.

Pistone, T., P. Guibert, F. Gay, D. Malvy, K. Ezzedine, M. C. Receveur, M. Siriwardana, B. Larouzé, and O. Bouchard. 2007. Malaria risk perception, knowledge and prophylaxis practices among travellers of African ethnicity living in Paris and visiting their country of origin in sub-Saharan Africa. *Transactions of the Royal Society of Tropical Medicine and Hygiene* 101(10):990-995.

Rao, C. Y., C. Kurukularatne, J. B. Garcia-Diaz, S. A. Kemmerly, D. Reed, S. K. Frickin, and J. Morgan. 2007. Implications of detecting the mold *Syncephalastrum* in clinical specimens of New Orleans residents after Hurricanes Katrina and Rita. *Journal of Occupational and Environmental Health* 49(4):411-416.

Rashid, H., I. M. Hunt, and W. Haider. 2007. Urban flood problems in Dhaka, Bangladesh: slum residents' choices for relocation to flood-free areas. *Environmental Management* 40(1):95-104.

Read, M. 2003. Journey of man. *National Geographic News*, http://news.nationalgeographic.com/news/2002/12/photogalleries/journey_of_man/photo2.html (accessed December 29, 2007).

Sanders, E. J., J. G. Rigau-Pérez, H. L. Smits, C. C. Deseda, V. A. Vorndam, T. Aye, R. A. Spiegel, R. S. Weyant, and S. L. Bragg. 1999. Increase of leptospirosis in dengue-negative patients after a hurricane in Puerto Rico in 1996 [correction of 1966]. *American Journal of Tropical Medicine and Hygiene* 61(3):399-404.

Sejvar, J., E. Bancroft, K. Winthrop, J. Bettinger, M. Bajani, S. Bragg, K. Shutt, R. Kaiser, N. Marano, T. Popovic, J. Tappero, D. Ashford, I. Mascola, D. Vugia, B. Perkins, N. Rosenstein, and Eco-Challenge Investigation Team. 2003. Leptospirosis in "Eco-Challenge" athletes, Malaysian Borneo, 2000. *Emerging Infectious Diseases* 9(6):702-707.

Sevbold, U., N. White, Y. F. Wang, J. S. Halvosa, and H. M. Blumberg. 2007. Colonization with multidrug-resistant organisms in evacuees after Hurricane Katrina. *Infection Control and Hospital Epidemiology* 28(6):726-729.

Smith, C. A. 1996. Plague in the ancient world: a study from Thucydides to Justinian. In *The Student Historical Journal*, vol. 28. Loyola University: New Orleans.

UNFCCC (United Nations Framework Convention on Climate Change). 2007. *Decision-/CP.13. Bali Action Plan.* United Nations Climate Change Conference, Bali, December 3-14, 2007, http://unfccc.int/files/meetings/cop_13/application/pdf/cp_bali_action.pdf (accessed December 31, 2007).

———. 2008. *Climate change information sheet 10: agriculture and food security*, http://unfccc.int/essential_background/background_publications_htmlpdf/climate_change_information_kit/items/288.php (accessed June 24, 2008).

UNHCR (United Nations High Commissioner for Refugees). 2007. *Protecting refugees and the role of the UNHCR*, http://www.unhcr.org/basics/BASICS/4034b6a34.pdf (accessed February 1, 2008).

WHA (World Health Assembly). 2005. *WHA 58.3 Revision of the International Health Regulations*, http://www.who.int/gb/ebwha/pdf_files/WHA58/WHA58_3-en.pdf (accessed November 25, 2007).

WHO (World Health Organization). 2003a. *Climate change and human health: risks and responses*, edited by A. J. McMichael, D. H. Campbell-Lendrum, C. F. Corvalán, K. L. Ebi, A. Githeko, J. D. Scheraga, and A. Woodward. Geneva, Switzerland: World Health Organization.

———. 2003b. *Areas with recent local transmission of SARS*, http://www.who.int/csr/sars/areas/en/ (accessed December 31, 2007).

———. 2006. Addressing the threat of tuberculosis caused by extensively drug-resistant *Mycobacterium tuberculosis. Weekly Epidemiological Record* 81(41):385-390, www.who.int/wer/2006/wer8141.pdf (accessed December 31, 2007).

———. 2008. *Situation updates—avian influenza*, http://www.who.int/csr/disease/avian_influenza/updates/en/ (accessed December 31, 2007).

Yee, E. L., H. Palacio, R. I. Almar, U. Shah, C. Kilborn, M. Faul, T. E. Gavagan, R. D. Feigin, J. Versalovic, F. H. Neil, A. L. Panililo, M. Miller, J. Spahr, and R. L. Glass. 2007. Widespread outbreak of norovirus gastroenteritis among evacuees of Hurricane Katrina residing in a large "megashelter" in Houston, Texas: lessons learned for prevention. *Clinical Infectious Diseases* 44(8):1032-1039.

Yew, E. 1980. Medical inspection of immigrants at Ellis Island, 1891-1924. *Bulletin of the New York Academy of Medicine* 56(5):488-510.

Zencovich, M., K. Kennedy, D. W. MacPherson, and B. D. Gushulak. 2006. Immigration medical screening and HIV infection in Canada. *International Journal of STDs and AIDS* 17(12):813-816.

Campbell-Lendrum References

Bailis, R., M. Ezzati, and D. M. Kammen. 2005. Mortality and greenhouse gas impacts of biomass and petroleum energy futures in Africa. *Science* 308(5718):98-103.

Black, R. E., L. H. Allen, Z. A. Bhutta, L. E. Caulfield, M. de Onis, M. Ezzati, C. Mathers, and J. Rivera. 2008. Maternal and child undernutrition: global and regional exposures and health consequences. *Lancet* 371(9608):243-260.

Bouma, M. J., C. Dye, and H. J. vanderKaay. 1996. Falciparum malaria and climate change in the Northwest Frontier Province of Pakistan. *American Journal of Tropical Medicine and Hygiene* 55(2):131-137.

Boykoff, M. T., and J. M. Boykoff. 2004. Balance as bias: global warming and the U.S. prestige press. *Global Environmental Change-Human and Policy Dimensions* 14(2):125-136.

Campbell-Lendrum, D., C. Corvalán, and M. Neira. 2007. Global climate change: implications for international public health policy. *Bulletin of the World Health Organization* 85(3):235-237.

Chan, M. 2007. *Climate change and health: preparing for unprecedented challenges*. The 2007 David E. Barmes Global Health Lecture. Bethesda, MD, December 10.

Checkley, W., L. D. Epstein, R. H. Gilman, D. Figueroa, R. I. Cama, J. A. Patz, and R. E. Black. 2000. Effects of El Niño and ambient temperature on hospital admissions for diarrhoeal diseases in Peruvian children. *Lancet* 355(9202):442-450.

Clark, J. S., S. R. Carpenter, M. Barber, S. Collins, A. Dobson, J. A. Foley, D. M. Lodge, M. Pascual, R. Pielke, W. Pizer, C. Pringle, W. V. Reid, K. A. Rose, O. Sala, W. H. Schlesinger, D. H. Wall, and D. Wear. 2001. Ecological forecasts: an emerging imperative. *Science* 293(5530):657-660.

Commission on Social Determinants of Health. 2007. *Achieving health equity: from root causes to fair outcomes*. Geneva, Switzerland: World Health Organization.

Confalonieri, U., B. Menne, R. Akhtar, K. L. Ebi, M. Hauengue, R. S. Kovats, B. Revich, and A. Woodward. 2007. Human health. In *Climate change 2007: impacts, adaptation and vulnerability*. Contribution of Working Group II to the Fourth Assessment Report of the Intergovernmental Panel on Climate Change, edited by M. L. Parry, O. F. Canziani, J. P. Palutikof, P. J. van der Linden, and C. E. Hanson. Cambridge, UK: Cambridge University Press.

Department of Health/Health Protection Agency. 2008. *Health effects of climate change in the UK 2008: an update of the Department of Health Report 2001/2002*. London, UK: Department of Health/Health Protection Agency.

Ezzati, M., A. D. Lopez, A. Rodgers, and C. J. L. Murray, eds. 2004. *Comparative quantification of health risks: global and regional burden of disease due to selected major risk factors*. Geneva, Switzerland: World Health Organization.

Hutton, G., and L. Haller. 2004. *Evaluation of the costs and benefits of water and sanitation improvements at the global level*. Geneva, Switzerland: World Health Organization.

IPCC (Intergovernmental Panel on Climate Change). 2007. *Climate change 2007: climate change impacts, adaptation and vulnerability*. Contribution of Working Group II to the Intergovernmental Panel on Climate Change Fourth Assessment Report. Cambridge, UK: Cambridge University Press.

Kovats, R. S., D. H. Campbell-Lendrum, A. J. McMichael, A. Woodward, and J. S. Cox. 2001. Early effects of climate change: do they include changes in vector-borne disease? *Philosophical Transactions of the Royal Society Series B* 356(1411):1057-1068.

McMichael, A., D. H. Campbell-Lendrum, C. Corvalán, K. Ebi, A. Githeko, J. Scheraga, and A. Woodward, eds. 2003. *Climate change and human health: risks and responses*. Geneva, Switzerland: World Health Organization.

McMichael, A. J., S. Friel, A. Nyong, and C. Corvalan. 2008. Global environmental change and health: impacts, inequalities, and the health sector. *British Medical Journal* 336(7637):191-194.

Millennium Ecosystem Assessment. 2005. *Ecosystems and human well being: health synthesis*. Geneva, Switzerland: World Health Organization.

NRC (National Research Council). 2001. *Under the weather: climate, ecosystems, and infectious disease*. Washington, DC: National Academy Press.

Pascual, M., J. A. Ahumada, L. F. Chaves, X. Rodo, and M. Bouma. 2006. Malaria resurgence in the East African highlands: temperature trends revisited. *Proceedings of the National Academy of Sciences* 103(15):5829-5834.

Pascual, M., B. Cazelles, M. J. Bouma, L. F. Chaves, and K. Koelle. 2008. Shifting patterns: malaria dynamics and rainfall variability in an African highland. *Proceedings. Biological Sciences* 275(1631):123-132.

Prince, R. 2008. Malaria warning as UK becomes warmer. *Daily Telegraph,* December 2.

Prüss-Üstün, A., D. Kay, F. Fewtrell, and J. Bartram. 2004. Unsafe water, sanitation and hygiene. In *Comparative quantification of health risks: global and regional burden of disease attributable to selected major risk factors*, edited by M. Ezzati, A. Lopez, A. Rodgers, and C. Murray. Geneva, Switzerland: World Health Organization.

Singh, R. B., S. Hales, N. de Wet, R. Raj, M. Hearnden, and P. Weinstein. 2001. The influence of climate variation and change on diarrheal disease in the Pacific Islands. *Environmental Health Perspectives* 109(2):155-159.

Smith, K. R., S. Mehta, and M. Meausezahl-Feuz. 2004. Indoor air pollution from household use of solid fuels. In *Comparative quantification of health risks: global and regional burden of disease due to selected major risk factors*, edited by M. Ezzati, A. Lopez, A. Rodgers, and C. Murray. Geneva, Switzerland: World Health Organization.

Stott, P. A., D. A. Stone, and M. R. Allen. 2004. Human contribution to the European heatwave of 2003. *Nature* 432(7017):610-614.

WHO (World Health Organization). 2002. *The world health report 2002*. Geneva, Switzerland: World Health Organization.

———. 2005. *Using climate to predict infectious disease epidemics*. Geneva, Switzerland: World Health Organization.

———. 2006a. *Burden of disease statistics*, http://www.who.int/healthinfo/bod/en/index.html (accessed April 18, 2008).

———. 2006b. *Fuel for life: household energy and health*. Geneva, Switzerland: World Health Organization.

———. 2006c. *WHO guidelines for the safe use of wastewater, excreta and greywater*, 3rd edition. Geneva, Switzerland: World Health Organization.

———. 2007. *International Health Regulations (IHR)*, http://www.who.int/csr/ihr/en/ (accessed April 18, 2008).

———. 2008a. *Protecting health from climate change: World Health Day 2008*. Geneva, Switzerland: World Health Organization.

———. 2008b. *Climate change and health: WHO publications 1990-2008*, http://www.who.int/globalchange/climate/publications/en/index.html (accessed April 18, 2008).

———. 2008c. *Documentation of the 122nd session of the executive board of the World Health Organization*, http://www.who.int/gb/e/e_eb122.html (accessed April 18, 2008).

Wilkinson, P., D. H. Campbell-Lendrum, and C. L. Bartlett. 2003. Monitoring the health effects of climate change. In *Climate change and human health: risks and responses*, edited by A. McMichael, D. H. Campbell-Lendrum, C. Corvalán, K. Ebi, A. Githeko, J. Scheraga, and A. Woodward. Geneva, Switzerland: World Health Organization.

Appendix A

Agenda

Global Climate Change and Extreme Weather Events: Understanding the Potential Contributions to the Emergence, Reemergence, and Spread of Infectious Disease

December 4-5, 2007
The Hotel Monaco, Paris Room
Washington, DC

DAY 1: DECEMBER 4, 2007

8:30-9:00	Registration and complimentary breakfast
9:00-9:15	Welcoming remarks Forum Leadership
9:15-9:40	Welcoming remarks Harvey Fineberg, M.D., Ph.D., President, Institute of Medicine
9:40-10:20	Keynote address Ecological, environmental, and infectious disease impacts of global climate change and extreme weather events Donald Burke, M.D. University of Pittsburgh
10:20-11:00	Questions from Forum members and audience
11:00-11:15	Break

Session I
The Direct and Indirect Influences of Extreme Weather Events and Climate Change on the Prevalence and Geographic Range of Vector- and Non-Vector-Borne Diseases

Moderator: Lonnie King, D.V.M., Centers for Disease Control and Prevention

11:15-12:15
- Sir Andrew Haines, M.D., M.B., B.Ch.
  London School of Hygiene and Tropical Medicine
- Paul Epstein, M.D., M.P.H.
  Harvard University

12:15-12:45 Questions from Forum members and audience

12:45-1:30 Lunch and continuation of Day 1 morning discussion

Session II
Environmental Trends and Their Influence on the Emergence, Reemergence, and Movement of Vector- and Non-Vector-Borne Infectious Diseases

Moderator: Ralph "Loren" Erickson, M.D., Dr.P.H., M.P.H., Department of Defense, Global Emerging Infections Surveillance and Response System

1:30-2:00 Climate change impacts on plant disease dynamics
Karen A. Garrett, Ph.D.
Kansas State University

2:00-2:30 Climate change: its effects on healthy aquatic and marine wildlife populations
Leslie Dierauf, V.M.D.
U.S. Geological Survey, National Wildlife Health Center

2:30-3:00 Sub-Saharan Africa—chikungunya and Rift Valley fever
Jean-Paul Chretien, M.D., Ph.D.
Department of Defense, Global Emerging Infections Surveillance and Response System

3:00-3:20 Break

3:20-3:50	Arctic environment Alan Parkinson, Ph.D. Centers for Disease Control and Prevention, Arctic Investigation Program
3:50-4:20	Plague and climate variation and change: a worldwide overview with a bias towards central Asia Nils Christian Stenseth, Ph.D. University of Oslo, Norway
4:20-5:20	Discussion of Session II
5:20-6:15	Open discussion of Day 1
6:15	Adjourn—Day 1
6:45-9:30	Dinner with speakers and Forum members and continuing discussion of Day 1

DAY 2: DECEMBER 5, 2007

8:30-9:00	Continental breakfast
9:00-9:15	Summary of Day 1 Margaret "Peggy" Hamburg, M.D. Vice Chair Forum on Microbial Threats

Session III
Scientific, Technological, and Policy Considerations Related to Prediction and Intervention in Vector- and Non-Vector-Borne Disease Outbreaks and Integration Strategies

Moderator: Patrick Fitch, Ph.D., National Biodefense Analysis and Countermeasures Center

9:15-9:45	The use of historical data sets in understanding ecosystem responses to climate change and the importance of long-term monitoring programs: drought Rodolfo Acuña-Soto, M.D. Universidad Nacional Autonoma de Mexico

9:45-10:15	Can on-the-ground monitoring of wildlife help detect the impacts of climate and environmental change on infectious disease epidemiology? William Karesh, D.V.M. Wildlife Conservation Society
10:15-10:30	Break
10:30-11:00	Marine environment Rita Colwell, Ph.D. University of Maryland
11:00-11:30	Use of remote sensing for detecting the impacts of climate and environmental change on infectious disease epidemiology Compton James Tucker, Ph.D. NASA Goddard Space Flight Center
11:30-12:00	The use of satellite-generated meteorological data to predict mosquito-borne encephalitis transmission William K. Reisen, Ph.D. University of California, Davis
12:00-12:45	Open discussion of Session III
12:45-1:30	Lunch and continuation of Day 2 discussion

Session IV
The International Public Health and Foreign Policy Implications of Drought, Flooding, and Extreme Weather Events on the Emergence and Spread of Infectious Diseases

Moderator: David Relman, M.D., Stanford University

1:30-2:00	Implications for international public health policy Diarmid Campbell-Lendrum, Ph.D. World Health Organization
2:00-2:30	Influences of migration and population mobility Douglas W. MacPherson, M.D. McMaster University

2:30-3:00	Global climate change: impacts on national security Maj. Gen. Richard L. Engel, USAF (Ret.) National Intelligence Council
3:00-3:30	Open discussion of Session IV
3:30-4:00	Wrap-up
4:00	Adjourn

Appendix B

Acronyms

AIDS	acquired immune deficiency syndrome
AVHRR	Advanced Very High Resolution Radiometer
CCF	Climate Change Futures
CDC	Centers for Disease Control and Prevention
CGIAR	Consultative Group for International Agricultural Research
CRED	Center for Research on the Epidemiology of Disasters
CRSP	Collaborative Research Support Program
DALY	disability adjusted life-year
DOD-GEIS	Department of Defense-Global Emerging Infections Surveillance and Response System
EID	emerging infectious disease
EIP	extrinsic incubation period
EMDAT	Emergency Events Database
ENSO	El Niño/Southern Oscillation
EU	European Union
FAO	Food and Agriculture Organization
FDA	Food and Drug Administration
GHG	greenhouse gas
GWP	gross world product

HAB	harmful algal bloom
HHS	U.S. Department of Health and Human Services
HIMAL	highland malaria project
HIV	human immunodeficiency virus
HPV	human papilloma virus
HTLV	human T-cell lymphotropic virus
IOM	Institute of Medicine
IPCC	Intergovernmental Panel on Climate Change
IPM	Integrated Pest Management
KEMRI	Kenya Medical Research Institute
LSQ	large square
MODIS	Moderate Resolution Imaging Spectroradiometer
NAS	National Academy of Sciences
NASA	National Aeronautics and Space Administration
NASA-GSFC	NASA Goddard Space Flight Center
NDVI	Normalized Difference Vegetation Index
NIA	National Intelligence Assessment
NIC	National Intelligence Council
NIE	National Intelligence Estimate
NIH	National Institutes of Health
NOAA	National Oceanic and Atmospheric Administration
NPP	net primary productivity
NRC	National Research Council
NSF	National Science Foundation
OFDA	Office of U.S. Foreign Disaster Assistance
OIRED	Office of International Research, Education, and Development
OLR	outgoing longwave radiation
PDO	Pacific Decadal Oscillation
PSQ	primary square
RSV	respiratory syncytial virus
RT-PCR	reverse transcriptase-polymerase chain reaction
RVF	Rift Valley fever
SANREM	Sustainable Agriculture and Natural Resource Management
SARS	severe acute respiratory syndrome

SRES	Special Report on Emissions Scenarios
SST	sea surface temperature
TBE	tick-borne encephalitis
TOPS	Terrestrial Observation and Prediction System
ULV	ultra low volume
UN	United Nations
UNDP	United Nations Development Programme
UNHCR	United Nations High Commissioner for Refugees
USAID	U.S. Agency for International Development
USAMRU	U.S. Army Medical Research Unit
VHS	viral hemorrhagic septicemia
WHO	World Health Organization
WNV	West Nile virus

Appendix C

Glossary

Abiotic: Nonliving chemical and physical factors in an environment.

Agent (of disease): Factor such as a microorganism whose presence is essential for the occurrence of a disease.

Anthropogenic: Caused or produced by humans.

Circumpolar region: The region that extends above 60°N latitude, borders the Arctic Ocean, and includes all, or the northern parts, of eight nations: the United States (Alaska), Canada, Greenland, Iceland, Norway, Finland, Sweden, and the Russian Federation.

Climate: Average meteorological conditions over a specified time period, usually at least a month, resulting from interactions among the atmosphere, oceans, and land surface. Climate variations occur over a wide range of spatial and temporal scales.

Climate change: A change of climate which is attributed directly or indirectly to human activity that alters the composition of the global atmosphere and which is in addition to natural climate variability observed over comparable time periods (http://www.ipcc.ch/pdf/glossary/ipcc-glossary.pdf).

Climate extremes: used to represent weather extremes (see definition below), but viewed over seasons (e.g., droughts), or longer periods (http://downloads.climatescience.gov/sap/sap3-3/sap3-3-final-FrontMaterials.pdf).

Climate variability: refers to variations or deviations from the mean state of the climate or temporal variations of the atmosphere-ocean system around a mean state measure over a long period of time. Typically, this term is used for timescales longer than those associated with synoptic weather events (i.e., months to millennia and longer). The term "natural climate variability" is further used to identify climate variations that are not attributable to or influenced by any activity related to humans. However it is recognized that such "internal or natural variability" could be affected by external factors driving climate change such as changes in the atmospheric concentration of greenhouse gases. The El Niño-Southern Oscillation (ENSO) phenomena is a good example of the variability in the coupled oceanic and atmosphere system that is a central factor in short-term climate variability and the interannual timescale (http://www.cpc.noaa.gov/products/analysis_monitoring/ensostuff/prelude_to_ensofaq.shtml; http://www.ncdc.noaa.gov/paleo/outreach/coral/coralenso.html; http://www.sws.uiuc.edu/atmos/statecli/Climate_change/glossary.htm).

Ecosystem: Mutually interrelated communities of species and abiotic components, existing as a system with specific interactions and exchange of matter, energy, and information.

El Niño: A warming of the surface waters of the tropical Pacific that occurs every 3 to 5 years, temporarily affecting weather worldwide.

Emerging infection: Either a newly recognized, clinically distinct infectious disease or a known infectious disease whose reported incidence is increasing in a given place or among a specific population.

Endemic: Restricted or peculiar to a locality or region. Endemic infection refers to a sustained, relatively stable pattern of infection in a specified population.

Epidemic: Appearance of an abnormally high number of cases of infection in a given population.

Epidemiology: Study of the distribution and determinants of health-related states or events in specified populations. Epidemiology is the basic quantitative science of public health.

Extreme weather: refers to weather phenomena that are at the extremes of the historical distribution and are rare for a particular place and/or time, especially severe or unseasonal weather. Such extremes include severe thunderstorms; severe snowstorms, ice storms, blizzards, flooding, hurricanes, and high winds, and heat waves. For example, although flooding is common in the United

States, the impacts of flooding are not consistent from year to year through time. Many years of small floods with little impact may be followed by a single large flood with a sizable loss (e.g., the June 2008 flooding in the midwestern United States) (http://www.greenhouse.gov.au/impacts/resources/glossary.html; http://en.wikipedia.org/wiki/Extreme_weather; http://www.sws.uiuc.edu/atmos/statecli/General/Illinois-climate-narrative.htm).

Extrinsic incubation period: Time required for the development of a disease agent in a vector from the time of uptake of the agent to the time the vector is infective.

Global warming: The gradual increase, observed or projected, in global surface temperature, as one of the consequences of radiative forcing caused by anthropogenic emissions (http://www.ipcc.ch/pdf/glossary/ar4-wg3.pdf).

Ground truth: information collected on location (http://en.wikipedia.org/wiki/Ground_truth).

Host (disease): Person or other living animal that affords subsistence or lodgment to an infectious agent under natural conditions.

Incidence: Number of cases of a disease commencing, or of persons falling ill, during a given period of time in a specified population. Incidence rate is the number of new cases of a specific disease diagnosed or reported during a defined interval of time divided by the number of all persons in a defined population during the same time.

Intertidal zone: The area that is exposed to the air at low tide and submerged at high tide (http://en.wikipedia.org/wiki/Intertidal_zone).

La Niña: Cooler than normal sea surface temperatures in the central and eastern tropical Pacific Ocean that impact global weather patterns. La Niña conditions recur every few years and can persist for as long as 2 years (http://www.elnino.noaa.gov/lanina_new_faq.html).

Mitigation: Initiatives that reduce the risk from natural and man-made hazards. With respect to climate change, mitigation usually refers to actions taken to reduce the emissions or enhance the sinks of greenhouse gases.

Nowcasts: Forecasts of events in the immediate future.

Ocean thermohaline circulation: The stabilizing ocean "conveyor belt."

Oomycete: A group of filamentous, unicellular heterokonts, physically resembling fungi (http://en.wikipedia.org/wiki/Oomycete).

Outbreak: Localized occurrence as opposed to a generalized epidemic.

Pandemic: Epidemic that occurs over a very wide area.

Pathogen: Organism capable of causing disease.

Permafrost: Permanently frozen land.

Phytohormone: Chemical that regulates plant growth (http://en.wikipedia.org/wiki/Plant_hormone).

Phytoplankton: Microscopic plants that live in the ocean (http://earth observatory.nasa.gov/Library/Phytoplankton/).

Poikilothermic: Any organism whose body temperature varies with the temperature of its surroundings.

Prevalence: Proportion of persons in a population currently affected by a particular disease. Prevalence rate is the number of cases of a specific disease at a particular time divided by the population at that time living in the same region.

Risk: Probability that an event will occur; a measure of the degree of loss expected by the occurrence of an event.

Seroconversion: Development of antibodies in blood serum as a result of infection.

Southern Oscillation: A large-scale atmospheric and hydrospheric fluctuation centered in the equatorial Pacific Ocean; it exhibits a nearly annual pressure anomaly, alternatively high over the Indian Ocean and high over the South Pacific; the variation in pressure is accompanied by variations in wind strengths, ocean currents, sea surface temperatures, and precipitation in the surrounding areas.

Transmission: Process by which a pathogen passes from a source of infection to a new host.

Vector: An organism, such as an insect, that transmits a pathogen from one host to another.

Vector-borne disease: (1) *Mechanical*: this includes simple mechanical carriage by a crawling or flying insect through soiling of its feet or proboscis or by passage of organisms through its gastrointestinal tract. This does not require multiplication or development of the organism. (2) *Biological*: propagation (multiplication), cyclic development, or a combination of these (cyclopropagative) is required before the arthropod can transmit the infective form of the agent to humans. An incubation period (extrinsic) is required following infection before the arthropod becomes infective. The infectious agent may be passed vertically to succeeding generations (transovarian transmission); transstadial transmission indicates its passage from one stage of the life cycle to another, as nymph to adult. Transmission may be by injection of salivary gland fluid during biting, or by regurgitation or deposition on the skin of feces or other material capable of penetrating the bite wound or an area of trauma from scratching or rubbing. This transmission is by an infected nonvertebrate host and not simple mechanical carriage by a vector or vehicle. However, an arthropod in either role is termed a vector.

Weather: Condition of the atmosphere at a particular place and time measured in terms of wind, temperature, humidity, atmospheric pressure, cloudiness, and precipitation. In most places, weather can change from hour to hour, from day to day, and from season to season.

Weather extremes (extreme weather events): Signifies individual weather events that are unusual in their occurrence (minimally, the event must lie in the upper or lower tenth percentile of the distribution) or have destructive potential, such as hurricanes and tornadoes (http://downloads.climatescience.gov/sap/sap3-3/sap3-3-final-FrontMaterials.pdf).

Zoonosis: Infection that causes disease in human populations but can be perpetuated solely in nonhuman host animals (e.g., bubonic plague); may be enzootic or epizootic.

Appendix D

Forum Member Biographies

David A. Relman, M.D. (*Chair*), is professor of medicine (infectious diseases and geographic medicine) and of microbiology and immunology at Stanford University School of Medicine, and chief of the infectious disease section at the Veterans Affairs (VA) Palo Alto Health Care System. Dr. Relman received his B.S. in biology from the Massachusetts Institute of Technology and his M.D. from Harvard Medical School. He completed his residency in internal medicine and a clinical fellowship in infectious diseases at Massachusetts General Hospital, Boston, after which he moved to Stanford for a postdoctoral fellowship in 1986 and joined the faculty there in 1994. His research focus is on understanding the structure and role of the human indigenous microbial communities in health and disease. This work brings together approaches from ecology, population biology, environmental microbiology, genomics, and clinical medicine. A second area of investigation explores the classification structure of humans and nonhuman primates with systemic infectious diseases, based on patterns of genome-wide gene transcript abundance in blood and other tissues. The goals of this work are to understand mechanisms of host-pathogen interaction, as well as predict clinical outcome early in the disease process. His scientific achievements include the description of a novel approach for identifying previously unknown pathogens; the characterization of a number of new human microbial pathogens, including the agent of Whipple's disease; and some of the most in-depth analyses to date of human indigenous microbial communities. Among his other activities, Dr. Relman currently serves as chair of the Board of Scientific Counselors of the National Institutes of Health (NIH) National Institute of Dental and Craniofacial Research, is a member of the National Science Advisory Board for Biosecurity, and advises a number of U.S. government departments and agencies on matters related to pathogen diversity, the future life sciences landscape, and the nature of present and future biological threats. He was co-chair of the Committee on Advances in Technology and the Prevention of Their Application to Next Generation Biowarfare Threats for the National Academy of Sciences (NAS). He received the Squibb Award from the Infectious Diseases Society of America (IDSA) in 2001, the Senior Scholar Award in Global Infectious Diseases from the Ellison Medical Foundation in 2002, an NIH Director's Pioneer Award in 2006, and a Doris Duke Distinguished Clinical Scientist Award in 2006. He is also a fellow of the American Academy of Microbiology.

Margaret A. Hamburg, M.D. (*Vice Chair*), was the founding vice president, Biological Programs, at the Nuclear Threat Initiative, a charitable organization working to reduce the global threat from nuclear, biological, and chemical weapons, and ran the program for many years. She currently serves as senior scientist for the organization. She completed her internship and residency in internal medicine at the New York Hospital-Cornell University Medical Center and is certified by the American Board of Internal Medicine. Dr. Hamburg is a graduate of Harvard College and Harvard Medical School. Before taking on her current position, she was the assistant secretary for planning and evaluation, U.S. Department of Health and Human Services (HHS), serving as a principal policy adviser to the secretary of health and human services, with responsibilities including policy formulation and analysis, the development and review of regulations and legislation, budget analysis, strategic planning, and the conduct and coordination of policy research and program evaluation. Prior to this, she served for nearly 6 years as the commissioner of health for the City of New York. As chief health officer in the nation's largest city, her many accomplishments included the design and implementation of an internationally recognized tuberculosis control program that produced dramatic declines in tuberculosis cases, the development of initiatives that raised childhood immunization rates to record levels, and the creation of the first public health bioterrorism preparedness program in the nation. She currently serves on the Harvard University Board of Overseers. She has been elected to membership in the Institute of Medicine (IOM), the New York Academy of Medicine, and the Council on Foreign Relations and is a fellow of the American Association for the Advancement of Science (AAAS) and the American College of Physicians.

David W. K. Acheson, M.D., F.R.C.P., is assistant commissioner for food protection in the U.S. Food and Drug Administration (FDA). Dr. Acheson graduated from the University of London Medical School in 1980 and, following training in internal medicine and infectious diseases in the United Kingdom, moved to the New England Medical Center and Tufts University in Boston in 1987. As an associate professor at Tufts University, he undertook basic molecular pathogenesis research on food-borne pathogens, especially Shiga toxin-producing *Escherichia* *coli*. In 2001, Dr. Acheson moved his laboratory to the University of Maryland Medical School in Baltimore to continue research on food-borne pathogens. In September 2002, Dr. Acheson accepted a position as chief medical officer at the FDA Center for Food Safety and Applied Nutrition (CFSAN). In January 2004, he also became the director of CFSAN's Food Safety and Security Staff, and in January 2005, the staff was expanded to become the Office of Food Safety, Defense and Outreach. In January 2007, the office was further expanded to become the Office of Food Defense, Communication and Emergency Response. On May 1, 2007, Dr. Acheson assumed the position of FDA assistant commissioner for food protection to provide advice and counsel to the commissioner on strategic and substantive food safety and food defense matters. Dr. Acheson has published extensively and is internationally recognized both for his public health expertise in food safety and for his research in infectious diseases. Additionally, Dr. Acheson is a fellow of both the Royal College of Physicians (London) and the IDSA.

Ruth L. Berkelman, M.D., is the Rollins Professor and director of the Center for Public Health Preparedness and Research at the Rollins School of Public Health, Emory University, in Atlanta. She received her A.B. from Princeton University and her M.D. from Harvard Medical School. Board certified in pediatrics and internal medicine, she began her career at the Centers for Disease Control and Prevention (CDC) in 1980 and later became deputy director of the National Center for Infectious Diseases (NCID). She also served as a senior adviser to the director of CDC and as assistant surgeon general in the U.S. Public Health Service. In 2001 she came to her current position at Emory University, directing a center focused on emerging infectious diseases and other urgent threats to health, including terrorism. She has also consulted with the biologic program of the Nuclear Threat Initiative and is most recognized for her work in infectious diseases and disease surveillance. She was elected to the IOM in 2004. Currently a member of the Board on Life Sciences of the National Academies, she also chairs the Board of Public and Scientific Affairs at the American Society for Microbiology (ASM).

Enriqueta C. Bond, Ph.D., is president of the Burroughs Wellcome Fund. She received her undergraduate degree from Wellesley College, her M.A. from the University of Virginia, and her Ph.D. in molecular biology and biochemical genetics from Georgetown University. She is a member of the IOM, AAAS, ASM, and the American Public Health Association. Dr. Bond chairs the Academies' Board on African Science Academy Development and serves on the Report Review Committee for the Academies. She serves on the board and executive committee of the Research Triangle Park Foundation, the board of the National Institute for Statistical Sciences, the board of the Northeast Biodefense Center and the New England Center of Excellence in Biodefense and Emerging Infectious Diseases, and the council of the National Institute of Child Health and Human Development. Prior to being named president of the Burroughs Wellcome Fund in 1994, Dr. Bond served on the staff of the IOM beginning in 1979, becoming its executive officer in 1989.

Roger G. Breeze, Ph.D., received his veterinary degree in 1968 and his Ph.D. in veterinary pathology in 1973, both from the University of Glasgow, Scotland. He was engaged in teaching, diagnostic pathology, and research on respiratory and cardiovascular diseases at the University of Glasgow Veterinary School from 1968 to 1977 and at Washington State University College of Veterinary Medicine from 1977 to 1987, where he was professor and chair of the Department of Microbiology and Pathology. From 1984 to 1987 he was deputy director of the Washington Technology Center, the state's high-technology sciences initiative, based in the College of Engineering at the University of Washington. In 1987, he was appointed director of the U.S. Department of Agriculture's (USDA's) Plum Island Animal Disease Center, a Biosafety Level 3 facility for research and diagnosis of the world's most dangerous livestock diseases. In that role he initiated research into the genomic and functional genomic basis of disease pathogenesis, diagnosis, and control of livestock RNA and DNA virus infections. This work became the basis of U.S. defense against natural and deliberate infection with these agents and led to his involvement in the early 1990s in biological weapons defense and proliferation prevention. From 1995 to 1998, he directed research programs in 20 laboratories in the Southeast for the USDA Agricultural Research Service before going to Washington, DC, to establish biological weapons defense research programs for USDA. He received the Distinguished Executive Award from President Clinton in 1998 for his work at Plum Island and in biodefense. Since 2004 he has been chief executive officer of Centaur Science Group, which provides consulting services in biodefense. His main commitment is to the Defense Threat Reduction Agency's Biological Weapons Proliferation Prevention Program in Europe, the Caucasus, and Central Asia.

Steven J. Brickner, Ph.D., is a research fellow in antibacterials chemistry at Pfizer Global Research and Development in Groton, Connecticut. He graduated from Miami University (Ohio) with a B.S. in chemistry with honors and received his M.S. and Ph.D. degrees in organic chemistry from Cornell University. He was an NIH postdoctoral research fellow at the University of Wisconsin-Madison. Dr. Brickner is a medicinal chemist with 25 years of research experience in the pharmaceutical industry, all focused on the discovery of novel antibacterial agents. He is an inventor or co-inventor on 21 U.S. patents and has published numerous scientific papers in the areas of oxazolidinones and novel azetidinones. Dr. Brickner has been a member of the IOM Forum on Microbial Threats since 1997 and is a member of the Editorial Advisory Board for *Current Pharmaceutical Design*. Dr. Brickner initiated the oxazolidinone research program at

Upjohn, led the team that discovered Zyvox® (linezolid), and is a co-inventor of this antibiotic used to treat multidrug-resistant Gram-positive infections. Zyvox is the first member of *any* entirely new class of antibiotics to reach the market in more than 35 years since the quinolones. He is a co-recipient of the 2007 American Chemical Society Team Innovation Award and the Pharmaceutical Research and Manufacturers of America's 2007 Discoverers Award. He was named the 2002-2003 Outstanding Alumni Lecturer, College of Arts and Science, Miami University (Ohio).

Gail H. Cassell, Ph.D., is currently vice president, Scientific Affairs, and Distinguished Lilly Research Scholar for Infectious Diseases, Eli Lilly and Company in Indianapolis, Indiana. She is the former Charles H. McCauley Professor and chairman of the Department of Microbiology at the University of Alabama Schools of Medicine and Dentistry at Birmingham, a department that ranked first in research funding from NIH during her decade of leadership. She obtained her B.S. from the University of Alabama in Tuscaloosa and in 1993 was selected as one of the top 31 female graduates of the twentieth century. She obtained her Ph.D. in microbiology from the University of Alabama at Birmingham and was selected as its 2003 Distinguished Alumnus. She is a past president of ASM (the oldest and single-largest life sciences organization with a membership of more than 42,000). She was a member of the NIH Director's Advisory Committee and a member of the Advisory Council of the National Institute of Allergy and Infectious Diseases of NIH. She was named to the original Board of Scientific Councilors of the CDC Center for Infectious Diseases and served as chair of the board. She recently served a 3-year term on the Advisory Board of the director of the CDC and as a member of the HHS secretary's Advisory Council of Public Health Preparedness. Currently she is a member of the Science Board of the FDA Advisory Committee to the Commissioner. Since 1996 she has been a member of the U.S.-Japan Cooperative Medical Science Program responsible for advising the respective governments on joint research agendas (U.S. State Department-Japan Ministry of Foreign Affairs). She has served on several editorial boards of scientific journals and has authored more than 250 articles and book chapters. Dr. Cassell has received national and international awards and an honorary degree for her research in infectious diseases. She is a member of the IOM and is currently serving a 3-year term on the IOM Council, its governing board. Dr. Cassell has been intimately involved in the establishment of science policy and legislation related to biomedical research and public health. For 9 years she was chairman of the Public and Scientific Affairs Board of ASM; she has served as an adviser on infectious diseases and indirect costs of research to the White House Office of Science and Technology Policy; and she has been an invited participant in numerous congressional hearings and briefings related to infectious diseases, antimicrobial resistance, and biomedical research. She has served two terms on the Liaison Committee for Medical Education (LCME), the accrediting body for

U.S. medical schools, as well as other national committees involved in establishing policies in training in the biomedical sciences. She has just completed a term on the Leadership Council of the School of Public Health of Harvard University. Currently she is a member of the Executive Committee of the Board of Visitors of Columbia University School of Medicine, the Board of Directors of the Burroughs Wellcome Fund, and the Advisory Council of the School of Nursing of Johns Hopkins.

Bill Colston, Ph.D., is division leader for the Chemical and Biological Countermeasures (CB) Division for the Global Security (GS) Principal Directorate at Lawrence Livermore National Laboratory. The newly formed CB Division is comprised of about 190 scientists from a variety of disciplines. The mission of this division is to provide national policy support, threat characterization, biological detection, chemical and explosives detection, instrumentation and systems development, decontamination and restoration, forensics and attribution, Biodefense Knowledge Center products, and incident response support operations. Prior to this assignment he held the positions of founding director of the Department of Homeland Security (DHS) Biodefense Knowledge Center (BKC) and deputy program leader for the Chemical and Biological Security Program. Dr. Colston holds a Ph.D. from the University of California, Davis, in biomedical engineering. He has published more than 40 publications in the scientific literature, holds more than 15 patents related to medical diagnostics and imaging devices, and has received three different R&D 100 Awards. His research interests are focused mainly on molecular characterization of infectious disease, with direct relevance to new diagnostic devices.

Col. Ralph (Loren) Erickson, M.D., M.P.H., Dr.P.H., is the director of the Department of Defense Global Emerging Infections Surveillance and Response System (DOD-GEIS) headquartered in Silver Spring, Maryland. He holds a B.S. degree in chemistry from the University of Washington, an M.D. from the Uniformed Services University of the Health Sciences, an M.P.H. from Harvard, and a Dr.P.H. from Johns Hopkins. Residency trained and board certified in preventive medicine, Dr. Erickson has held a number of leadership positions within the Army Medical Department, including director of the General Preventive Medicine Residency Program, Walter Reed Army Institute of Research; director, Epidemiology and Disease Surveillance, U.S. Army Center for Health Promotion and Preventive Medicine; commander of the U.S. Army Center for Health Promotion and Preventive Medicine (Europe); and specialty leader for all U.S. Army preventive medicine physicians.

Mark Feinberg, M.D., Ph.D., is vice president for medical affairs and policy in global vaccine and infectious diseases at Merck & Co., Inc., and is responsible for global efforts to implement vaccines to achieve the greatest health benefits, including efforts to expand access to new vaccines in the developing world. Dr. Feinberg received a bachelor's degree magna cum laude from the University of Pennsylvania in 1978 and his M.D. and Ph.D. degrees from Stanford University School of Medicine in 1987. His Ph.D. research at Stanford was supervised by Dr. Irving Weissman and included time spent studying the molecular biology of the human retroviruses—HTLV-I (human T-cell lymphotrophic virus, type I) and HIV—as a visiting scientist in the laboratory of Dr. Robert Gallo at the National Cancer Institute. From 1985 to 1986, Dr. Feinberg served as a project officer for the IOM Committee on a National Strategy for AIDS. After receiving his M.D. and Ph.D. degrees, Dr. Feinberg pursued postgraduate residency training in internal medicine at the Brigham and Women's Hospital of Harvard Medical School and postdoctoral fellowship research in the laboratory of Dr. David Baltimore at the Whitehead Institute for Biomedical Research. From 1991 to 1995, Dr. Feinberg was an assistant professor of medicine and microbiology and immunology at the University of California, San Francisco (UCSF), where he also served as an attending physician in the AIDS-oncology division and as director of the virology research laboratory at San Francisco General Hospital. From 1995 to 1997, Dr. Feinberg was a medical officer in the Office of AIDS Research in the Office of the Director of NIH, the chair of the NIH Coordinating Committee on AIDS Etiology and Pathogenesis Research, and an attending physician at the NIH Clinical Center. During this period, he also served as executive secretary of the NIH Panel to Define Principles of Therapy of HIV Infection. Prior to joining Merck in 2004, Dr. Feinberg served as professor of medicine and microbiology and immunology at the Emory University School of Medicine, as an investigator at the Emory Vaccine Center, and as an attending physician at Grady Memorial Hospital. At UCSF and Emory, Dr. Feinberg and colleagues were engaged in the preclinical development and evaluation of novel vaccines for HIV and other infectious diseases and in basic research studies focused on revealing fundamental aspects of the pathogenesis of AIDS. Dr. Feinberg also founded and served as the medical director of the Hope Clinic of the Emory Vaccine Center—a clinical research facility devoted to the clinical evaluation of novel vaccines and to translational research studies of human immune system biology. In addition to his other professional roles, Dr. Feinberg has also served as a consultant to, and a member of, several IOM and NAS committees. Dr. Feinberg currently serves as a member of the National Vaccine Advisory Committee (NVAC) and is a member of the Board of Trustees of the National Foundation for Infectious Diseases (NFID). Dr. Feinberg has earned board certification in internal medicine; he is a fellow of the American College of Physicians, a member of the Association of American Physicians, and the recipient of an Elizabeth Glaser Scientist Award from the Pediatric AIDS Foundation and an Innovation in Clinical Research Award from the Doris Duke Charitable Foundation.

J. Patrick Fitch, Ph.D., is laboratory director for the National Biodefense Analysis and Countermeasures Center (NBACC) and president of Battelle National Biodefense Institute, LLC (BNBI). BNBI manages and operates the NBACC national laboratory for DHS as a Federally Funded Research and Development Center (FFRDC) established in 2006. NBACC's mission is to provide the nation with the scientific basis for awareness of biological threats and attribution of their use against the American public. Dr. Fitch joined Battelle in 2006 as vice president for Biodefense Programs after more than 20 years of experience leading multidisciplinary applied science teams at the University of California's Lawrence Livermore National Laboratory (LLNL). From 2001 to 2006, he led the LLNL Chemical and Biological National Security Program (CBNP), with applied science programs from pathogen biology to deployed systems. CBNP accomplishments include performing more than 1 million assays on national security samples; setting up and operating 24/7 reach-back capabilities; setting up a nationwide bioalert system; receiving three R&D 100 awards; designing signatures for validated assays in the CDC Laboratory Response Network and the National Animal Health Laboratory Network; and designing. His advisory board activities have included the U.S. Animal Health Association, Texas A&M University DHS Center of Excellence, Central Florida University (College of Engineering), Colorado State University (College of Engineering), California State Breast Cancer Research Program, and *Biomolecular Engineering*. Dr. Fitch was a fellow of the American Society for Laser Medicine and Surgery and an associate editor of *Circuits, Systems and Signal Processing*. He has received two national awards for medical devices, a technical writing award for an article in *Science*, and an international best paper award from the Institute of Electrical and Electronics Engineers. He also co-invented the technology, developed the initial business plan, and successfully raised venture investments for a medical device start-up company. Dr. Fitch received his Ph.D. from Purdue University and B.S. from Loyola College of Maryland.

Capt. Darrell R. Galloway, M.S.C., Ph.D., is chief of the Medical Science and Technology Division for the Chemical and Biological Defense Directorate at the Defense Threat Reduction Agency. He received his baccalaureate degree in microbiology from California State University in Los Angeles in 1973. After completing military service in the U.S. Army as a medical corpsman from 1969 to 1972, Captain Galloway entered graduate school and completed a doctoral degree in biochemistry in 1978 from the University of California, followed by 2 years of postgraduate training in immunochemistry as a fellow of the National Cancer Institute (NCI) at the Scripps Clinic and Research Foundation in La Jolla, California. Captain Galloway began his Navy career at the Naval Medical Research Institute in Bethesda, Maryland, where he served as a research scientist working on vaccine development from 1980 to 1984. In late 1984, Captain Galloway left active service to pursue an academic appointment at Ohio State University, where he is now a tenured faculty member in the Department of Microbiology. He also holds appointments at the University of Maryland Biotechnology Institute and the Uniformed Services University of the Health Sciences. He has an international reputation in the area of bacterial toxin research and has published more than 50 research papers on various studies of bacterial toxins. In recent years, Captain Galloway's research has concentrated on anthrax and the development of DNA-based vaccine technology. His laboratory has contributed substantially to the development of a new DNA-based vaccine against anthrax that has completed the first phase of clinical trials. Captain Galloway is a member of the ASM and has served as president of the Ohio branch of that organization. He received an NIH Research Career Development Award. In 2005, Captain Galloway was awarded the Joel M. Dalrymple Award for significant contributions to biodefense vaccine development.

S. Elizabeth George, Ph.D., is deputy director, Biological Countermeasures Portfolio Science and Technology Directorate, Department of Homeland Security. Until merging into the new department in 2003, she was program manager of the Chemical and Biological National Security Program in the Department of Energy's National Nuclear Security Administration's Office of Nonproliferation Research and Engineering. Significant accomplishments include the design and deployment of BioWatch, the nation's first civilian biological threat agent monitoring system, and PROTECT, the first civilian operational chemical detection and response capability deployed in the Washington, DC.area subway system. Previously, she spent 16 years at the U.S. Environmental Protection Agency (EPA), Office of Research and Development, National Health and Ecological Effects Research Laboratory, Environmental Carcinogenesis Division, where she was branch chief of the Molecular and Cellular Toxicology Branch. She received her B.S. in biology in 1977 from Virginia Polytechnic Institute and State University and her M.S. and Ph.D. in microbiology in 1979 and 1984, respectively, from North Carolina State University. From 1984 to 1986, she was an NRC fellow in the laboratory of Dr. Larry Claxton at EPA. Dr. George is the 2005 chair of the Chemical and Biological Terrorism Defense Gordon Research Conference. She has served as councilor for the Environmental Mutagen Society and president and secretary of the Genotoxicity and Environmental Mutagen Society. She holds memberships in the ASM and the AAAS and is an adjunct faculty member in the School of Rural Public Health, Texas A&M University. She is a recipient of the EPA Bronze Medal and Scientific and Technological Achievement Awards and the DHS Under Secretary's Award for Science and Technology. She is author of numerous journal articles and has presented her research at national and international meetings.

Jesse L. Goodman, M.D., M.P.H., is director of the FDA's Center for Biologics Evaluation and Research (CBER), which oversees medical, public health, and policy activities concerning the development and assessment of vaccines, blood products, tissues, and related devices and novel therapeutics, including cellular and gene therapies. He moved to the FDA full-time in 2001 from the University of Minnesota, where he was professor of medicine and director of the Division of Infectious Diseases. A graduate of Harvard College, he received his M.D. at the Albert Einstein College of Medicine; did residency and fellowship training at the Hospital of the University of Pennsylvania and at the University of California, Los Angeles (UCLA), where he was also chief medical resident; and is board certified in internal medicine, oncology, and infectious diseases. He trained in the virology laboratory of Jack Stevens at UCLA and has had an active laboratory program in the molecular pathogenesis of infectious diseases. In 1995, his laboratory isolated the etiologic agent of human granulocytic ehrlichiosis (HGE) and subsequently characterized fundamental events involved in the infection of leukocytes, including their cellular receptors. He is editor of the book *Tick Borne Diseases of Humans* published by ASM Press in 2005 and is a staff physician and infectious diseases consultant at the NIH Clinical Center and the National Naval Medical Center-Walter Reed Army Medical Center, as well as adjunct professor of medicine at the University of Minnesota. He is active in a wide variety of clinical, public health, and product development issues, including pandemic and emerging infectious disease threats; bioterrorism preparedness and response; and blood, tissue, and vaccine safety and availability. In these activities, he has worked closely with CDC, NIH, and other HHS components, academia, and the private sector, and he has put into place an interactive team approach to emerging threats. This model was used in the collaborative development and rapid implementation of nationwide donor screening of the U.S. blood supply for West Nile virus. He has been elected to the American Society for Clinical Investigation (ASCI) and to the IOM.

Eduardo Gotuzzo, M.D., is principal professor and director at the Instituto de Medicina Tropical Alexander von Humbolt, Universidad Peruana Cayetan Heredia in Lima, Peru, as well as chief of the Department of Infectious and Tropical Diseases at the Cayetano Heredia Hospital. He is also an adjunct professor of medicine at the University of Alabama, Birmingham, School of Medicine. Dr. Gotuzzo is an active member of numerous international societies and has been president of the Latin America Society of Tropical Disease (2000-2003), the IDSA Scientific Program (2000-2003), the International Organizing Committee of the International Congress of Infectious Diseases (1994 to present), president-elect of the International Society for Infectious Diseases (1996-1998), and president of the Peruvian Society of Internal Medicine (1991-1992). He has published more than 230 articles and chapters as well as six manuals and one book. Recent honors and awards include being named an honorary member of the American Society of Tropical Medicine and Hygiene in 2002, an associate member of the National Academy of Medicine in 2002, an honorary member of the Society of

Internal Medicine in 2000, and a distinguished visitor at the Faculty of Medical Sciences, University of Cordoba, Argentina, in 1999. In 1988 he received the Golden Medal for Outstanding Contribution in the Field of Infectious Diseases awarded by Trnava University, Slovakia.

Jo Handelsman, Ph.D., received her Ph.D. in molecular biology from the University of Wisconsin, Madison (UW-M), in 1984 and joined the faculty of the UW-M Department of Plant Pathology in 1985, where she is currently a Howard Hughes Medical Institute (HHMI) professor. Her research focuses on the genetic and functional diversity of microorganisms in soil and insect gut communities. The Handelsman lab has concentrated on the discovery and biological activity of novel antibiotics from cultured and uncultured bacteria and has contributed to the pioneering of a new technique called metagenomics that facilitates the genomic analysis of assemblages of uncultured microorganisms. Handelsman is studying the midgut of the gypsy moth to understand the basis for the resistance and susceptibility of microbial communities to invasion, developing it as a model for the microbial community in the human gut. In addition to her passion for understanding the secret lives of bacteria, Dr. Handelsman is dedicated to improving science education and to the advancement of women in research universities. She is director of the HHMI New Generation Program for Scientific Teaching, which is dedicated to teaching graduate and postdoctoral students the principles and practices of teaching and mentoring. She is co-director of the National Academies Summer Institute for Undergraduate Education in Biology, a collaborative venture between HHMI and the National Academies that aims to train a nationwide network of faculty who are outstanding teachers and mentors. Dr. Handelsman is co-director of the Women in Science and Engineering Leadership Institute at UW-M, whose mission is to understand the impediments to the successful recruitment and advancement of women faculty in the sciences and to develop and study interventions intended to reduce those barriers.

Carole A. Heilman, Ph.D., is the director of the Division of Microbiology and Infectious Diseases (DMID), at the National Institute of Allergy and Infectious Diseases (NIAID), a component of NIH-HHS. As director of DMID she has responsibility for scientific direction, oversight, and management of all extramural research programs on infectious diseases (except AIDS) within NIH. In addition, since 2001 Dr. Heilman has played a critical role in launching and directing NIAID's extramural biodefense research program. Previously, Dr. Heilman served as deputy director of NIAID's Division of AIDS for 3 years. Dr. Heilman has a Ph.D. in microbiology from Rutgers University. She did her postdoctoral work in molecular virology at the National Cancer Institute and continued at the NCI as a senior staff fellow in molecular oncology. She moved into health science administration in 1986, focusing on respiratory pathogens, particularly vaccine development. She has received numerous awards for scientific management and leadership, including three HHS Secretary's Awards for Distinguished Service for her contributions to developing pertussis, biodefense, and AIDS vaccines.

David L. Heymann, M.D., is currently executive director of the World Health Organization (WHO) Communicable Diseases Cluster. From October 1995 to July 1998, he was director of the WHO Programme on Emerging and Other Communicable Diseases Surveillance and Control. Prior to becoming director of this program, he was the chief of research activities in the Global Programme on AIDS. From 1976 to 1989, before joining WHO, Dr. Heymann spent 13 years working as a medical epidemiologist in sub-Saharan Africa (Cameroon, Ivory Coast, the former Zaire, and Malawi) on assignment from CDC in CDC-supported activities aimed at strengthening capacity in the surveillance and control of infectious diseases, with special emphasis on childhood immunizable diseases, African hemorrhagic fevers, pox viruses, and malaria. While based in Africa, he participated in the investigation of the first outbreak of Ebola in Yambuku in the former Zaire in 1976, then investigated the second outbreak of Ebola in 1977 in Tandala; in 1995, he directed the international response to the Ebola outbreak in Kikwit. Prior to 1976, Dr. Heymann spent 2 years in India as a medical officer in the WHO Smallpox Eradication Programme. He holds a B.A. from Pennsylvania State University, an M.D. from Wake Forest University, and a diploma in tropical medicine and hygiene from the London School of Hygiene and Tropical Medicine. He has also completed practical epidemiology training in CDC's Epidemic Intelligence Service training program. He has published 131 scientific articles on infectious diseases in peer-reviewed medical and scientific journals.

Phil Hosbach is vice president, New Products and Immunization Policy, at Sanofi Pasteur. The areas under his supervision are new product marketing, state and federal government policy, business intelligence, bids and contracts, medical communications, public health sales, and public health marketing. His current responsibilities include oversight of immunization policy development. He acts as Sanofi Pasteur's principal liaison with CDC. Mr. Hosbach graduated from Lafayette College in 1984 with a degree in biology. He has 20 years of pharmaceutical industry experience, including the past 17 years focused solely on vaccines. He began his career at American Home Products in clinical research in 1984. He joined Aventis Pasteur (then Connaught Labs) in 1987 as clinical research coordinator and has held research and development positions of increasing responsibility, including clinical research manager and director of clinical operations. Mr. Hosbach also served as project manager for the development and licensure of Tripedia, the first diphtheria, tetanus, and acellular pertussis (DTaP) vaccine approved by the FDA for use in U.S. infants. During his clinical research career at Aventis Pasteur, he contributed to the development and licensure of seven vaccines and has authored or coauthored several clinical research articles. From 2000 through 2002, Mr. Hosbach served on the board of directors for Pocono Medical

Center in East Stroudsburg, Pennsylvania. Since 2003 he has served on the board of directors of Pocono Health Systems, which includes Pocono Medical Center.

James M. Hughes, M.D., is professor of medicine and public health at Emory University's School of Medicine and Rollins School of Public Health, serving as director of the Emory Program in Global Infectious Diseases, associate director of the Southeastern Center for Emerging Biological Threats, and senior advisor to the Emory Center for Global Safe Water. He also serves as senior scientific advisor for infectious diseases to the International Association of National Public Health Institutes funded by the Bill and Melinda Gates Foundation. Prior to joining Emory in June 2005, Dr. Hughes served as director of the National Center for Infectious Diseases (NCID) at the CDC. Dr. Hughes received his B.A. and M.D. degrees from Stanford University and completed postgraduate training in internal medicine at the University of Washington, infectious diseases at the University of Virginia, and preventive medicine at the CDC. After joining the CDC as an EIS officer in 1973, Dr. Hughes worked initially on food-borne and waterborne diseases and subsequently on infection control in healthcare settings. He served as director of CDC's Hospital Infections Program from 1983 to 1988, as deputy director of NCID from 1988 to 1992, and as director of NCID from 1992 to 2005. A major focus of Dr. Hughes' career has been on building partnerships among the clinical, research, public health, and veterinary communities to prevent and respond to infectious diseases at the national and global levels. His research interests include emerging and reemerging infectious diseases; antimicrobial resistance; food-borne diseases; healthcare-associate infections; vector-borne and zoonotic diseases; rapid detection of and response to infectious diseases and bioterrorism; strengthening public health capacity at the local, national, and global levels; and prevention of water-related diseases in the developing world. Dr. Hughes is a fellow of the American Association for the Advancement of Science, the American College of Physicians, and the Infectious Diseases Society of America, a member of IOM, and a Councillor of the American Society of Tropical Medicine and Hygiene.

Stephen A. Johnston, Ph.D., is currently director of the Center for Innovations in Medicine in the Biodesign Institute at Arizona State University. His center focuses on formulating and implementing disruptive technologies for basic problems in health care. The center has three divisions: Genomes to Vaccines, Cancer Eradication, and DocInBox. Genomes to Vaccines has developed high-throughput systems to screen for vaccine candidates and is applying them to predict and produce chemical vaccines. The Cancer Eradication group is working on formulating a universal prophylactic vaccine for cancer. DocInBox is developing technologies to facilitate presymptomatic diagnosis. Dr. Johnston founded the Center for Biomedical Inventions (a.k.a. Center for Translation Research) at the University of Texas-Southwestern, the first center of its kind in the medical arena. He and his colleagues have developed numerous inventions and innovations, including the gene gun, genetic immunization, TEV protease system, organelle transformation, digital optical chemistry arrays, expression library immunization, linear expression elements, and others. He also was involved in transcription research for years, first cloning *Gal4*, then later discovering functional domains in transcription factors and the connection of the proteasome to transcription. He has been professor at the University of Texas Southwestern Medical Center at Dallas and associate and assistant professor at Duke University. He has been involved in several capacities as an adviser on biosecurity since 1996 and is a member of the WRCE SAB and a founding member of BioChem 20/20.

Gerald T. Keusch, M.D., is associate provost and associate dean for global health at Boston University and Boston University School of Public Health. He is a graduate of Columbia College (1958) and Harvard Medical School (1963). After completing a residency in internal medicine, fellowship training in infectious diseases, and 2 years as an NIH research associate at the Southeast Asia Treaty Organization (SEATO) Medical Research Laboratory in Bangkok, Thailand, Dr. Keusch joined the faculty of the Mt. Sinai School of Medicine in 1970, where he established a laboratory to study the pathogenesis of bacillary dysentery and the biology and biochemistry of Shiga toxin. In 1979 he moved to Tufts Medical School and New England Medical Center in Boston to found the Division of Geographic Medicine, which focused on the molecular and cellular biology of tropical infectious diseases. In 1986 he integrated the clinical infectious diseases program into the Division of Geographic Medicine and Infectious Diseases, continuing as division chief until 1998. He has worked in the laboratory and in the field in Latin America, Africa, and Asia on basic and clinical infectious diseases and HIV/AIDS research. From 1998 to 2003, he was associate director for international research and director of the Fogarty International Center at NIH. Dr. Keusch is a member of ASCI, the Association of American Physicians, the ASM, and the IDSA. He has received the Squibb (1981), Finland (1997), and Bristol (2002) awards of the IDSA. In 2002 he was elected to the IOM.

Rima F. Khabbaz, M.D., is director of the National Center for Preparedness, Detection, and Control of Infectious Diseases at CDC. She became director of the National Center for Infectious Diseases at CDC in December 2005 and led its transition to the current centers. She is a graduate of the American University of Beirut, Lebanon, where she obtained both her bachelor's degree in science and her medical doctorate degree. She trained in internal medicine and completed a fellowship in infectious diseases at the University of Maryland in Baltimore. She is also a clinical associate professor of medicine (infectious diseases) at Emory University. She began her CDC career in 1980 as an epidemic intelligence service officer in the Hospital Infections Program. She later served as a medical epidemiologist in CDC's Retrovirus Diseases Branch, where she made major contributions to defining the epidemiology of non-HIV retroviruses (HTLV-I and II) in the United States and developing guidance for counseling HTLV-infected persons. Following the hantavirus pulmonary syndrome outbreak in the southwestern United States in 1993, she led CDC's efforts to set up national surveillance for the syndrome. Prior to becoming director of NCID, she was acting deputy director and, before that, associate director for epidemiologic science, NCID. Additional positions held at CDC include associate director for science and deputy director of the Division of Viral and Rickettsial Diseases. She played a leading role in developing CDC's blood safety programs and its food safety programs related to viral diseases. She also had a key role in CDC's responses to outbreaks of new and/or reemerging viral infections including Nipah, Ebola, West Nile, SARS (severe acute respiratory syndrome), and monkeypox. She led CDC's field team to the nation's capital during the public health response to the anthrax attack of 2001. She is a fellow of IDSA, a member of the American Epidemiologic Society, ASM, and the Council of State and Territorial Epidemiologists. She served on FDA's Blood Product Advisory Committee and on its Transmissible Spongiform Encephalopathy Advisory Committee. She also served on IDSA's Annual Meeting Scientific Program Committee and serves on the society's National and Global Public Health Committee. She is a graduate of the National Preparedness Leadership Initiative at Harvard University and of the Public Health Leadership Institute at the University of North Carolina.

Lonnie J. King, D.V.M., is currently director of CDC's new National Center for Zoonotic, Vector-Borne, and Enteric Diseases (NCZVED). Dr. King leads the center's activities for surveillance, diagnostics, disease investigations, epidemiology, research, public education, policy development, and disease prevention and control programs. NCZVED also focuses on water-borne, food-borne, vector-borne, and zoonotic diseases of public health concern, which also include most of CDC's select and bioterrorism agents, neglected tropical diseases, and emerging zoonoses. Before serving as director, he was the first chief of the agency's Office of Strategy and Innovation. In 1996, Dr. King was appointed dean of the College of Veterinary Medicine, Michigan State University. He served for 10 years as dean of the college. As dean, he was the chief executive officer for academic programs, research, the teaching hospital, diagnostic center for population and animal health, basic and clinical science departments, and outreach and continuing education programs. As dean and professor of large animal clinical sciences, Dr. King was instrumental in obtaining funds for construction of the $60 million Diagnostic Center for Population and Animal Health, initiated the Center for Emerging Infectious Diseases in the college, served as the campus leader in food safety, and had oversight for the National Food Safety and Toxicology Center. He brought the Center for Integrative Toxicology to the college and was the university's designated leader for counterbioterrorism activities for his college.

Prior to this, Dr. King was administrator for USDA's Animal and Plant Health Inspection Service (APHIS). Dr. King served as the country's chief veterinary officer for 5 years and worked extensively in global trade agreements within the North American Free Trade Agreement and the World Trade Organization. Before beginning his government career in 1977, he was in private veterinary practice for 7 years in Dayton, Ohio, and in Atlanta, Georgia. He received his B.S. and D.V.M. from Ohio State University in 1966 and 1970, respectively. He earned his M.S. in epidemiology from the University of Minnesota while on special assignment with the U.S. Department of Agriculture in 1980. He received his master's in public administration from the American University in Washington, DC, in 1991. Dr. King has a broad knowledge of animal agriculture and the veterinary profession through his work with other government agencies, universities, major livestock and poultry groups, and private practitioners. Dr. King is a board-certified member of the American College of Veterinary Preventive Medicine and has completed the senior executive fellowship program at Harvard University. He served as president of the Association of American Veterinary Medical Colleges from 1999 to 2000 and was vice chair for the National Commission on Veterinary Economic Issues from 2000 to 2004. Dr. King helped start the National Alliance for Food Safety, served on the Governor's Task Force on Chronic Wasting Disease for the State of Michigan, and was a member of four NAS committees; most recently he chaired the National Academies Committee on Assessing the Nation's Framework for Addressing Animal Diseases. Dr. King is one of the developers of the Science, Politics, and Animal Health Policy Fellowship Program, and he lectures extensively on the future of animal health, emerging zoonoses, and veterinary medicine. He served as a consultant and member of the Board of Scientific Counselors to CDC's National Center for Infectious Diseases and is a member of the IOM's Forum on Microbial Threats. Dr. King was an editor for the OIE (World Organisation for Animal Health) *Scientific Review on Emerging Zoonoses*, is a current member of FDA's Board of Scientific Advisors, and is president of the American Veterinary Epidemiology Society. Dr. King was elected to the IOM in 2004.

Col. George W. Korch, Ph.D., is commander, U.S. Army Medical Research Institute for Infectious Diseases, Ft. Detrick, Maryland. Dr. Korch attended Boston University and earned a B.S. in biology in 1974, followed by postgraduate study in mammalian ecology at the University of Kansas from 1975 to 1978. He earned his Ph.D. from the Johns Hopkins School of Hygiene and Public Health in immunology and infectious diseases in 1985, followed by postdoctoral experience at Johns Hopkins from 1985 to 1986. His areas of training and specialty are the epidemiology of zoonotic viral pathogens and medical entomology. For the past 15 years, he has also been engaged in research and program management for medical defense against biological pathogens used in terrorism or warfare.

Joshua Lederberg, Ph.D.,[1] is professor emeritus of molecular genetics and informatics and Sackler Foundation Scholar at the Rockefeller University in New York City. His lifelong research, for which he received the Nobel Prize in 1958, has been in genetic structure and function in microorganisms. He has a keen interest in international health and from 1990 to 1992 was co-chair of a previous IOM Committee on Emerging Microbial Threats to Health. Currently he is co-chair of the Committee on Emerging Microbial Threats to Health in the Twenty-First Century. He has been a member of the NAS since 1957 and is a charter member of the IOM.

Stanley M. Lemon, M.D., is the John Sealy Distinguished University Chair and director of the Institute for Human Infections and Immunity at the University of Texas Medical Branch (UTMB) at Galveston. He received his undergraduate A.B. degree in biochemical sciences from Princeton University summa cum laude and his M.D. with honors from the University of Rochester. He completed postgraduate training in internal medicine and infectious diseases at the University of North Carolina at Chapel Hill and is board certified in both. From 1977 to 1983 he served with the U.S. Army Medical Research and Development Command, followed by a 14-year period on the faculty of the University of North Carolina School of Medicine. He moved to UTMB in 1997, serving first as chair of the Department of Microbiology and Immunology, then as dean of the School of Medicine from 1999 to 2004. Dr. Lemon's research interests relate to the molecular virology and pathogenesis of the positive-stranded RNA viruses responsible for hepatitis. He has had a long-standing interest in antiviral and vaccine development and has served as chair of FDA's Anti-Infective Drugs Advisory Committee. He is the past chair of the Steering Committee on Hepatitis and Poliomyelitis of the WHO Programme on Vaccine Development. He is past chair of the NCID-CDC Board of Scientific Councilors and currently serves as a member of the U.S. Delegation to the U.S.-Japan Cooperative Medical Sciences Program. He was co-chair of the NAS Committee on Advances in Technology and the Prevention of Their Application to Next Generation Biowarfare Threats, and he recently chaired an IOM study committee related to vaccines for the protection of the military against naturally occurring infectious disease threats.

Lynn Marks, M.D., is senior vice president of the Infectious Diseases Medicine Development Center at GlaxoSmithKline. Dr. Marks received his medical degree from the University of South Alabama College of Medicine and is board certified in internal medicine and infectious diseases. He joined SmithKline Beecham in 1993 as associate director and later director, Anti-Infectives Clinical Research, Development, and Medical Affairs. He then moved to the Consumer Healthcare Division where he held the positions of worldwide medical director, Rx to OTC

[1]Deceased February 2, 2008.

Switch, and then vice president and director, Worldwide Medical, Regulatory, and Toxicology. Later he returned to Pharma as vice president, Global Commercial Strategy, Infectious Diseases, and he subsequently became senior vice president, Infectious Diseases, Medicine Development Center. Prior to joining industry, Dr. Marks was with the University of South Alabama College of Medicine, where he held the positions of assistant professor of medicine in the Division of Infectious Diseases and adjunct assistant professor in the Department of Microbiology and immunology as well as the Department of Pharmacology. His NIH-supported research centered on the molecular genetics of *Rickettsia*.

Edward McSweegan, Ph.D., is a program officer at the National Institute of Allergy and Infectious Diseases. He graduated from Boston College with a B.S. in biology in 1978. He has an M.S. in microbiology from the University of New Hampshire and a Ph.D. in microbiology from the University of Rhode Island. He was an NRC associate from 1984 to 1986 and did postdoctoral research at the Naval Medical Research Institute in Bethesda, Maryland. Dr. McSweegan served as a AAAS diplomacy fellow in the U.S. State Department from 1986 to 1988 where he helped to negotiate science and technology agreements with Poland, Hungary, and the former Soviet Union. After moving to NIH, he continued to work on international health and infectious disease projects in Egypt, Israel, India, and Russia. Currently, he manages NIAID's bilateral program with India, the Indo-U.S. Vaccine Action Program, and he represents NIAID in the HHS Biotechnology Engagement Program with Russia and related countries. He is a member of AAAS, the ASM, and the National Association of Science Writers. He is the author of numerous journal and freelance articles.

Stephen S. Morse, Ph.D., is professor of epidemiology and founding director of the Center for Public Health Preparedness at the Mailman School of Public Health of Columbia University. He returned to Columbia in 2000 after 4 years in government service as program manager at the Defense Advanced Research Projects Agency (DARPA), where he co-directed the Pathogen Countermeasures Program and subsequently directed the Advanced Diagnostics Program. Before coming to Columbia, he was assistant professor of virology at the Rockefeller University in New York, where he remains an adjunct faculty member. He is the editor of two books, *Emerging Viruses* (Oxford University Press, 1993; paperback, 1996), which was selected by American Scientist for its list of 100 Top Science Books of the 20th Century, and *The Evolutionary Biology of Viruses* (Raven Press, 1994). He was a founding section editor of the CDC journal *Emerging Infectious Diseases* and was formerly an editor-in-chief of the Pasteur Institute's journal *Research in Virology*. Dr. Morse was chair and principal organizer of the 1989 NIAID-NIH Conference on Emerging Viruses, for which he originated the term and concept of emerging viruses-infections. He has served as a member of the IOM-NAS Committee on Emerging Microbial Threats to Health, chaired its Task Force on Viruses, and was a contributor to the resulting report *Emerging Infections* (1992). He was a member of the IOM's Committee on Xenograft Transplantation, and he currently serves on the Steering Committee of the IOM's Forum on Emerging Infections (now the Forum on Microbial Threats). Dr. Morse also served as an adviser to WHO and several government agencies. He is a fellow of the New York Academy of Sciences and a past chair of its microbiology section, a fellow of the American Academy of Microbiology of the American College of Epidemiology, and an elected life member of the Council on Foreign Relations. He was the founding chair of ProMED, the nonprofit international Program to Monitor Emerging Diseases, and was one of the originators of ProMED-mail, an international network inaugurated by ProMED in 1994 for outbreak reporting and disease monitoring using the Internet. Dr. Morse received his Ph.D. from the University of Wisconsin, Madison.

Michael T. Osterholm, Ph.D., M.P.H., is director of the Center for Infectious Disease Research and Policy and director of the NIH-sponsored Minnesota Center for Excellence in Influenza Research and Surveillance at the University of Minnesota. He is also professor at the School of Public Health and adjunct professor at the Medical School. Previously, Dr. Osterholm was the state epidemiologist and chief of the acute disease epidemiology section for the Minnesota Department of Health. He has received numerous research awards from NIAID and CDC. He served as principal investigator for the CDC-sponsored Emerging Infections Program in Minnesota. He has published more than 300 articles and abstracts on various emerging infectious disease problems and is the author of the best-selling book *Living Terrors: What America Needs to Know to Survive the Coming Bioterrorist Catastrophe*. He is past president of the Council of State and Territorial Epidemiologists. He currently serves on the IOM Forum on Microbial Threats. He has also served on the IOM Committee to Ensure Safe Food from Production to Consumption, on the IOM Committee on the Department of Defense Persian Gulf Syndrome Comprehensive Clinical Evaluation Program, and as a reviewer for the IOM report, *Chemical and Biological Terrorism: Research and Development to Improve Civilian Medical Response*.

George Poste, Ph.D., D.V.M., is director of the Biodesign Institute and Del E. Webb Distinguished Professor of Biology at Arizona State University. From 1992 to 1999, he was chief science and technology officer and president, Research and Development, of SmithKline Beecham (SB). During his tenure at SB, he was associated with the successful registration of 29 drug, vaccine, and diagnostic products. He is chairman of Orchid Cellmark. He serves on the board of directors of Monsanto and Exelixis. He is a distinguished fellow at the Hoover Institution at Stanford University. He is a member of the Defense Science Board of the U.S. Department of Defense and of the IOM Forum on Microbial Threats. Dr. Poste is a board-certified pathologist, a fellow of the

Royal Society, and a fellow of the Academy of Medical Sciences. He was awarded the rank of Commander of the British Empire by Queen Elizabeth II in 1999 for services to medicine and for the advancement of biotechnology. He has published more than 350 scientific papers; has coedited 15 books on cancer, biotechnology, and infectious diseases; and serves on the editorial board of several technical journals.

Gary A. Roselle, M.D., received his medical degree from the Ohio State University School of Medicine in 1973. He served his residency at the Northwestern University School of Medicine and his infectious diseases fellowship at the University of Cincinnati School of Medicine. He is program director for infectious diseases for the Department of Veterans Affairs Central Office in Washington, DC, as well as the chief of the medical service at the Cincinnati VA Medical Center. He is a professor of medicine in the Department of Internal Medicine, Division of Infectious Diseases, at the University of Cincinnati College of Medicine. Dr. Roselle serves on several national advisory committees. In addition, he is currently heading the Emerging Pathogens Initiative for the VA. He has received commendations from the under secretary for health for the VA and the secretary of veterans affairs for his work in the Infectious Diseases Program for the VA. He has been an invited speaker at several national and international meetings and has published more than 90 papers and several book chapters.

Janet Shoemaker is director of the ASM's Public Affairs Office, a position she has held since 1989. She is responsible for managing the legislative and regulatory affairs of this 42,000-member organization, the largest single biological science society in the world. She has served as principal investigator for a project funded by the National Science Foundation (NSF) to collect and disseminate data on the job market for recent doctorates in microbiology and has played a key role in ASM projects, including production of the ASM *Employment Outlook in the Microbiological Sciences* and *The Impact of Managed Care and Health System Change on Clinical Microbiology*. Previously, she held positions as assistant director of public affairs for ASM; as ASM coordinator of the U.S.-U.S.S.R. Exchange Program in Microbiology, a program sponsored and coordinated by the NSF and the U.S. Department of State; and as a freelance editor and writer. She received her baccalaureate, cum laude, from the University of Massachusetts and is a graduate of the George Washington University programs in public policy and in editing and publications. She has served as commissioner to the Commission on Professionals in Science and Technology and as ASM representative to the ad hoc Group for Medical Research Funding, and she is a member of Women in Government Relations, the American Society of Association Executives, and AAAS. She has coauthored published articles on research funding, biotechnology, biological weapons control, and public policy issues related to microbiology.

P. Frederick Sparling, M.D., is the J. Herbert Bate Professor Emeritus of Medicine, Microbiology, and Immunology at the University of North Carolina (UNC) at Chapel Hill, and professor of medicine, Duke University. He is director of the North Carolina Sexually Transmitted Infections Research Center and also the Southeast Regional Centers of Excellence in Biodefense and Emerging Infections. Previously he served as chair of the Department of Medicine and chair of the Department of Microbiology and Immunology at UNC. He was president of IDSA from 1996 to 1997. He was also a member of the IOM Committee on Microbial Threats to Health (1990-1992) and the IOM Committee on Emerging Microbial Threats to Health in the 21st Century (2001-2003). Dr. Sparling's laboratory research has been on the molecular biology of bacterial outer membrane proteins involved in pathogenesis, with a major emphasis on gonococci and meningococci. His work helped to define the genetics of antibiotic resistance in gonococci and the role of iron-scavenging systems in the pathogenesis of human gonorrhea.

Brian Staskawicz, Ph.D., is professor and chair, Department of Plant and Microbial Biology, University of California, Berkeley. Dr. Staskawicz received his B.A. in biology from Bates College in 1974 and his Ph.D. from the University of California, Berkeley, in 1980. Dr. Staskawicz's work has contributed greatly to understanding the molecular interactions between plants and their pathogens. He was elected to the NAS in 1998 for elucidating the mechanisms of disease resistance, as his lab was the first to clone a bacterial effector gene from a pathogen and among the first to clone and characterize plant disease resistance genes. Dr. Staskawicz's research focuses on the interaction of the bacteria *Pseudomonas* and *Xanthomonas* with *Arabidopsis*, tomato, and pepper. He has published extensively in this area and is one of the leading scientists in the world working on elucidating the molecular basis of plant innate immunity.

Terence Taylor is director of the Global Health and Security Initiative and president and director of the International Council for the Life Sciences (ICLS). He is responsible for the overall direction of the ICLS and its programs, which have the goal of enhancing global biosafety and biosecurity. From 1995 to 2005, he was assistant director of the International Institute for Strategic Studies (IISS), a leading independent international institute, and president and executive director of its U.S. office (2001-2005). He studies international security policy, risk analysis, and scientific and technological developments and their impact on political and economic stability worldwide. At IISS he was one of the Institute's leading experts on issues associated with nuclear, biological, and chemical weapons and their means of delivery. In his previous appointments, he has had particular responsibilities for issues affecting public safety and security in relation to biological risks and advances in the life sciences. He was one of the commissioners to the United Nations Special Commission on Iraq, for which he also conducted missions as a chief inspector. He was a science fellow at the Center for International Security and Cooperation at Stanford University, where he carried out, among other subjects, studies of the implications for government and industry of the weapons of mass destruction treaties and agreements. He has also carried out consultancy work for the International Committee of the Red Cross (ICRC) on the implementation and development of the laws of armed conflict and serves as a member of the Editorial Board of the *ICRC Review*. He has served as chairman of the World Federation of Scientists' Permanent Monitoring Panel on Risk Analysis. He was a career officer in the British Army on operations in many parts of the world, including counterterrorist operations and UN peacekeeping. His publications include monographs, book chapters, and articles for, among others, Stanford University, the World Economic Forum, Stockholm International Peace Research Institute (SIPRI), the Crimes of War Project, the *International Herald Tribune*, the *Wall Street Journal*, the *International Defence Review*, the *Independent* (London), *Tiempo* (Madrid), the *International and Comparative Law Quarterly*, the *Washington Quarterly*, and other scholarly journals, including unsigned contributions to IISS publications.

Murray Trostle, Dr.P.H., is a foreign service officer with the U.S. Agency for International Development (USAID) presently serving as the deputy director of the Avian and Pandemic Influenza Preparedness and Response Unit. Dr. Trostle attended Yale University where he received a master's in public health in 1978, focusing on health services administration. In 1990, he received his doctorate in public health from UCLA. His research involved household survival strategies during famine in Kenya. Dr. Trostle has worked in international health and development for approximately 38 years. He first worked overseas in the Malaysian national malaria eradication program in 1968 and has since focused on health development efforts in the former Soviet Union, Africa, and Southeast Asia. He began his career with USAID in 1992 as a postdoctoral fellow with AAAS. During his career he has worked with a number of development organizations such as the American Red Cross, Project Concern International, and the Center for Development and Population Activities. With USAID, Dr. Trostle has served as director of the child immunization cluster, where he was chairman of the European Immunization Interagency Coordinating Committee and the USAID representative to the Global Alliance on Vaccines and Immunization. Currently, Dr. Trostle leads the USAID Infectious Disease Surveillance Initiative as well as the Avian Influenza Unit.